Advanced renewable energy systems

Advanced renewable energy systems

Part – I

S. C. Bhatia

WOODHEAD PUBLISHING INDIA PVT LTD

New Delhi

Published by Woodhead Publishing India Pvt. Ltd.
Woodhead Publishing India Pvt. Ltd., 303, Vardaan House, 7/28, Ansari Road, Daryaganj, New Delhi - 110002, India
www.woodheadpublishingindia.com

First published 2014, Reprinted 2017

Woodhead Publishing India Pvt. Ltd. ISBN: 978-93-80308-43-2
Woodhead Publishing India Pvt. Ltd. e-ISBN: 978-93-80308-73-9

Typeset by Mind Box Solutions, New Delhi
Printed and bound by Replika Press Pvt. Ltd.

Contents

Part – II

Preface

Renewable energy is a natural energy which does not have a limited supply. Renewable energy can be used again and again and will never run out. Renewable energy is derived from natural processes that are replenished constantly. In its various forms, it derives directly from the sun or from heat generated deep within the earth. Included in the definition is electricity and heat generated from solar, wind, ocean, hydropower, biomass, geothermal resources, and biofuels and hydrogen derived from renewable resources.

Renewable energy replaces conventional fuels in four distinct areas: power generation, hot water/space heating, transport fuels, and rural (off-grid) energy services. Airflows can be used to run wind turbines. Energy in water can be harnessed and used. Since water is about 800 times denser than air, even a slow flowing stream of water or moderate sea swell can yield considerable amount of energy. Solar energy is the energy derived from the sun in the form of solar radiation. Solar-powered electrical generation relies on photovoltaics and heat engines. Solar technologies are broadly characterised as either passive solar or active solar depending on the way they capture, convert and distribute solar energy. Active solar techniques include the use of photovoltaic panels and solar thermal collectors to harness the energy. Biomass (plant material) is a renewable energy source because the energy it contains comes from the sun. Through the process of photosynthesis, plants capture the sun's energy. When the plants are burnt, they release the sun's energy they contain. In this way, biomass functions as a sort of natural battery for storing solar energy. As long as biomass is produced sustainably, with only as much used as is grown, the battery will last indefinitely. Biofuels include a wide range of fuels which are derived from biomass. The term covers solid biomass, liquid fuels and various biogases. Liquid biofuels include bioalcohols, such as bioethanol, and oils, such as biodiesel. Gaseous biofuels include biogas, landfill gas and synthetic gas. Geothermal energy is energy obtained by trapping the heat of the earth itself, both from kilometres deep into the earth's crust in volcanically active locations of the globe or from shallow depths, as in geothermal heat pumps in most locations of the planet. It is expensive to build a power station but operating costs are low resulting in low energy costs for suitable sites. Ultimately, this energy derives from heat in the earth's core.

This book on renewable energy sources summarises the various aspects of renewable energy and is divided into 26 chapters.

Chapter 1 is devoted to energy resources and their utilisation. The chapter gives brief outline of non-conventional resources along with India's and world energy resources. Chapter 2 deals with solar radiations which are becoming increasingly appreciated because of their influence on living matter and the feasibility of their application for useful purposes. Chapter 3 concentrates on solar devices such as solar photovoltaic modules, silicon solar cells, solar lanterns, solar lights, solar water pumps and heaters, etc. Chapter 4 focuses on solar thermal energy which can be utilised through two routes—solar thermal route and solar photovoltaic route. Chapter 5 acquaints the readers with solar photovoltaic systems which convert sunlight directly into electricity without creating any air or water pollution. Chapter 6 is devoted to energy-consuming and -converting equipments. Nanotechnology has the potential for significant impact at all stages of the energy value chain. Considering this, Chapter 7 focuses on nanotechnology and solar power. Chapter 8 deals with wind energy which is a form of solar energy and caused by the uneven heating of atmosphere by sun, the irregularities of the earth's surface and rotation of earth. Chapter 9 acquaints the readers with status of wind power in India. Chapter 10 discusses hydroelectric power which comes from flowing water. Water, when it is falling by force of gravity, can be used to turn turbines and generators produce heat. Chapter 11 is devoted to small hydroelectric power on a scale serving a small community or industrial plant. Chapter 12 acquaints the readers with hydropower development in India. Chapter 13 concentrates on tide, wave and ocean energy. These are the form of hydropower that converts the energy of tides into useful forms of power—mainly electricity. Chapter 14 focuses on geothermal power generation which means energy or power extracted from beneath the earth. Geothermal energy is called renewable source of energy because heat is continuously produced inside the earth. Chapter 15 acquaints the readers with geothermal energy resources and its potential in India. Chapter 16 is devoted to biofuels: a review. Biofuels are the types of fuels whose energy is derived from biological carbon fixation. Biofuels include fuels derived from biomass conversion as well as solid biomass, liquid fuels and various biogases. Chapter 17 concentrates on biogas which refers to gas produced by biological breakdown or organic matter in the absence of oxygen. Chapter 18 discusses biomass gasification which is incomplete combustion of biomass resulting in production of combustible gases such as producer gas, hydrogen, traces of methane, etc. Chapter 19 deals with cogeneration of power and low-pressure stream by way of expansion through a turbine.

Chapter 20 is devoted to photosynthesis which is a chemical process that converts carbon dioxide into organic compounds, especially sugars, using energy from sunlight. Chapter 21 concentrates on ethanol which is most often used as a motor fuel in the United States and Brazil. Chapter 22 focuses on biodiesel which refers to a vegetable-oil- or animal-fat-based diesel fuel consisting of long-chain alkyl esters. Chapter 23 deals with biohydrogen which is produced biologically, most commonly by algae and bacteria. Biohydrogen is a potential biofuel obtainable from both cultivation and waste organic materials. Chapter 24 is devoted to algae fuel for future. Chapter 25 deals with nanotech biofuels and fuels additives. Chapter 26 discusses issues related to biofuels which include the effect of moderating oil prices, the 'food vs. fuel debate', carbon emission levels, deforestation and soil erosion.

Such wide coverage makes this book a treatise on the subject. Glossary, references and index have been provided at the end for quick reference. Diagrams, figures, tables and index supplement the text. All the topics have been covered in a cogent and lucid style to help the reader grasp the information quickly and easily.

It will not be wrong to hold that this book on renewable energy sources is essential reading for all students pursuing B. Tech (Civil/Chemical/Mechanical/Environmental Engineering). Besides students, the book is useful to consultants, industrialists and researchers in these and allied fields.

The book also caters to the requirement of the syllabus prescribed by various Indian universities for undergraduate student pursuing undergraduate courses in engineering. It has been prepared with meticulous care, aiming at making the book error-free. Constructive suggestions are always welcome from users of this book.

S. C. Bhatia

1

Energy resources and their utilisation

1.1 Introduction

Renewable energy is energy which comes from natural resources such as sunlight, wind, rain, tides, and geothermal heat, which are renewable (naturally replenished). About 16 per cent of global final energy consumption comes from renewables, with 10 per cent coming from traditional biomass, which is mainly used for heating, and 3.4 per cent from hydroelectricity. New renewables (small hydro, modern biomass, wind, solar, geothermal, and biofuels) accounted for another 2.8 per cent and are growing very rapidly. The share of renewables in electricity generation is around 19 per cent, with 16 per cent of global electricity coming from hydroelectricity and 3 per cent from new renewables.

While many renewable energy projects are large-scale, renewable technologies are also suited to rural and remote areas, where energy is often crucial in human development. As of 2011, small solar PV systems provide electricity to a few million households, and micro-hydro configured into mini-grids serves many more. Over 44 million households use biogas made in household-scale digesters for lighting and/or cooking, and more than 166 million households rely on a new generation of more-efficient biomass cookstoves.

Climate change concerns, coupled with high oil prices, peak oil, and increasing government support, are driving increasing renewable energy legislation, incentives and commercialisation. New government spending, regulation and policies helped the industry weather the global financial crisis better than many other sectors. According to a 2011 projection by the International Energy Agency, solar power generators may produce most of the world's electricity within 50 years, dramatically reducing the emissions of greenhouse gases that harm the environment.

1.2 India's energy resources

India's need to increase energy provision for its population and fast growing economy poses a formidable challenge which is perceived as both a great opportunity as well as a necessity for the country to increase the share of renewables in the overall energy mix. India has been making continuous progress in conventional as well as renewable power generation. India's approach is to meet its energy needs in a responsible, sustainable and ecofriendly manner.

The country has made a remarkable growth in last 2–3 years in the field of renewable energy power generation. The past few years saw a record addition of 2332 MW of renewable energy sources i.e. solar, wind, biomass, geothermal and hydro, etc. which could make important contributions to sustainable development. All these development have taken place in 11th plan period. The bulk addition is in wind generation at 1565 MW, small hydro power segment recorded an addition of 305 MW, cogeneration 295 MW and biomass 153 MW. The lowest additions were in the solar at 8 MW and waste-to-energy segments at 4.7 MW. A target of 14,000 MW capacity addition has been set for the 11th Plan (2007–08 to 2011–12).

The need for renewable energy arises as energy Security is important to our country. As per World Energy Outlook report, India will become the third largest net importer of oil before 2025 after the United States and China. This will not be sustainable in the long run given the high volatility of international crude oil prices. Coal imports are also likely to increase from 12 per cent in 2005 to 28 per cent in 2030. In order to insulate itself from any future supply disruption and price shocks of fossil fuels and furthermore to achieve energy security and also meet global climate change objectives, renewables are a must and has indeed caught the imagination of India.

Riding on the crest of a high growth trajectory and in pursuit of sustainable sources to meet its rising domestic energy demand and access to energy for overall development, the country is focusing on harnessing renewable energy production through maximising the utilisation of renewable energy.

Currently, their exploitation in commercial markets is low, being constrained by costs and uncompensated benefits, as well as intermittent supplies and other technical and institutional considerations. But they have hold on social consideration for:

1. Enhanced energy security by providing supplies that are abundant, diverse and indigenous.
2. Reduced global and local atmospheric emissions when used in place of fossil fuels.
3. Improved options in rural areas and in newly industrialising and developing countries.
4. Increased local and regional employment opportunities in energy infrastructure manufacturing, installation and maintenance for developed and developing countries.

1.3 Renewable energy sources

Conventionally, natural resources, like, wind, solar and hydro are termed as renewable energy sources. These terms can be easily understood. Bioenergy, which is another important renewable energy source covers a wide spectrum of energy activities from direct production heat through combustion of fuel wood and other biomass residues, to the generation of electricity and the production of gases and liquid fuel and chemicals. It is widely used globally. The geothermal resources are the internal heat of the earth. Its use covers a range of option from power generation to space heating and/or air conditioning.

1.3.1 Growth of renewable power in India

The renewable energy potential in India is as below:

1. Wind power 48,000 MW
2. Small hydro 15,000 MW
3. Biomass power 16,700 MW
4. Bagasse cogeneration 5000 MW
5. Solar energy 20 MW/sq. km
6. Waste to energy 3800 MW

India has been making continuous progress in conventional as well as renewable power generation. The trajectory of growth of installed capacity since year 2002 (start of 10th Plan), 2007 (start of 11th Plan), and at present as on 30-9-2010, is given in Table 1.1.

It may be observed that renewable grid capacity has increased more than 5 times, from 2 per cent to around 11 per cent in only 8 years.

Table 1.1 Source-wise contribution to installed power generation capacity (MW).

Time period	Thermal (%)	Hydro (>25 MW) (%)	Nuclear (%)	Renewable power (%)
1-4-2002	59% (74429)	26% (26, 269)	2% (2720)	2% (1628)
31-9-2007	65% (87015)	26% (34, 654)	3% (3900)	6% (10, 258)
1-4-2010	64% (106518)	22.4% (37, 328)	2.7% (4560)	10.90% (18, 155)

Total installed capacity: India's total installed capacity of electricity generation has expanded from 1,05,046 MW at the end of 2,00,102 to 1,66,561 MW at the end of September, 2010. In fact, India ranks sixth globally in terms of total electricity generation.

Thermal power: 64.0 per cent of the total installed capacity, producing 1,06,518 MW.

Hydel power: Hydel power plants come next with 22.4 per cent of the total an installed capacity of 37,328 MW.

Renewable energy: Renewable energy sources contribute around 10.9 per cent to the total power generation in the country producing 18,155 MW (as on 31.9.2010).

1.4 Renewable energy capacity in India

All sectors of renewable energy are being developed and promoted. However, the wind power program has been the fastest growing contributing to around 75 per cent of the grid-connected renewable energy power installed capacity.

Wind power: Wind power has seen a phenomenal growth of around 33 per cent CAGR (compound annual growth rate) in the last 5 years and the total capacity at end of 2010 was 12129 MW (as on 31-7-2010) with most of the capacity installed in the state of Tamil Nadu in India.

Solar energy: Among various renewable energy resources, India possesses a very large solar energy resource which is seen as having the highest potential for the future. The first, recently announced, very ambitious Jawaharlal Nehru National Solar Mission with a target of 20,000 MW grid solar power, 2000 MW of off-grid capacity including 20 million solar lighting systems and 20 million sq. m. solar thermal collector area by 2022 is under implementation, likely to be implemented in three phases. The first phase will be of three years (up to March, 2013), the second till March 2017 and the third phase will continue till March 2022. The target for phase-I is to set up 1100 MW grid connected solar plants including 100 MW of roof top and small solar plants and 200 MW capacity equivalent off-grid solar applications and 7 million sq.m. solar thermal collector area. The main objectives of the mission are to help reach grid parity by 2022 and help set up indigenous manufacturing capacity.

Mini hydro plants: Mini hydro plants are supposed to have a 15 GW potential out of which 2.5 GW has been realised.

Biomass energy: Biomass energy has 25 GW of potential out of which around 2.1 GW has been realised.

Rural electrification to meet unmet demand through renewable energy is a priority by itself. 150 villages have been covered in last 2 years through mini grid by rice-husk based gasification systems in Bihar. There are plans to cover about 10,000 villages from biomass-based systems and over 1000 villages from solar power by 2022.

The grid interactive projects (up to 2 MW) at the tail end of the grid at 11 kV to reduce the transmission and distribution losses and stabilise grid voltages are now being piloted. So far, about 112 MW equivalent biomass gasifier systems have been set up in industries for captive power and thermal applications.

1.5 Renewables and the environment

Renewable energy is usually more environmental friendly than alternative energy sources. Life-cycle emissions from renewable energy use are small compared with those from fossil fuel plants. Nuclear power generation does have a major environmental impact, it releases no sulphur dioxide (SO_2) or nitrogen oxides (NO_x) and little carbon dioxide (CO_2). Its life-cycle emissions of these gases fall within the ranges shown for non-hydroelectric renewable energy.

Renewable energy entails a number of other potential environmental impacts. On the negative side, renewable energy can make large tracts of land unusable for competing uses, disrupt marine life, bird life and flora/fauna, and produce visual and noise pollution. Generally though, these potential environmental impacts are site-specific and there are a number of ways to minimise the effects, which are usually small and reversible. There are environmental benefits from renewable other than reduction of greenhouse gas and other air emissions. For example, hydroelectric schemes can improve water supplies and facilitate reclamation of degraded land and habitat.

The use of bioenergy can have many environmental benefits if the resource is produced and used in a sustainable way. If the land from which bioenergy is produced is replanted, bioenergy is used sustainably and the carbon released will be recycled into the next generation of growing plants. Bioenergy plants have lower emissions of SO_2 than coal and oil plants, but they may produce more particulate matter. These emissions are controllable but they increase generating costs.

The environmental and social effects of large-scale hydropower are site specific and are the subject of much controversy. Large-scale projects may disturb local ecosystems, reduce biological diversity or modify water quality. They may also cause socio-economic damage by displacing local populations. Mini- and micro-hydro systems have relatively modest and localised effects on the environment, but their kWh cost is generally higher. Hydropower emits some greenhouse gases on a life-cycle basis (especially methane generated by decaying bioenergy in reservoirs), but in most cases far less than the burning of fossil fuels.

Geothermal plants may release gaseous emissions into the atmosphere during their operation. These gases are mainly carbon dioxide and hydrogen sulphide with traces of ammonia, hydrogen, nitrogen, methane, radon, and the volatile species of boron, arsenic and mercury. This could slow the future development of geothermal resources. Emissions can be managed through strict regulations and by control methods used by the geothermal industry to meet these regulatory requirements. Hydrogen sulphide abatement systems reduce environmental damage but are costly to install.

Wind-power generation has very low emissions on a life-cycle basis, but has a number of environmental effects that may limit its potential. The most important effects on the environment are:

1. Visual effects: Wind turbines must be in exposed areas and are therefore highly visible. They are considered unsightly by some people, and concerns have increased with the larger size of new generation turbines.
2. Noise: Wind turbines produce aerodynamic noise, from air passing over the blades and mechanical noise from the moving parts of the turbine, especially the gearbox. Better designs have reduced noise, and research continues. Wind farms developed far from highly populated areas are, by definition, less offensive.
3. Electromagnetic interference: Wind turbines may scatter electromagnetic signals causing interference to communication systems. Appropriate siting (avoiding military zones or airports) can minimise this impact.
4. Bird safety: Birds get killed when they collide with the rotating blades of a turbine. Migratory species are at higher risk than resident species. Siting the turbines away from migratory routes reduces the impact.

1.6 Renewable potential vs installed capacity in India

A report of MNRE, shows the total potential of renewable energy that could be harnessed in the country. Table 1.2 shows the break-up of the potentials of different sources.

Table 1.2 Renewable potential and installed capacity.

Source	Potential (MW)	Installed capacity (MW) (June 2010 as per MNRE)
Wind	45,195	12,009.00
Biomass	16,881	901.10
Small hydro	15,000	2767.00
Bagasse-based cogeneration	5000	1412.00
Waste-to-energy	2700	72.46
Solar	50 MW/sq. km	12.00
Total		17,174.00

1.7 Total generation installed capacity in India

Table 1.3 shows the generation installed capacity as on June 2010.

Table 1.3 Installed capacity.

Fuel	Capacity (MW)	Share (%)
Thermal	1,05,646.98	64.26
Coal and Lignite	87093.85	52.97
Gas	17353.85	10.55
Diesel and liquid fuels	1199.75	0.73
Hydro	37033.40	22.52
Renewable	17173.90	10.45
Wind	12009.48	7.30
Biomass	901.10	1.68
Small hydro	2767.05	0.55
Bagasse-based cogeneration	1411.53	0.86
Waste-to-energy	72.46	0.04
Solar	12.38	0.01
Nuclear	4560.00	2.77
Total	164414.28	100

1.8 World energy resources and consumption

World energy consumption in 2010: Over 5 per cent growth energy markets have combined crisis recovery and strong industry dynamism. Energy consumption in the G20 soared by more than 5 per cent in 2010, after the slight decrease of 2009. This strong increase is the result of two converging trends. On the one-hand, industrialised countries, which experienced sharp decreases in energy demand in 2009, recovered firmly in 2010, almost coming back to historical trends. Oil, gas, coal, and electricity markets followed the same trend. On the other hand, China and India, which showed no signs of slowing down in 2009, continued their intense demand for all forms of energy.

In 2009, world energy consumption decreased for the first time in 30 years (–1.1 per cent) or 130 Mtoe (Megaton oil equivalent), as a result of the financial and economic crisis (GDP drop by 0.6 per cent in 2009). This evolution is the result of two contrasting trends. Energy consumption growth remained vigorous in several developing countries, specifically in Asia (+4 per cent). Conversely, in OECD, consumption was severely cut by 4.7 per cent in 2009 and was thus almost down to its 2000 levels. In North America, Europe and Commonwealth of Independent States (CIS), consumptions shrank by 4.5 per cent, 5 per cent and 8.5 per cent respectively due to the slowdown in economic activity. China became the world's largest energy consumer (18 per cent of the total) since its consumption surged by 8 per cent during 2009 (up from 4 per cent in 2008). Oil remained the largest energy source (33 per cent) despite the fact that its share has been decreasing over time. Coal posted a growing role in the world's energy consumption in 2009, it accounted for 27 per cent of the total.

In 2008, total worldwide energy consumption was 474 exajoules (474×10^{18} J = 1,32,000 TWh). This is equivalent to an average energy consumption rate of 15 terawatts (1.504×10^{13} W). The potential for renewable energy is: solar energy 1600 EJ (4,44,000 TWh), wind power 600 EJ (1,67,000 TWh), geothermal energy 500 EJ (1,39,000 TWh), biomass 250 EJ (70,000 TWh), hydropower 50 EJ (14,000 TWh) and ocean energy 1 EJ (280 TWh).

More than half of the energy has been consumed in the last two decades since the industrial revolution, despite advances in efficiency and sustainability. According to IEA world statistics in four years (2004–2008) the world population increased 5 per cent, annual CO_2 emissions increased 10 per cent and gross energy production increased 10 per cent. Most energy is used in the country of origin, since it is cheaper to transport final products than raw materials. In 2008 the share export of the total energy production by fuel was: oil 50 per cent (1952/3941 MT), gas 25 per cent (800/3149 bcm), hard coal 14 per cent (793/5845 MT) and electricity 1 per cent (269/20,181 TWh).

Most of the world's energy resources are from the sun's rays hitting earth. Some of that energy has been preserved as fossil energy, some is directly or indirectly usable; for example, via wind, hydro- or wave power. The term solar constant is the amount of incoming solar electromagnetic radiation per unit area, measured on the outer surface of earth's atmosphere, in a plane perpendicular to the rays. The solar constant includes all types of solar radiation, not just visible light. It is measured by satellite to be roughly 1366 watts per square meter, though it fluctuates by about 6.9 per cent during a year—from 1412 Wm^{-2} in early January to 1321 Wm^{-2} in early July, due to the earth's varying distance from the sun, and by a few parts per thousand from day-to-day. For the whole earth, with a cross section of 12,74,00,000 km^2, the total energy rate is 174 petawatts (1.740×10^{17} W), plus or minus 3.5 per cent. This value is the total rate of solar energy received by the planet; about half, 89 PW, reaches the earth's surface.

The estimates of remaining non-renewable worldwide energy resources vary, with the remaining fossil fuels totalling an estimated 0.4 YJ (1 YJ = 1024 J) and the available nuclear fuel such as uranium exceeding 2.5 YJ. Fossil fuels range from 0.6 to 3 YJ if estimates of reserves of methane clathrates are accurate and become technically extractable. The total energy flux from the sun is 3.8 YJ/year, dwarfing all non-renewable resources.

From 1990 to 2008 the average use of energy per person as IEA data increased 10 per cent and the world population increased 27 per cent. Regional energy use grew from 1990 to 2008: Middle East 170 per cent, China 146 per cent, India 91 per cent, Africa 70 per cent, Latin America 66 per cent, USA 20 per cent, EU-27 7 per cent and world 39 per cent.

1.9 Consumption of energy

1.9.1 Primary energy

The United States Energy Information Administration regularly publishes a report on world consumption for most types of primary energy resources.

Fuel type	**Average power in TW**		
	1980	**2004**	**2007**
Oil	4.38	5.58	5.74
Gas	1.80	3.45	3.61
Coal	2.34	3.87	4.27
Hydroelectric	0.60	0.93	1.00
Nuclear power	0.25	0.91	0.93
Geothermal, wind, solar energy, wood	0.02	0.13	0.16
Total	9.48	15.0	15.8

1.9.2 Renewable energy

Renewable energy comes from natural resources such as sunlight, wind, rain, tides, and geothermal heat, which are renewable (naturally replenished). As of 2010, about 16 per cent of global final energy consumption comes from renewables, with 10 per cent coming from traditional biomass, which is mainly used for heating, and 3.4 per cent from hydroelectricity. New renewables (small hydro, modern biomass, wind, solar, geothermal, and biofuels) accounted for another 2.8 per cent and are growing very rapidly. The share of renewables in electricity generation is around 19 per cent, with

16 per cent of global electricity coming from hydroelectricity and 3 per cent from new renewables.

1.9.3 Hydropower

Worldwide hydroelectricity installed capacity reached 816 GW in 2005, consisting of 750 GW of large plants, and 66 GW of small hydro installations. Large hydro capacity totalling 10.9 GW was added by China, Brazil, and India during the year, but there was a much faster growth (8 per cent) small hydro, with 5 GW added, mostly in China where some 58 per cent of the world's small hydro plants are now located. China is the largest hydropower producer in the world, and continues to add capacity. In the Western world, although Canada is the largest producer of hydroelectricity in the world, the construction of large hydro plants has stagnated due to environmental concerns.

The trend in both Canada and the United States has been to micro hydro because it has negligible environmental impacts and opens up many more locations for power generation. In British Columbia alone, the estimates are that micro hydro will be able to more than double electricity production in the province.

1.9.4 Biomass and biofuels

Until the beginning of the nineteenth century biomass was the predominant fuel, today it has only a small share of the overall energy supply. Electricity produced from biomass sources was estimated at 44 GW for 2005. Biomass electricity generation increased by over 100 per cent in Germany, Hungary, the Netherlands, Poland, and Spain. A further 220 GW was used for heating (in 2006), bringing the total energy consumed from biomass to around 264 GW. The use of biomass fires for cooking is excluded.

World production of bioethanol increased by 8 per cent in 2007 to reach 33 billion litres (8.72 billion US gallons), with most of the increase in the United States, bringing it level to the levels of consumption in Brazil. Biodiesel increased by 85 per cent to 3.9 billion litres (1.03 billion US gallons), making it the fastest growing renewable energy source in 2007. Over 50 per cent is produced in Germany.

1.9.5 Wind power

Wind power is growing at the rate of 30 per cent annually, with a worldwide installed capacity of 198 gigawatts (GW) in 2010, and is widely used in

Europe, Asia, and the United States. Wind power accounts for approximately 19 per cent of electricity use in Denmark, 9 per cent in Spain and Portugal, and 6 per cent in Germany and the Republic of Ireland. The United States is an important growth area and installed US wind power capacity reached 25,170 MW at the end of 2008.

1.9.6 Solar power

The available solar energy resources are 3.8 YJ/year (1,20,000 TW). Less than 0.02 per cent of available resources are sufficient to entirely replace fossil fuels and nuclear power as an energy source. Assuming that our rate of usage in 2007 remains constant, estimated reserves are accurate, and no new unplanned reserves are found, we will run out of conventional oil in 2045, and coal in 2159. In practice, neither will actually run out as natural constraints will force production to decline as the remaining reserves dwindle. The rate at which demand increases and reserves dwindle has been increasing dramatically because the rate of consumption is not constant. For example, if demand for oil doubled, reserves would not last as long. In addition, the cost of fossil fuels continues to rise while solar power becomes more economically viable.

In 2007 grid-connected photovoltaic electricity was the fastest growing energy source, with installations of all photovoltaics increasing by 83 per cent in 2009 to bring the total installed capacity to 15 GW. Nearly half of the increase was in Germany, which is now the world's largest consumer of photovoltaic electricity (followed by Japan). Solar cell production increased by 50 per cent in 2007, to 3800 MW, and has been doubling every two years.

The consumption of solar hot water and solar space heating was estimated at 88 GW (gigawatts of thermal power) in 2006. The heating of water for unglazed swimming pools is excluded.

1.9.7 Geothermal

Geothermal energy is used commercially in over 70 countries. In the year 2006, 200 PJ (57 TWh) of electricity was generated from geothermal resources, and an additional 270 PJ of geothermal energy was used directly, mostly for space heating. In 2007, the world had a global capacity for 10 GW of electricity generation and an additional 28 GW of direct heating, including extraction by geothermal heat pumps. Heat pumps are small and widely distributed, so estimates of their total capacity are uncertain and range up to 100 GW.

1.9.8 Consumption by country

Energy consumption is loosely correlated with gross national product and climate, but there is a large difference even between the most highly developed countries, such as Japan and Germany with 6 kWh per person and United States with 11.4 kWh per person. In developing countries, particularly those that are subtropical or tropical such as India, the per person energy use is closer to 0.7 kWh. Bangladesh has the lowest consumption with 0.2 kWh per person.

The US consumes 25 per cent of the world's energy with a share of global GDP at 22 per cent and a share of the world population at 4.59 per cent. The most significant growth of energy consumption is currently taking place in China, which has been growing at 5.5 per cent per year over the last 25 years. Its population of 1.3 billion people (19.6 per cent of the world population) is consuming energy at a rate of 1.6 kWh per person.

One measurement of efficiency is energy intensity. This is a measure of the amount of energy it takes a country to produce a dollar of gross domestic product.

1.9.9 Consumption by sector

Industrial users (agriculture, mining, manufacturing, and construction) consume about 37 per cent of the total 15 TW. Personal and commercial transportation consumes 20 per cent; residential heating, lighting, and appliances use 11 per cent; and commercial uses (lighting, heating and cooling of commercial buildings, and provision of water and sewer services) amount to 5 per cent of the total.

The other 27 per cent of the world's energy is lost in energy transmission and generation. In 2007, global electricity consumption averaged 2 TW. The energy rate used to generate 2 TW of electricity is approximately 5 TW, as the efficiency of a typical existing power plant is around 38 per cent. The new generation of gas-fired plants reaches a substantially higher efficiency of 55 per cent. Coal is the most common fuel for the world's electricity plants.

Total world energy use per sector was in 2008 industry 28 per cent, transport 27 per cent and residential and service 36 per cent.

1.10 Renewable resources

Renewable resources are available each year, unlike non-renewable resources, which are eventually depleted. A simple comparison is a coal mine and a forest. While the forest could be depleted, if it is managed it represents a continuous supply of energy, vs. the coal mine, which once has been exhausted is gone. Most of earth's available energy resources are renewable resources. Renewable resources account for more than 93 per cent of total US energy reserves. Annual renewable resources were multiplied times thirty years for comparison with non-renewable resources. In other words, if all non-renewable resources were uniformly exhausted in 30 years, they would only account for 7 per cent of available resources each year, if all available renewable resources were developed.

1.10.1 Solar energy

Renewable energy sources are even larger than the traditional fossil fuels and in theory can easily supply the world's energy needs. 89 PW of solar power falls on the planet's surface. While it is not possible to capture all, or even most, of this energy, capturing less than 0.02 per cent would be enough to meet the current energy needs. Barriers to further solar generation include the high price of making solar cells and reliance on weather patterns to generate electricity. Also, current solar generation does not produce electricity at night, which is a particular problem in high northern and southern latitude countries; energy demand is highest in winter, while availability of solar energy is lowest. This could be overcome by buying power from countries closer to the equator during winter months, and may also be addressed with technological developments like Infrared Solar Panel systems, which can capture energy in cloudy conditions and even at night. Globally, solar generation is the fastest growing source of energy, seeing an annual average growth of 35 per cent over the past few years. Japan, Europe, China, US and India are the major growing investors in solar energy.

1.10.2 Wind power

The available wind energy estimates range from 300 TW to 870 TW. Using the lower estimate, just 5 per cent of the available wind energy would supply the current worldwide energy needs. Most of this wind energy is available over the open ocean. The oceans cover 71 per cent of the planet and wind tends to blow more strongly over open water because there are fewer obstructions.

1.10.3 Wave and tidal power

At the end of 2006, 0.3 GW of electricity was produced by tidal power. Due to the tidal forces created by the moon (68 per cent) and the sun (32 per cent), and the earth's relative rotation with respect to moon and sun, there are fluctuating tides. These tidal fluctuations result in dissipation at an average rate of about 3.7 TW.

Another physical limitation is the energy available in the tidal fluctuations of the oceans, which is about 0.6 EJ (exajoule). Note this is only a tiny fraction of the total rotational energy of the earth. Without forcing, this energy would be dissipated (at a dissipation rate of 3.7 TW) in about four semi-diurnal tide periods. So, dissipation plays a significant role in the tidal dynamics of the oceans. Therefore, this limits the available tidal energy to around 0.8 TW (20 per cent of the dissipation rate) in order not to disturb the tidal dynamics too much. Waves are derived from wind, which is in turn derived from solar energy, and at each conversion there is a drop of about two orders of magnitude in available energy. The total power of waves that wash against our shores add up to 3 TW.

1.10.4 Geothermal

Estimates of exploitable worldwide geothermal energy resources vary considerably, depending on assumed investments in technology and exploration and guesses about geological formations. According to a 2007 study, it was thought that this might amount to between 65 and 138 GW of electrical generation capacity 'using enhanced technology'. Other estimates range from 35 to 2000 GW of electrical generation capacity, with a further potential for 140 EJ/year of direct use.

A 2006 report by MIT that took into account the use of enhanced geothermal systems (EGS) concluded that it would be affordable to generate 100 GWe (gigawatts of electricity) or more by 2050, just in the United States, for a maximum investment of 1 billion US dollars in research and development over 15 years. The MIT report calculated the world's total EGS resources to be over 13 YJ, of which over 200 ZJ would be extractable, with the potential to increase this to over 2 YJ with technology improvements—sufficient to provide all the world's energy needs for several millennia. The total heat content of the earth is 1,30,00,000 YJ.

1.10.5 Biomass

Production of biomass and biofuels are growing industries as interest in sustainable fuel sources is growing. Utilising waste products avoids a food vs. fuel trade-off, and burning methane gas reduces greenhouse gas emissions, because even though it releases carbon dioxide, carbon dioxide is 23 times less of a greenhouse gas than is methane. Biofuels represent a sustainable partial replacement for fossil fuels, but their net impact on greenhouse gas emissions depends on the agricultural practices used to grow the plants used as feedstock to create the fuels. While it is widely believed that biofuels can be carbon-neutral, there is evidence that biofuels produced by current farming methods are substantial net carbon emitters. Geothermal and biomass are the only two renewable energy sources that require careful management to avoid local depletion.

1.10.6 Hydropower

In 2006, hydroelectric power supplied 16.4 per cent of world electricity, down from 21.0 per cent in 1973, but only 2.2 per cent of the world's energy.

1.10.7 Alternative energy paths

Denmark and Germany have started to make investments in solar energy, despite their unfavourable geographic locations. Germany is now the largest consumer of photovoltaic cells in the world. Denmark and Germany have installed 3 GW and 17 GW of wind power respectively. In 2006, wind generated 18.5 per cent of all the electricity in Denmark. Brazil invests in ethanol production from sugar cane, which is now a significant part of the transportation fuel in that country. Starting in 1965, France made large investments in nuclear power and to this date three quarters of its electricity comes from nuclear reactors. Switzerland is planning to cut its energy consumption by more than half to become a 2000-watt society by 2050 and the United Kingdom is working towards a zero energy building standard for all new housing by 2016.

1.11 Energy planning

Energy planning has a number of different meanings. However, one common meaning of the term is the process of developing long-range policies to help guide the future of a local, national, regional or even the global energy system. Energy planning is often conducted within governmental organisations but may also be carried out by large energy companies such as electric utilities or oil and gas producers. Energy planning may be carried out with input from different stakeholders drawn from government agencies, local utilities, academia and other interest groups. Energy planning is often conducted using integrated approaches that consider both the provision of energy supplies and the role of energy efficiency in reducing demands. Energy planning should always reflect the outcomes of population growth.

Energy planning has traditionally played a strong role in setting the framework for regulations in the energy sector (for example, influencing what type of power plants might be built or what prices were charged for fuels). But in the past two decades many countries have deregulated their energy systems so that the role of energy planning has been reduced, and decisions have increasingly been left to the market. This has arguably led to increased competition in the energy sector, although there is little evidence that this has translated into lower energy prices for consumers. Indeed in some cases, deregulation has led to significant concentrations of 'market power' with large very profitable companies having a large influence as price setters.

This trend now seems to be reversing as concerns grow over the environmental impacts of energy consumption and production, particularly in light of the threat of global climate change, which is caused largely by emissions of greenhouse gases from the world's energy systems.

Sustainable energy planning is particularly appropriate for communities who want to develop their own energy security, while employing best available practice in their planning processes.

1.12 Energy intensity

Energy intensity is a measure of the energy efficiency of a nation's economy. It is calculated as units of energy per unit of GDP.

1. High energy intensities indicate a high price or cost of converting energy into GDP.
2. Low energy intensity indicates a lower price or cost of converting energy into GDP.

Energy Intensity as defined here is not to be confused with energy use intensity (EUI), a measure of building energy use per unit area.

Many factors influence an economy's overall energy intensity. It may reflect requirements for general standards of living and weather conditions in an economy. It is not a typical for particularly cold or hot climates to require greater energy consumption in homes and workplaces for heating (furnaces or electric heaters) or cooling (air conditioning, fans, refrigeration). A country with an advanced standard of living is more likely to have a wider prevalence of such consumer goods and thereby be impacted in its energy intensity than one with a lower standard of living. Energy efficiency of appliances and buildings (through use of building materials and methods, such as insulation), fuel economy of vehicles, vehicular distances travelled (frequency of travel or larger geographical distances), better methods and patterns of transportation, capacities and utility of mass transit, energy rationing or conservation efforts, 'off-grid' energy sources, and stochastic economic shocks such as disruptions of energy due to natural disasters, wars, massive power outages, unexpected new sources, efficient uses of energy or energy subsidies may all impact overall energy intensity of a nation.

Thus, a nation that is highly economically productive, with mild and temperate weather, demographic patterns of work places close to home, and uses fuel efficient vehicles, supports carpools, mass transportation or walks or rides bicycles, will have a far lower energy intensity than a nation that is economically unproductive, with extreme weather conditions requiring heating and cooling, long commutes, and extensive use of generally poor fuel economy vehicles. Paradoxically, some activities that may seem to promote high energy intensities, such as long commutes, could in fact result in lower energy intensities by causing a disproportionate increase in GDP output. Figures of energy consumption used in statistics are energy sources marketed through major energy industries. Therefore some small scale but frequent consumption of energy source like firewood, charcoal peat, water wheel, and wind mill are not in its count. The countries that do not have such developed energy industries or people with highly self-energy efficient life style report smaller energy consumption figures. Energy intensity can be used as a comparative measure between countries; whereas the change in energy consumption required to raise GDP in a specific country over time is described as its energy elasticity.

1.13 Energy development

Energy development is the effort to provide sufficient primary energy sources and secondary energy forms for supply, cost, impact on air pollution and water pollution, mitigation of climate change with renewable energy.

Technologically advanced societies have become increasingly dependent on external energy sources for transportation, the production of many manufactured goods, and the delivery of energy services. This energy allows people who can afford the cost to live under otherwise unfavourable climatic conditions through the use of heating, ventilation, and/or air conditioning. Level of use of external energy sources differs across societies, as do the climate, convenience, levels of traffic congestion, pollution and availability of domestic energy sources. All terrestrial energy sources except nuclear, geothermal and tidal are from current solar insolation or from fossil remains of plant and animal life that relied directly and indirectly upon sunlight, respectively. Ultimately, solar energy itself is the result of the sun's nuclear fusion. Geothermal power from hot, hardened rock above the magma of the earth's core is the result of the decay of radioactive materials present beneath the earth's crust, and nuclear fission relies on man-made fission of heavy radioactive elements in the earth's crust; in both cases these elements were produced in supernova explosions before the formation of the solar system.

1.14 Fossil fuels

Fossil fuels sources burn coal or hydrocarbon fuels, which are the remains of the decomposition of plants and animals. There are three main types of fossil fuels: coal, petroleum, and natural gas. Another fossil fuel, liquefied petroleum gas (LPG), is principally derived from the production of natural gas. Heat from burning fossil fuel is used either directly for space heating and process heating or converted to mechanical energy for vehicles, industrial processes or electrical power generation.

Pros

1. The technology and infrastructure already exist for the use of the fossil fuels.
2. Petroleum energy density in terms of volume (cubic space) and mass (weight) is superior to some alternative energy sources [or energy storage devices, like a battery (electricity)].
3. Fossil fuels are currently more economical, and more suitable for decentralised energy use.

Cons

1. Petroleum-powered vehicles are very inefficient. Only about 30 per cent of the energy from the fuel they consume is converted into mechanical energy. The rest of the fuel-source energy is inefficiently expanded as waste heat. The heat and gaseous pollution emissions harm our environment.
2. The inefficient atmospheric combustion (burning) of fossil fuels in vehicles, buildings, and power plants contributes to urban heat islands.
3. The combustion of fossil fuels leads to the release of pollution into the atmosphere.
4. Dependence on fossil fuels from volatile regions or countries creates energy security risks for dependent countries.
5. Fossil fuels are non-renewable, unsustainable resources, which will eventually decline in production and become exhausted, with dire consequences to societies that remain highly dependent on them.
6. Extraction of fossil fuels results in extensive environmental degradation, such as the strip mining and mountaintop removal of coal.

1.14.1 Nuclear energy

Nuclear fission

Nuclear power stations use nuclear fission to generate energy by the reaction of uranium-235 inside a nuclear reactor. The reactor uses uranium rods, the atoms of which are split in the process of fission, releasing a large amount of energy. The process continues as a chain reaction with other nuclei. The energy heats water to create steam, which spins a turbine generator, producing electricity.

Nuclear fusion

Fusion power could solve many of the problems of fission power (the technology mentioned above) but, despite research having started in the 1950s, no commercial fusion reactor is expected before 2050. Many technical problems remain unsolved. Proposed fusion reactors commonly use deuterium, an isotope of hydrogen, as fuel and in most current designs also lithium. Assuming a fusion energy output equal to the current global output and that this does not increase in the future, then the known current lithium reserves would last 3000 years, lithium from sea water would last 60 million years, and a more complicated fusion process using only deuterium from sea water would have fuel for 150 billion years.

Fuel cell

A fuel cell is a device that converts the chemical energy from a fuel into electricity through a chemical reaction with oxygen or another oxidising agent. Hydrogen is the most common fuel, but hydrocarbons such as natural gas and alcohols like methanol are sometimes used. Fuel cells are different from batteries in that they require a constant source of fuel and oxygen to run, but they can produce electricity continually for as long as these inputs are supplied.

Welsh Physicist William Grove developed the first crude fuel cells in 1839. The first commercial use of fuel cells was in NASA space programs to generate power for probes, satellites and space capsules. Since then, fuel cells have been used in many other applications. Fuel cells are used for primary and backup power for commercial, industrial and residential buildings and in remote or inaccessible areas. They are used to power fuel cell vehicles, including automobiles, buses, forklifts, airplanes, boats, motorcycles and submarines.

There are many types of fuel cells, but they all consist of an anode (negative side), a cathode (positive side) and an electrolyte that allows charges to move between the two sides of the fuel cell. Electrons are drawn from the anode to the cathode through an external circuit, producing direct current electricity. As the main difference among fuel cell types is the electrolyte, fuel cells are classified by the type of electrolyte they use.

Fuel cells come in a variety of sizes. Individual fuel cells produce very small amounts of electricity, about 0.7 volts, so cells are 'stacked' or placed in series or parallel circuits, to increase the voltage and current output to meet an application's power generation requirements. In addition to electricity, fuel cells produce water, heat and, depending on the fuel source, very small amounts of nitrogen dioxide and other emissions. The energy efficiency of a fuel cell is generally between 40–60 per cent or up to 85 per cent efficient if waste heat is captured for use.

Fuel cells come in many varieties; however, they all work in the same general manner. They are made up of three segments which are sandwiched together: the anode, the electrolyte, and the cathode. Two chemical reactions occur at the interfaces of the three different segments. The net result of the two reactions is that fuel is consumed, water or carbon dioxide is created, and an electric current is created, which can be used to power electrical devices, normally referred to as the load. Figure 1.1 shows a block diagram of a fuel cell.

At the anode a catalyst oxidises the fuel, usually hydrogen, turning the fuel into a positively charged ion and a negatively charged electron. The electrolyte is a substance specifically designed so ions can pass through it, but the electrons cannot. The freed electrons travel through a wire creating the electric current. The ions travel through the electrolyte to the cathode. Once reaching the cathode, the ions are reunited with the electrons and the two react with a third chemical, usually oxygen, to create water or carbon dioxide.

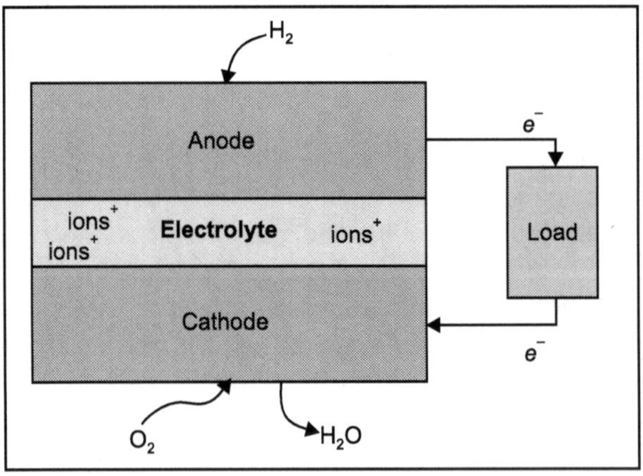

Figure 1.1 A block diagram of a fuel cell

A typical fuel cell produces a voltage from 0.6 V to 0.7 V at full rated load. Voltage decreases as current increases, due to several factors:

1. Activation loss.

2. Ohmic loss (voltage drop due to resistance of the cell components and interconnects).
3. Mass transport loss (depletion of reactants at catalyst sites under high loads, causing rapid loss of voltage).

To deliver the desired amount of energy, the fuel cells can be combined in series and parallel circuits, where series yields higher voltage, and parallel allows a higher current to be supplied. Such a design is called a fuel cell stack. The cell surface area can be increased, to allow stronger current from each cell.

1.15 Renewable sources

Renewable energy is energy which comes from natural resources such as sunlight, wind, rain, tides, and geothermal heat, which are renewable (naturally replenished). Renewable energy is an alternative to fossil fuels and nuclear power, and was commonly called alternative energy in the 1970s and 1980s.

1.15.1 Biomass, biofuels and vegetable oil

Biomass production involves using garbage or other renewable resources such as corn or other vegetation to generate electricity. When garbage decomposes, the methane produced is captured in pipes and later burned to produce electricity. Vegetation and wood can be burned directly to generate energy, like fossil fuels or processed to form alcohols.

Vegetable oil is generated from sunlight, H_2O, and CO_2 by plants. It is safer to use and store than gasoline or diesel as it has a higher flash point. Straight vegetable oil works in diesel engines if it is heated first. Vegetable oil can also be transesterified to make biodiesel, which burns like normal diesel.

Pros

1. Biomass production can be used to burn organic by-products resulting from agriculture.
2. Biomass is abundant on earth and is generally renewable. In theory, we will never run. Biomass is found throughout the world, a fact that should alleviate energy pressures in third world nations.
3. When methods of biomass production other than direct combustion of plant mass are used, such as fermentation and pyrolysis, there is little effect on the environment. Alcohols and other fuels produced by these alternative methods are clean burning and are feasible replacements to fossil fuels.

4. Since CO_2 is first taken out of the atmosphere to make the vegetable oil and then put back after it is burned in the engine, there is no net increase in CO_2.
5. By combining the use of biomass with geo-sequestration of CO_2 then this could result in a net decrease of CO_2 in the atmosphere.
6. Vegetable oil has a higher flash point and therefore is safer than most fossil fuels.
7. Transitioning to vegetable oil could be relatively easy as biodiesel works where diesel works, and straight vegetable oil takes relatively minor modifications.
8. Infrastructure for biodiesel around the World is significant and growing.

Cons

1. Direct combustion of any carbon-based fuel leads to air pollution similar to that from fossil fuels.
2. Direct competition with land use for food production and water use. As this decreases food supply, the price of food increases worldwide.
3. Current production methods would require enormous amounts of land to replace all gasoline and diesel.

1.15.2 Geothermal energy

Geothermal energy harnesses the heat energy present underneath the earth. Two wells are drilled. One well injects water into the ground to provide water. The hot rocks heat the water to produce steam. The steam that shoots back up the other hole(s) is purified and is used to drive turbines, which power electric generators. When the water temperature is below the boiling point of water a binary system is used. A low boiling point liquid is used to drive a turbine and generator in a closed system similar to a refrigeration unit running in reverse.

Pros

1. Geothermal energy is base load power.
2. Economically feasible in high grade areas now.
3. Low deployment costs.
4. Geothermal power plants have a high capacity factor; they run continuously day and night with an uptime typically exceeding 95 per cent.
5. Once a geothermal power station is implemented, the energy produced from the station is practically free, minus maintenance costs. A small

amount of energy is required in order to run a pump, although this pump can be powered by excess energy generated at the plant.

6. Geothermal power stations are relatively small, and have a lesser impact on the environment than tidal or hydroelectric plants. Because geothermal technology does not rely on large bodies of water, but rather, small, but powerful jets of water, like geysers, large generating stations can be avoided without losing functionality.
7. Geothermal is now feasible in areas where the earth's crust is thicker. Using enhanced geothermal technology, it is possible to drill deeper and to inject water to generate geothermal power.
8. Geothermal energy does not produce air or water pollution if performed correctly.

Cons

1. Geothermal power extracts small amounts of minerals such as sulphur that are removed prior to feeding the turbine and re-injecting the water back into the injection well.
2. Geothermal power requires locations that have suitable subterranean temperatures within 5 km of surface.
3. Some geothermal stations have created geological instability, even causing earthquakes strong enough to damage buildings.

1.15.3 Hydroelectric energy

Energy in water can be harnessed and used. Since water is about 800 times denser than air, even a slow flowing stream of water or moderate sea swell, can yield considerable amounts of energy. In hydro energy, the gravitational descent of a river is compressed from a long run to a single location with a dam or a flume. This creates a location where concentrated pressure and flow can be used to turn turbines or water wheels, which drive a mechanical mill or an electric generator.

Pros

1. Hydroelectric power stations can promptly increase to full capacity, unlike other types of power stations. This is because water can be accumulated above the dam and released to coincide with peak demand.
2. Electricity can be generated constantly, so long as sufficient water is available.

3. Hydroelectric power produces no primary waste or pollution.
4. Hydropower is a renewable resource.
5. Hydroelectricity assists in securing a country's access to energy supplies.
6. Much hydroelectric capacity is still undeveloped, such as in Africa.

Cons

1. The construction of a dam can have a serious environmental impact on the surrounding areas. The amount and the quality of water downstream can be affected, which affects plant life both aquatic, and land-based. Because a river valley is being flooded, the local habitat of many species is destroyed, while people living nearby may have to relocate their homes.
2. Hydroelectricity can only be used in areas where there is a sufficient and continuing supply of water.
3. Flooding submerges large forests (if they have not been harvested). The resulting anaerobic decomposition of the carboniferous materials releases methane, a greenhouse gas.
4. Dams can contain huge amounts of water. As with every energy storage system, failure of containment can lead to catastrophic results, e.g. flooding.
5. Dams create large lakes that may have adverse effects on earth tectonic system causing intense earthquakes.
6. Hydroelectric plants rarely can be erected near load centers, requiring long transmission lines.

1.15.4 Solar power

Solar power involves using solar cells to convert sunlight into electricity, using sunlight hitting solar thermal panels to convert sunlight to heat water or air, using sunlight hitting a parabolic mirror to heat water (producing steam) or using sunlight entering windows for passive solar heating of a building. It would be advantageous to place solar panels in the regions of highest solar radiation.

Pros

1. Solar power imparts no fuel costs.
2. Solar power is a renewable resource. As long as the sun exists, its energy will reach earth.

3. Solar power generation releases no water or air pollution, because there is no combustion of fuels.
4. In sunny countries, solar power can be used in remote locations, like a wind turbine. This way, isolated places can receive electricity, when there is no way to connect to the power lines from a plant.
5. Solar energy can be used very efficiently for heating (solar ovens, solar water and home heaters) and day lighting.
6. Coincidentally, solar energy is abundant in regions that have the largest number of people living off grid—in developing regions of Africa, Indian subcontinent and Latin America. Hence cheap solar, when available, opens the opportunity to enhance global electricity access considerably, and possibly in a relatively short time period.
7. Passive solar building design and zero energy buildings are demonstrating significant energy bill reduction, and some are cost-effectively off the grid.
8. Photovoltaic equipment cost has been steadily falling, the production capacity is rapidly rising.
9. Distributed point-of-use photovoltaic systems eliminate expensive long-distance electric power transmission losses.
10. Photovoltaics are much more efficient in their conversion of solar energy to usable energy than biofuel from plant materials.

Cons

1. Solar electricity is currently more expensive than grid electricity.
2. Solar heat and electricity are not available at night and may be unavailable because of weather conditions; therefore, a storage or complementary power system is required for off-the-grid applications.
3. Solar cells produce DC which must be converted to AC (using a grid tie inverter) when used in currently existing distribution grids. This incurs an energy loss of 4–12 per cent.
4. The energy payback time—the time necessary for producing the same amount of energy as needed for building the power device—for photovoltaic cells is about 1–5 years, depending primarily on location.

1.15.5 Tidal power generation

Tidal power can be extracted from moon-gravity-powered tides by locating a water turbine in a tidal current or by building impoundment pond dams that admit-or-release water through a turbine. The turbine can turn an electrical generator or a gas compressor, that can then store energy until needed. Coastal tides are a source of clean, free, renewable, and sustainable energy.

Pros

1. Tidal power is free once the dam is built. This is because tidal power harnesses the natural power of tides and does not consume fuel. In addition, the maintenance costs associated with running a tidal station are relatively inexpensive.
2. Tides are very reliable because it is easy to predict when high and low tides will occur. The tide goes in and out twice a day usually at the predicted times. This makes tidal energy easy to maintain, and positive and negative spikes in energy can be managed.
3. Tidal energy is renewable, because nothing is consumed in the rising of tides. Tidal power relies on the gravitational pull of the moon and sun, which pull the sea backwards and forwards, generating tides.

Cons

1. Tidal power is not currently economically feasible, because the initial costs of building a dam are tremendous. Furthermore, it only provides power for around 10 hours each day, when the tide is moving in or out of the basin.
2. The barrage construction can affect the transportation system in water. Boats may not be able to cross the barrage, and commercial ships, used for transport or fishery, need to find alternative routes or costly systems to go through the barrage.
3. The erection of a barrage may affect the aquatic ecosystems surrounding it. The environment affected by the dam is very wide, altering areas numerous miles upstream and downstream. For example, many birds rely on low tides to unearth mud flats, which are used as feeding areas.
4. Maximum energy production is limited to 2.5 terawatts. This is the total amount of tidal dissipation or the friction measured by the slowing of the lunar orbit.

1.15.6 Wind power

This type of energy harnesses the power of the wind to propel the blades of wind turbines. These turbines cause the rotation of magnets, which creates electricity. Wind towers are usually built together on wind farms.

Pros

1. Wind power produces no water or air pollution that can contaminate the environment, because there are no chemical processes involved in wind power generation. Hence, there are no waste by-products, such as carbon dioxide.
2. Power from the wind does not contribute to global warming because it does not generate greenhouse gases.
3. Wind generation is a renewable source of energy, which means that we will never run out of it.
4. Wind towers can be beneficial for people living permanently or temporarily, in remote areas. It may be difficult to transport electricity through wires from a power plant to a faraway location and thus, wind towers can be set up at the remote setting.
5. Farming and grazing can still take place on land occupied by wind turbines.
6. Those utilising wind power in a grid-tie configuration will have backup power in the event of a power outage.
7. Because of the ability of wind turbines to coexist within agricultural fields, siting costs are frequently low.

Cons

1. Wind is unpredictable; therefore, wind power is not predictably available. When the wind speed decreases less electricity is generated. This makes wind power unsuitable for base load generation.
2. Wind farms may be challenged in communities that consider them an eyesore or obstruction.
3. Wind farms, depending on the location and type of turbine, may negatively affect bird migration patterns, and may pose a danger to the birds themselves (primarily an issue with older/smaller turbines).
4. Windfarms may interfere with radar creating a hole in radar coverage and so affect national security.
5. Tall wind turbines have been proven to impact Doppler radar towers and affect weather forecasting in a negative way. This can be prevented by not having the wind turbines in the radar's line of sight.

1.15.7 Photovoltaics

Photovoltaics (PV) is a method of generating electrical power by converting solar radiation into direct current electricity using semiconductors that exhibit the photovoltaic effect. Photovoltaic power generation employs solar panels composed of a number of solar cells containing a photovoltaic material. Materials presently used for photovoltaics include monocrystalline silicon, polycrystalline silicon, amorphous silicon, cadmium telluride, and copper indium gallium selenide/sulphide. Due to the growing demand for renewable energy sources, the manufacturing of solar cells and photovoltaic arrays has advanced considerably in recent years.

Photovoltaics are best known as a method for generating electric power by using solar cells to convert energy from the sun into a flow of electrons. The photovoltaic effect refers to photons of light exciting electrons into a higher state of energy, allowing them to act as charge carriers for an electric current. The term photovoltaic denotes the unbiased operating mode of a photodiode in which current through the device is entirely due to the transduced light energy. Virtually all photovoltaic devices are some type of photodiode.

Solar cells produce direct current electricity from sun light, which can be used to power equipment or to recharge a battery. The first practical application of photovoltaics was to power orbiting satellites and other spacecraft, but today the majority of photovoltaic modules are used for grid connected power generation. In this case an inverter is required to convert the DC to AC. There is a smaller market for off-grid power for remote dwellings, boats, recreational vehicles, electric cars, roadside emergency telephones, remote sensing, and cathodic protection of pipelines.

In buildings: Photovoltaic arrays are often associated with buildings: either integrated into them, mounted on them or mounted nearby on the ground. Building-integrated photovoltaics (BIPV) are increasingly incorporated into new domestic and industrial buildings as a principal or ancillary source of electrical power.

In transport: PV has traditionally been used for electric power in space. PV is rarely used to provide motive power in transport applications, but is being used increasingly to provide auxiliary power in boats and cars.

Rural electrification: Unlike the past decade, which saw solar solutions purchased mainly by international donors, it is now the locals who are increasingly opening their wallets to make the switch from their traditional energy means. That is because solar products prices in recent years have declined to become cheaper than kerosene and batteries. Developing countries where many villages are often more than five kilometres away from grid power are increasingly using photovoltaics. In remote locations in India a rural lighting program has been providing solar powered LED lighting to replace kerosene lamps.

Advantages and disadvantages of solar photovoltaics

Advantages

Solar power is pollution-free during use. Production end-wastes and emissions are manageable using existing pollution controls. End-of-use recycling technologies are under development and policies are being produced that encourage recycling from producers.

PV installations can operate for many years with little maintenance or intervention after their initial set-up, so after the initial capital cost of building any solar power plant, operating costs are extremely low compared to existing power technologies. As of 2010, the price of PV modules per MW has fallen by 60 per cent since the summer of 2008, according to Bloomberg New Energy Finance estimates, putting solar power for the first time on a competitive footing with the retail price of electricity in a number of sunny countries. There has been fierce competition in the supply chain, and further improvements in the levelised cost of energy for solar lie ahead, posing a growing threat to the dominance of fossil fuel generation sources in the next few years.

When grid-connected, solar electric generation replaces some or all of the highest-cost electricity used during times of peak demand (in most climatic regions). This can reduce grid loading, and can eliminate the need for local battery power to provide for use in times of darkness. These features are enabled by net metering. Time-of-use net metering can be highly favourable, but requires newer electronic metering, which may still be impractical for some users.

Disadvantages

Photovoltaics are costly to install. Solar electricity is not produced at night and is greatly reduced in cloudy conditions requiring alternate sources of power. While many buildings with photovoltaic arrays are tied into the power grid which absorbs any excess electricity generated throughout the day and provides electricity in the evening, such systems use a grid tie inverter to convert direct current (DC) alternating current (AC) incurring an energy loss of 4–12 per cent. Off-grid systems use either storage batteries which also incur significant energy losses and require regular maintenance or engine-generators which consume costly fuel.

Solar electricity production depends on the limited power density of the location's insolation. Average daily output of a flat plate collector at latitude tilt in the contiguous US is 3–7 kilowatt·h/m^2/day and on average lower in Europe.

Solar electricity is more expensive than most other forms of small-scale alternative energy production. Without governments mandating 'feed-in-tariffs' for green solar energy, solar PV is less affordable to homeowners than solar hot water or solar space heating.

Photovoltaic panels are specifically excluded in Europe from RoHS (Restriction on Hazardous Substances) since 2003 and were again excluded in 2011. California has largely adopted the RoHS standard through EWRA. Therefore, PV panels may legally in Europe and California contain lead, mercury and cadmium which are forbidden or restricted in all other electronics. Much of the investment in a home-mounted system may be lost if the homeowner moves and the buyer puts less value on the system than the seller.

1.16 Environmental impacts of hydrogen-based energy systems

There is increasing interest in the role that hydrogen-based energy systems may play in the future, especially in the transport sector. They appear to be an attractive alternative to current fossil fuel-based energy systems in the future, since these have been proven to affect climate due to greenhouse gases emissions. Nevertheless, any future hydrogen-based economy would need to assess the possible global environmental impacts of such alternative energy production.

British scientists have recently reviewed current understanding of the fate and behaviour of hydrogen in the atmosphere and characterised its major sources and sinks. They show that, contrary to most expectations, hydrogen is an indirect greenhouse gas with a potential global warming effect. They then quantified the global warming potential of hydrogen in comparison to CO_2 and the global warming consequences of replacing the current fossil fuel-based economy with one based on hydrogen.

The results suggest that because hydrogen reacts in the atmosphere with tropospheric OH radicals, the emission of hydrogen into the atmosphere would disrupt the distribution of methane and ozone, the second and third most important greenhouse gases. Emissions of hydrogen lead to increased burdens of methane and ozone and hence to an increase in global warming. Therefore, hydrogen can be considered as an indirect greenhouse gas with the potential to increase global warming.

The scientists have estimated that the potential effects on climate from hydrogen-based energy systems would be much lower than those from fossil fuel-based energy systems. However, such impacts will depend on the rate of hydrogen leakage during its synthesis, storage and use. The researchers have calculated that a global hydrogen economy with a leakage rate of 1 per cent of the produced hydrogen would produce a climate impact of 0.6 per cent of the fossil fuel system it replaces. If the leakage rate was 10 per cent, then the climate impact would be 6 per cent of that of the fossil fuel system.

The current study suggests that a future hydrogen-based economy would not be free from climate disturbance, although this may be considerably less pronounced that caused by the current fossil fuel energy systems. Careful attention would have to be paid to reducing hydrogen leakage to a minimum if the potential climate benefits of a future global hydrogen economy are to be realised.

2

Solar radiations

2.1 Introduction

Solar radiations are becoming increasingly appreciated because of their influence on living matter and the feasibility of its application for useful purposes. It is a perpetual source of natural energy that, along with other forms of renewable energy, has a great potential for a wide variety of applications because it is abundant and accessible. Solar radiation is rapidly gaining ground as a supplement to the nonrenewable sources of energy, which have a finite supply. The electromagnetic radiation emitted by the sun covers a very large range of wavelengths, from radiowaves through the infrared, visible and ultraviolet to X-rays and gamma rays. However, 99 per cent of the energy of solar radiation is contained in the wavelength band from 0.15 to 4 μm, comprising the near ultraviolet, visible and near infrared regions of the solar spectrum, with a maximum at about 0.5 μm. About 40 per cent of the solar radiation received at the earth's surface on clear days is visible radiation within the spectral range 0.4 to 0.7 μm, while 51 per cent is infrared radiation in the spectral region 0.7 to 4 μm.

The total radiation emitted by the sun in unit time remains practically constant. The variations actually observed in association with solar phenomena like sunspots, prominences and solar flares are mainly confined to the extreme ultraviolet end of the solar spectrum and to the radiowaves. The contribution of these variations to the total energy emitted is extremely small and can be neglected in solar energy applications. The planet earth revolves around the sun in an elliptical orbit of very small eccentricity with the sun at one of the foci, completing one revolution in one year. The axis of rotation of the earth is inclined at about 23½ degrees with respect to the plane of orbital revolution and is directed always to a fixed point in space. As a consequence of this geometry of the sun and the earth, large seasonal variations occur in the amount of solar radiation received at different latitudes of the earth. The largest annual variations occur near the two poles and the smallest near the equator. During the course of its annual motion around the sun in an elliptical orbit, the earth comes nearest to the sun each year around January 5 (perihelion) and farthest around July 5 (aphelion). The sun-earth distance at perihelion is

1.471×10^8 km and at aphelion 1.521×10^8 km. The mean distance is 1.496×10^8 km, which is known as 1 Astronomical Unit. Due to the variations in the sun-earth distance, the solar radiation intercepted by the earth varies by ±3.3 per cent around the mean value, being maximum at the beginning of January and minimum at the beginning of July.

Of the entire quantity of radiant energy emitted by the sun's spherical surface, only a small fraction (4.5×10^{-10}) is actually intercepted by the planet earth. The amount of solar energy falling in unit time on unit area, held normal to the sun's rays outside the earth's atmosphere when the earth is at the mean distance from the sun, is called the solar constant. According to the latest measurements, the solar constant has a value of 136 mW/cm^2 or 1.36 kW/m^2 or 1.95 calories/cm^2 per minute.

2.2 Interaction of the sun's radiation with the earth's atmosphere

Since the earth is surrounded by an atmosphere which contains various gaseous constituents, suspended dust and other minute solid and liquid particulate matter and clouds of various types, marked depletion of solar energy takes place during its passage through the atmosphere to the surface of the earth. In a cloud-free atmosphere, the depletion occurs by three distinct physical processes operating simultaneously: (i) selective absorption by molecular oxygen, ozone, carbon dioxide, and water vapour in certain specific wavelengths, (ii) Rayleigh scattering by molecules of the different gases that constitute the atmosphere, and (iii) Mie scattering. In Rayleigh scattering, where the size of the scattering particles is small compared with the wavelength of light, the intensity of scattering follows the well-known λ–4 law, as a consequence of which a spatial redistribution of the energy of the incoming radiation takes place in the scattered light, the shorter wavelengths predominating and giving rise to the blue of the sky.

Roughly one half of the scattered radiation is lost to space and the remaining half is directed downwards to the earth's surface from different directions as diffuse radiation. In Mie scattering where the sizes of the scattering particles, mostly dust, are comparable with the wavelength of visible and infrared radiation, a depletion in the solar radiation occurs both by true scattering (involving a redistribution of incident energy) as well as by absorption by the particles, wherein a part of the radiant energy is transformed into heat. This type of scattering also leads to a fraction of the solar radiation being lost to space and another fraction being directed downwards as diffuse radiation. Unlike Rayleigh scattering, however, the dust-scattering process is more complicated and the redistribution of energy is much more asymmetrical with respect to the different directions when compared to Rayleigh scattering. Because of absorption by oxygen and ozone at high levels of the atmosphere, the short-wavelength limit of solar radiation received at the earth's surface is approximately 0.29 μ.

In a cloudy atmosphere, considerable depletion of the direct solar radiation takes place. A large fraction is reflected back to space from the tops of clouds, another part transmitted downwards to the earth as diffuse radiation and a small fraction absorbed by the clouds. Dense dark clouds of appreciable vertical depth can cut off as much as 80 per cent of the incident energy, but thinner clouds deplete only 20–50 per cent of the radiant energy, depending on their depth and liquid water content. The effect of an increase in dust in the atmosphere is to decrease the direct solar radiation and increase the diffuse radiation. When the cloud amount increases from 0 to about 4 or 5 octas of the sky, there is a corresponding increase in diffuse radiation. However, when the cloud amount increases still further to overcast conditions, a decrease in diffuse radiation takes place, largely because of increased absorption and reflection from cloud tops. The fraction of the total solar radiant energy reflected back to space through reflection from the tops of clouds, scattering by atmospheric gases and dust particles and by reflection at the earth's surface is called the albedo of the earth-atmosphere system and has a value of about 0.30 for the earth as a whole. The mean monthly value of the intensity of direct solar radiation normal to the solar beam actually received at the earth's surface at noon time in India varies from 0.51 to 1.05 kW/m^2, depending on latitude, altitude of the station, and season.

The solar radiant energy in the shortwave spectrum falling on the earth's surface as well as that absorbed by the atmosphere is re-emitted back into space as longwave radiation.

This longwave radiation has the characteristics of blackbody radiation in the temperature range of about –60° to +30°C. Since, in the long run, the earth's mean surface temperature as well as the mean temperature of the atmosphere remains unchanged, it follows that the net heat absorbed by the earth–atmosphere system from the shortwave solar radiation is equal to the net heat emitted as longwave radiation by the planet earth and its atmosphere.

2.3 Terminology of radiation parameters

For practical purposes it is convenient to divide the entire radiation regime within the earth's atmosphere into two parts, the solar or shortwave radiation and the terrestrial or longwave radiation. This is possible because solar radiation corresponds approximately to blackbody radiation at about 6000°K, the energy of which is almost entirely (99 per cent) confined to wavelengths less than 4 μ, while the radiation emitted by the earth and its atmosphere, which corresponds approximately to blackbody radiation at temperatures less than 300°K, has, for all practical purposes, its energy between the wavelength limits 4–100 μm. Thus, a division of the spectrum at about 4 μm effectively separates the solar and terrestrial radiation as shortwave and longwave radiation. The different radiation parameters are defined below:

1. Direct solar radiation at normal incidence, IN, is the quantity of shortwave solar radiant energy emitted by the solid angle subtended by the visible disc of the sun and passing through a unit area held normal to the solar beam at the earth's surface in unit time.
2. Global solar radiation, G, is the total quantity of shortwave radiant energy emitted by the sun's disc as well as that scattered diffusively by the atmosphere and clouds passing through a unit area in the horizontal in unit time. The global solar radiation is also referred to as incoming total shortwave radiation, K↓.
3. Direct solar radiation on a horizontal surface, IH, is the quantity of solar radiant energy emitted from the solid angle subtended by the visible disc of the sun and passing through a unit area in the horizontal in unit time. This is also called the vertical component of the direct solar radiation.
4. Diffuse solar radiation, D, is that part of shortwave radiation scattered by the atmosphere reflected diffusely and transmitted by clouds and passing through unit horizontal area in unit time. This radiation comes from a solid angle of 2π with the exception of the solid angle subtended by the sun's disc. There is a simple relation between G, D and IH and IN, given by:

$$I_H = (G - D) \text{ and } I_N = \frac{(G - D)}{\sin h}$$

   where, h is the angle of elevation of the sun.
5. Reflected solar radiation, K↑, is that part of the incoming shortwave radiation which is reflected by the earth's surface and diffused by the atmospheric layer between the ground and the point of observation. The albedo of the earth's surface, A, is expressed as the ratio of the reflected solar radiation to the global solar radiation,

$$A = \frac{K\uparrow}{G}.$$

6. Upward longwave radiation, L↑, is the quantity of longwave radiation emitted upwards by the earth's surface and the shallow layer of atmosphere below the level of the measuring instrument, and passing through unit area in the horizontal in unit time.

7. Downward longwave radiation, L↓, is the longwave radiation emitted downwards by the earth's atmosphere and passing through a unit horizontal area at the level of the instrument in unit time.
8. Net longwave radiation, L*, is the difference between the outgoing and incoming longwave radiation and is given by L* = L↑ – L↓.
9. Net radiation, Q*, is the difference between the total incoming short and longwave radiation and the outgoing short and longwave radiation and is given by Q* = K↓ – K↑ + L↓ – L↑.

Other radiation parameters of interest are direct, global or diffuse solar radiation in restricted portions of the solar spectrum and various atmospheric turbidity factors.

2.3.1 Latitude

The latitude of a location on the earth is the angular distance of that location south or north of the equator. The latitude is an angle, and is usually measured in degrees (marked with °). The equator has a latitude of 0°, the North pole has a latitude of 90° north (written 90°N or +90°), and the South pole has a latitude of 90° south (written 90°S or –90°). Together, latitude and longitude can be used as a geographic coordinate system to specify any location on the globe. Curves of constant latitude on the earth (running east-west) are referred to as lines of latitude or parallels. Each line of latitude is actually a circle on the earth parallel to the equator, and for this reason lines of latitude are also known as circles of latitude. In spherical geometry, lines of latitude are examples of circles of a sphere, with the equator being a great circle.

Latitude [usually denoted by the Greek letter phi (φ)] is often measured in degrees, with minutes and seconds for finer measurements. For example, the Eiffel Tower has a latitude of 48° 51′ 29″ N—that is, 48 degrees plus 51 minutes plus 29 seconds. Alternatively, latitude may be measured entirely in degrees, e.g. 48.85806° N. Besides the equator, four other lines of latitude are commonly used to mark maps of the earth. Each of these lines plays an important role in the geometrical relationship between the earth and the sun:

1. Arctic circle: 66° 33′ 39″ N.
2. Tropic of Cancer: 23° 26′ 21″ N.
3. Equator: 0° Latitude.
4. Tropic of Capricorn: 23° 26′ 21″ S.
5. Antarctic Circle: 66° 33′ 39″ S.

Effect of latitude: A region's latitude has a great effect on its climate and weather. Latitude more loosely determines prevailing winds, and other physical characteristics of geographic locations. Auroral activity is common at high geomagnetic latitudes. Some observers have noted a correlation between the latitude of a country and its economic activity.

2.3.2 Declination

Declination is one of the two coordinates of the equatorial coordinate system, the other being either right ascension or hour angle. Declination in astronomy is comparable to geographic latitude, but projected onto the celestial sphere. Declination is measured in degrees north and south of the celestial equator. Points north of the celestial equator have positive declinations, while those to the south have negative declinations (Fig. 2.1).

1. An object on the celestial equator has a declination of 0°.
2. An object at the celestial north pole has a declination of +90°.
3. An object at the celestial south pole has a declination of –90°.

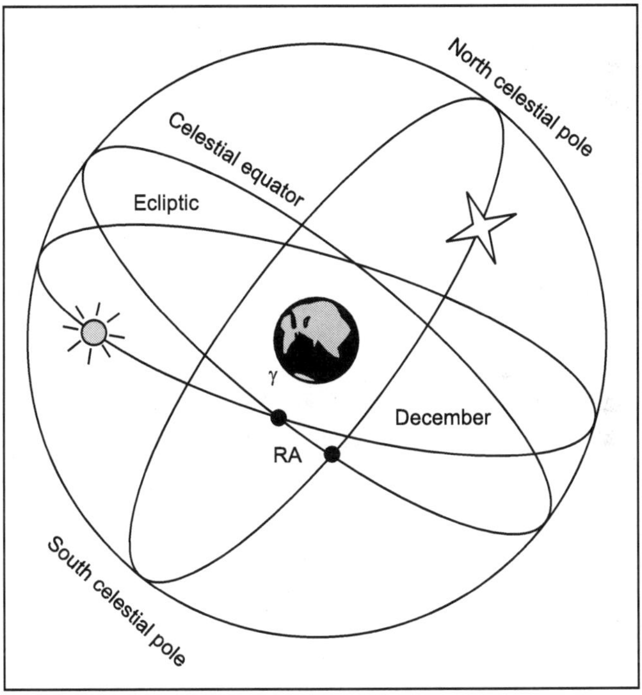

Figure 2.1 Declination

The sign is customarily included even if it is positive. Any unit of angle can be used for declination, but it is often expressed in degrees, minutes, and seconds of arc.

A celestial object directly overhead (at the zenith) has a declination very nearly equal to the observer's latitude. A pole star therefore has the declination near to +90° or –90°. At northern latitudes φ (φ = observer's latitude) >0, celestial objects with a declination greater than 90° – φ are always visible. Such stars are called circumpolar stars, while the phenomenon of the sun not setting is called midnight sun.

When an object is directly overhead its declination is almost always within 0.01 degree of the observer's latitude; it would be exactly equal except for two complications. The first complication applies to all celestial objects: the object's declination equals the observer's astronomic latitude, but the term 'latitude' ordinarily means geodetic latitude, which is the latitude on maps and GPS devices. The difference usually doesn't exceed a few thousandths of a degree, but in a few places (such as the big island of Hawaii) it can exceed 0.01 degree.

For practical purposes the second complication only applies to solar system objects: 'declination' is ordinarily measured at the center of the earth, which is not quite spherical, so a line from the center of the earth to the object is not quite perpendicular to the earth's surface.

It turns out that when the moon is directly overhead its geocentric declination can differ from the observer's astronomic latitude by up to 0.005 degree. The importance of this complication is inversely proportional to the object's distance from the earth, so for most practical purposes it is only a concern with the moon.

The declination of the sun, dO·, is the angle between the rays of the sun and the plane of the earth's equator. The earth's axial tilt (called the obliquity of the ecliptic by astronomers) is the angle between the earth's axis and a line perpendicular to the earth's orbit. The earth's axial tilt changes gradually over thousands of years, but its current value is about ε = 23°26′. Because this axial tilt is nearly constant, solar declination (dO·) varies with the seasons and its period is one year.

2.3.3 Azimuth

An azimuth is an angular measurement in a spherical coordinate system. The vector from an observer (origin) to a point of interest is projected perpendicularly onto a reference plane; the angle between the projected vector and a reference vector on the reference plane is called the azimuth (Fig. 2.2).

An example of an azimuth is the measurement of the position of a star in the sky. The star is the point of interest, the reference plane is the horizon or the surface of the sea, and the reference vector points to the north. The azimuth is the angle between the north point and the perpendicular projection of the star down onto the horizon. Azimuth is usually measured in degrees (°). The concept is used in many practical applications including navigation, astronomy, engineering, mapping, mining and artillery.

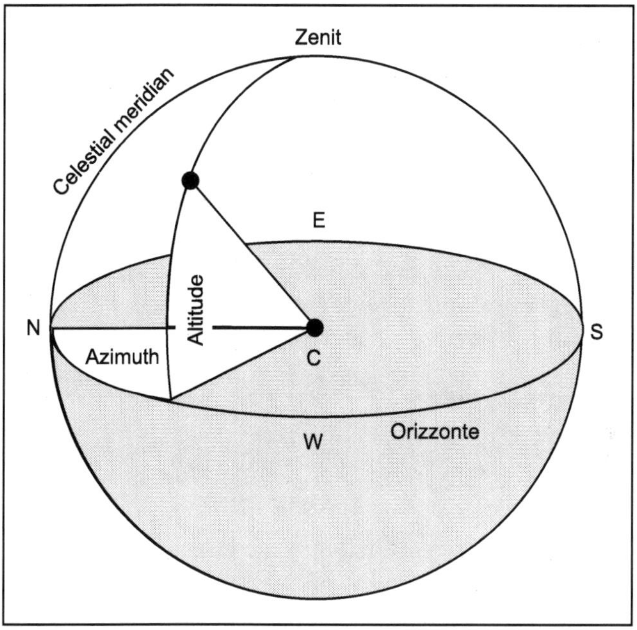

Figure 2.2 The azimuth is the angle that start from North direction, indicated here as 45 degrees

In land navigation, azimuth is usually denoted as alpha (α), and defined as a horizontal angle measured clockwise from a north base line or meridian. Azimuth has also been more generally defined as a horizontal angle measured clockwise from any fixed reference plane or easily established base direction line. Today, the reference plane for an azimuth in a general navigational context is typically true north, measured as a 0° azimuth, though other angular units (grad, mil) can also be employed. In any event, the azimuth cannot exceed the highest number of units in a circle—for a 360° circle, this is 359 degrees, 59 arcminutes, 59 arcseconds (359° 59′ 59″).

For example, moving clockwise on a 360° degree circle, a point due east would have an azimuth of 90°, south 180°, and west 270°. However, there are exceptions: some navigation systems use geographic south as the reference plane. Any direction can potentially serve as the plane of reference, as long as it is clearly defined for everyone using that system.

2.3.4 Solar azimuth angle

The solar azimuth angle is the azimuth angle of the sun. It is most often defined as the angle from due north in a clockwise direction. It can be calculated, to a good approximation, using the following formula, however angles should be interpreted with care due to the inverse sine, i.e. $x = \sin^{-1}(y)$ has more than one solution, only one of which will be correct.

$$\sin \phi_s = \frac{-\sin h \cos \delta}{\cos \theta_s}$$

The following two formulas can also be used to approximate the solar azimuth angle, however because these formulas utilize cosine, the azimuth angle will always be positive, and therefore, should be interpreted as the angle less than 180 degrees when the hour angle, h, is negative (morning) and the angle greater than 180 degrees when the hour angle, h, is positive (afternoon).

$$\cos \phi_s = \frac{\sin \delta \cos \Phi - \cos h \cos \delta \sin \Phi}{\cos \theta_s}$$

$$\cos \phi_s = \frac{\sin \delta - \sin \theta_s \sin \Phi}{\cos \theta_s \cos \Phi}$$

The previous formulas use the following terminology:
ϕ_s is the solar azimuth angle.
θ_s is the solar elevation angle.
h is the hour angle of the present time.
δ is the current sun declination.
Φ is the local latitude.

2.3.5 Hour angle

The hour angle is one of the coordinates used in the equatorial coordinate system to give the position of a point on the celestial sphere. The hour angle of a point is the angle between two planes: one containing the earth's axis and the zenith (the meridian plane), and the other containing the earth's axis and the given point. The angle is negative east of the meridian plane and positive west of the meridian plane or it can be positive westward zero to 360 degrees. The angle may be measured in degrees or in time, with 24 hours equalling 360 degrees exactly. The hour angle is paired with the declination to fully specify the position of a point on the celestial sphere. The hour angle (HA) of an object is equal to the difference between the current local sidereal time (LST) and the right ascension (α) of that object:

$$HA_{object} = LST - \alpha_{object}$$

Thus, the object's hour angle indicates how much sidereal time has passed since the object was on the local meridian. It is also the angular distance between the object and the meridian, measured in sidereal hours (1 hour = 15 degrees). For example, if an object has an hour angle of 2.5 hours, it crossed the local meridian 2.5 sidereal hours ago (i.e. hours measured using sidereal time), and is currently 37.5 degrees west of the meridian. Negative hour angles indicate the time until the next transit across the local meridian. Of course, an hour angle of zero means the object is currently on the local meridian.

2.3.6 Solar hour angle

The hour angle (h or ha) of a point on the earth's surface is the angle through which the earth would turn to bring the meridian of the point directly under the sun. The earth is rotating, so this angular displacement represents time. So in observing the sun from earth, the solar hour angle is an expression of time, expressed in angular measurement, usually degrees, from the solar noon.

At solar noon, at the observer's longitude on earth, the hour angle is 0.000 degrees with the time before solar noon expressed as negative degrees, and the local time after solar noon expressed as positive degrees. The hour angle is the angular displacement of the sun east or west of the local meridian due to rotation of the earth on its axis at 15° per hour with morning being negative and afternoon being positive. For example, at 10:30 a.m. local apparent time the hour angle is –22.5° (15° per hour times 1.5 hours before noon).

The cosine of the hour angle [cos(h)] becomes an easy computation tool in determining the cosine term for the computation of the angle of the sun's altitude (or the complementary zenith angle) at any time during the day. At solar noon, h = 0.000 so of course, cos(h) = 1, and before and after solar noon the cos(± h) term = the same value for morning (negative hour angle) or afternoon (positive hour angle), i.e. the sun is at the same altitude in the sky at both 11:00 am and 1:00 pm solar time, etc.

2.3.7 Local mean time

Local mean time is a form of solar time that corrects the variations of local apparent time, forming a uniform time scale at a specific longitude. Its uniformity depends only on the accuracy of the clocks used to measure it. Local mean time was used from the early 19th century, when local solar time or sundial time was last used, until standard time was adopted on various dates in the several countries. Standard time means that the same time is used throughout some region—usually, it was either offset from Greenwich Mean

Time or was the local mean time of the capital of the region. The difference between local mean time and local apparent time is the equation of time.

2.3.8 Solar time

Time of day as determined by the position of the sun in the sky. Apparent solar time, the time given by a sundial, is not uniform because of the varying speed of the earth in its elliptical orbit. Mean solar time is a uniform time that coincides with apparent solar time at four instants through the year. The difference between them is known as the equation of time, and is greatest in early November when the sun is more than 16 minutes fast on mean solar time. Mean solar time on the Greenwich meridian is known as Greenwich Mean Time and is the basis of civil timekeeping.

2.4 Apparent motions of the sun

During the year the earth moves around the sun. As a result, the sun appears to move around the sky once with respect to the stars as seen from earth. Earth takes about 365 and a quarter days to travel once around the sun. Since the earth moves around the sun and we are resident on the earth, it means that the sun appears to shift in the sky about 1 degree per day. This number is calculated by taking the total degrees in a circle (360 degrees, since the earth moves 360 degrees around the sun) and dividing that amount by the total number of days in a year (365.2422). The result is 360 degree/365.2422 days to equal about 1 degree per day.

The plane of the earth's orbit is called the ecliptic. Since the earth orbits the sun, the sun is also on the ecliptic. As a result, the sun appears to us to move around the sky on the ecliptic.

Please note that the ecliptic is not the same thing as the earth's extended equator, the celestial equator. The earth's axis (and hence also the celestial equator) is actually tilted by 23.5 degrees with respect to the plane of the earth's orbit or the ecliptic. If we consider the celestial equator as the reference circle, the ecliptic appears to be inclined or tilted, 23.5 degrees from the plane of the earth's equator, as demonstrated above. This orientation means that the sun appears to transit the sky north of the celestial equator during part of the year and south of the celestial equator the other part of the year.

2.5 Day length

Day length or length of day or length of daytime, refers to the time each day from the moment the upper limb of the sun's disk appears above the horizon during sunrise to the moment when the upper limb disappears below the horizon during sunset. Due to the diffusion and refraction of sunlight by the atmosphere, there is actually daylight even when the sun is slightly below the horizon. The period when it is still somewhat light even though the sun is below the horizon is called twilight. In general, the length of a day varies throughout the year, and depends upon latitude. This variation is caused by the tilt of the earth's axis of rotation with respect to the ecliptic plane of the earth around the sun. At the solstice occurring about June 20–22, the north pole is tilted toward the sun, and therefore the northern hemisphere has days ranging in duration from just over 12 hours in the southern portion of the Tropic of Cancer to 24 hours in the Arctic Circle, while the southern hemisphere has days ranging in duration from just under 12 hours in the northern portion of the Tropic of Capricorn to zero in the Antarctic Circle.

At the equinox occurring about September 22–23, the poles are neither tilted toward nor away from the sun, and the duration of a day is generally about 12 hours all over the earth. At the solstice occurring about December 20–22, the south pole is tilted toward the sun, and therefore the southern hemisphere has days ranging in duration from just over 12 hours in the northern portion of the Tropic of Capricorn to 24 hours in the Antarctic Circle, whereas the northern hemisphere has days ranging in duration from just under 12 hours in the southern portion of the Tropic of Cancer to zero in the Arctic Circle. At the equinox occurring about March 19–21, the poles are again aligned so that the duration of a day is generally about 12 hours all over the earth.

In each hemisphere, the higher the latitude, the shorter the day during winter. Between winter and summer solstice, the day's duration increases, and the rate of increase is larger the higher the latitude. A fast increase of day length is what allows a very short day on winter solstice at 60 degrees latitude (either north or south) to reach about 12 hours by the spring equinox, while a slower increase is required for a much longer day on winter solstice at 20 degrees latitude (again, either north or south) to reach 12 hours by the spring equinox. The rate of change of day duration is generally fastest at the equinoxes, although at high latitudes the change is similar for several weeks before and after the equinoxes. The rate of change of day duration at each solstice is zero as the change goes from positive to negative or vice versa. Some interesting facts are as follows:

1. On the equator, the duration of daylight is not exactly 12 hours all the year round, but rather—due to atmospheric refraction and the size of the Sun—exceeds 12 hours by about 7 minutes each day.
2. Because the sun is north of the equator for almost 4 days more than half the year, the duration of the average day at a given latitude in the northern hemisphere exceeds the duration of the average day at the same latitude in the southern hemisphere by a few minutes.

3. During a few days around the equinoxes—about March 19–22 and September 21–24—both poles experience simultaneously 24 hours of daytime, due mainly to atmospheric refraction.
4. Each pole has only one sunrise and one sunset per year, around the time of the equinoxes. Each pole's sunrise is nearly coincident with the other's sunset, with minor differences due mainly to atmospheric refraction.

2.6 Solar energy reaching the earth's surface

2.6.1 Solar Constant

The sun is considered to produce a constant amount of energy. At the surface of the Sun the intensity of the solar radiation is about 6.33×10^7 W/m^2 (note that this is a power, in watts, per unit area in metres).

As the sun's rays spread out into space the radiation becomes less intense and by the time the rays reach the edge of the earth's atmosphere they are considered to be parallel.

The solar constant (I_{SC}) is the average radiation intensity falling on an imaginary surface, perpendicular to the sun's rays and at the edge of the earth's atmosphere (Fig. 2.3). The word 'constant' is a little misleading since, because of the earth's elliptical orbit the intensity of the solar radiation falling on the earth changes by about 7 per cent between January 1st, when the earth is nearest the sun, and July 3rd, when the earth is furthest from the sun. A yearly average value is thus taken and the solar constant equals 1367 W/m^2. Even this value is inaccurate since the output of the sun changes by about ±0.25 per cent due to sun spot cycles.

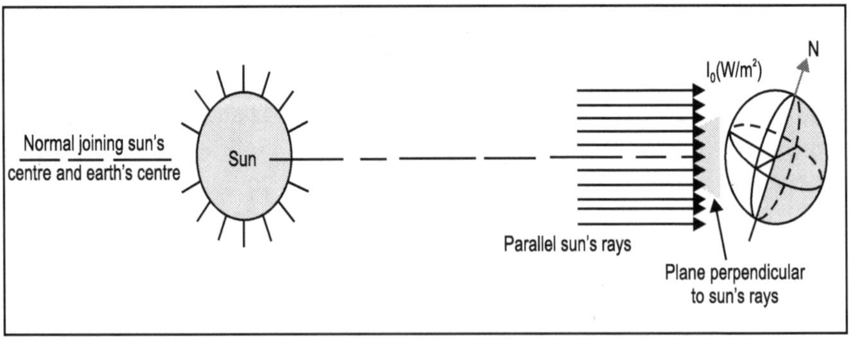

Figure 2.3 The sun's rays incident on the earth. I0 = irradiance on a plane perpendicular to the sun's rays

The solar radiation intensity falling on a surface is called irradiance or insolation and is measured in W/m^2 or kW/m^2.

The solar constant can be used to calculate the irradiance incident on a surface perpendicular to the sun's rays outside and the earth's atmosphere (Fig. 2.3) on any day of the year (i.e. as the distance between the sun and earth changes throughout the year):

$$I_0 = I_{SC}\left[1 + 0.034\cos\left(2\pi\frac{n}{365.25}\right)\right] \quad \text{... (2.1)}$$

where,

I_0 = extraterrestrial (outside the atmosphere) irradiance on a plane perpendicular to the sun's rays (W/m^2).

I_{SC} = the solar constant (1367 W/m^2).

n = the day of the year such that for January the 1st n = 1.

Figure 2.4 shows the variation in I_0 over the course of a year. Most solar power calculations use I_0 as a starting point because, for any given day of the year it is the maximum possible energy obtainable from the sun at the edge of the earth's atmosphere.

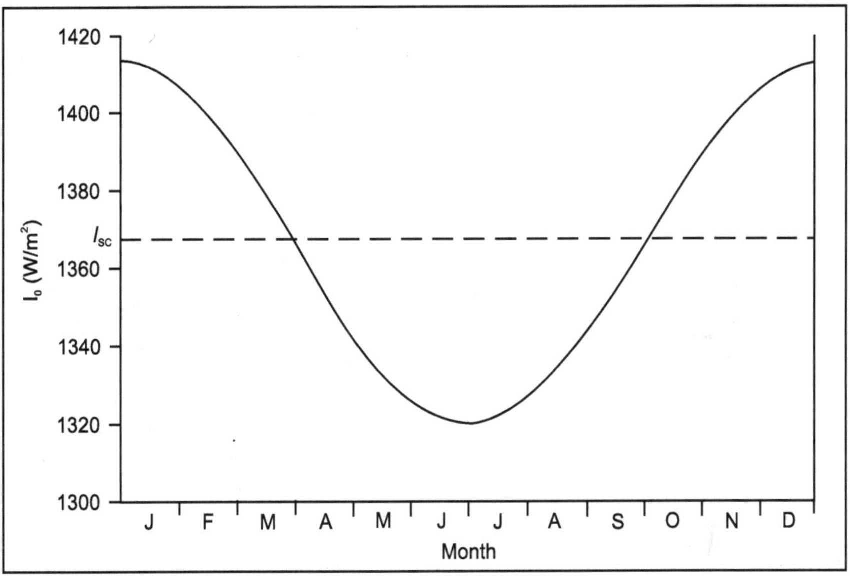

Figure 2.4 The variation in I_0 over the course of a year. The dashed line shows the value of the solar constant (ISC)

2.6.2 Cosine Effect

The value of I0 is the same no matter where you are on the earth's surface, however not all points on the earth's surface are perpendicular to the sun's rays. A useful quantity to calculate is the solar irradiance incident on an imaginary surface that is parallel to a horizontal plane on the earth's surface (Fig. 2.5). The irradiance on such a surface is smaller than I_0 because of the cosine effect and is the maximum amount of solar energy that could be collected on a horizontal plane at the earth's surface if the atmosphere did not scatter and absorb any radiation.

Figure 2.5 shows three plane surfaces:

1. Plane A, a horizontal plane at the point P on the earth's surface.
2. Plane B, a surface parallel to plane A but on the edge of the earth's atmosphere, often referred to as the horizontal plane.
3. Plane C, a surface perpendicular to the sun's rays, often referred to as the normal plane.

I_0 is the irradiance intensity on the normal plane and the irradiance intensity on the horizontal plane can be calculated from:

$$I_{0h} = \cos \theta_z \qquad \text{... (2.2)}$$

where θ_z is the solar zenith angle described already and I_{0h} is the extraterrestrial irradiance intensity on the horizontal plane. It can be seen in Fig. 2.2, the qz is also the angle of incidence of the sun's rays on a horizontal plane. Note that since cosine values fall between 1 and –1, I_{0h} will never be greater than I_0, and I_{0h} = I0 at point P′ where $\cos \theta_z = 1$ ($\theta_z = 0°$).

2.6.3 Irradiation

Just to be confusing the intensity of solar radiation is called irradiance and is measures in the units of power per unit area (W/m^2 or kW/m^2) however, the total amount of solar radiation energy is called irradiation and is measures in the units of energy per unit area (J/m^2). Irradiation is given the symbol H, so that:

1. H_0 is the total daily amount of extraterrestrial radiation on a plane perpendicular to the sun's rays.
2. H_{0h} is the total daily amount of extraterrestrial radiation on a plane horizontal to the earth's surface.

Note that these planes are considered to rotate with the earth so that H_0 and H_{0h} are daily values, and the planes are shaded at night. Figures 2.6 and 2.7 show how the values of H_0 and H_{0h} vary throughout the year in the northern hemisphere. Note that for any given day the value of H_0 changes from latitude to latitude despite the value of I_0 being constant for all latitudes. This occurs because the length of the days changes and the effects is most obvious inside the Arctic circle where much of the year is either 24 hours of darkness or 24 hours of daylight.

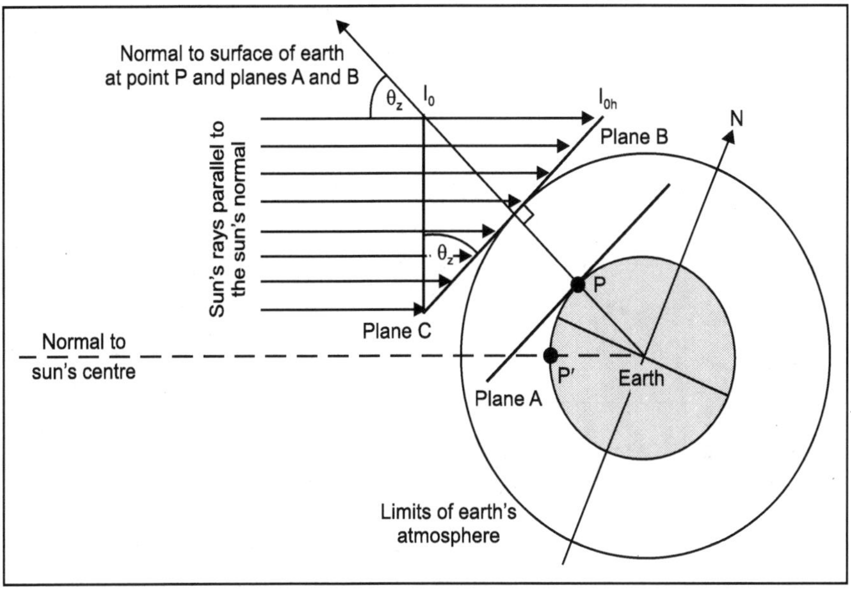

Figure 2.5 The cosine effect

2.6.4 Solar spectrum

The sun's radiation is a good approximation of black body radiation (a continuous distribution of wavelengths with no wavelengths missing) with wavelengths in the range of about 0.2 μm to 2.6 μm (Fig. 2.8). The solar spectrum consists of ultraviolate rays in the range of 200–400 nm, visible light in the range 390 nm (violet) to 740 nm (red) and the infra red in the range 700 nm to 1 mm. Table 2.1 shows the subdivisions of the ultraviolate range and Table 2.2 shows the distribution of extraterrestrial solar radiation.

Table 2.1 Subdivision of ultraviolate radiation.

UV-A	320–400 nm	Not harmful in normal doses, vitamin D production
UV-B	290–320 nm	Tanning, can burn
UV-C	230–290 nm	Causes skin cancer

Table 2.2 Distribution of extraterrestrial solar radiation.

Ultraviolet	200–400 nm	8.7%
Visible	400–700 nm	38.3%
Near infrared	700–3500 nm	51.7%

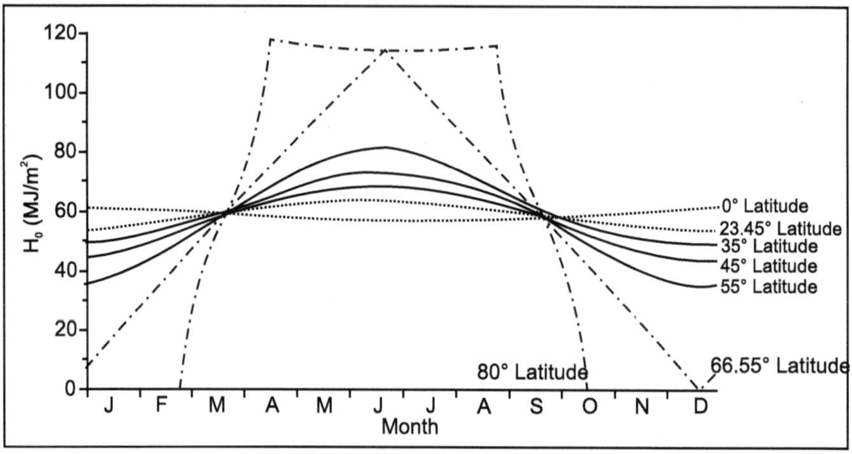

Figure 2.6 The total daily amount of extraterrestrial irradiation on a plane perpendicular to the sun's rays (H_0) for different latitudes

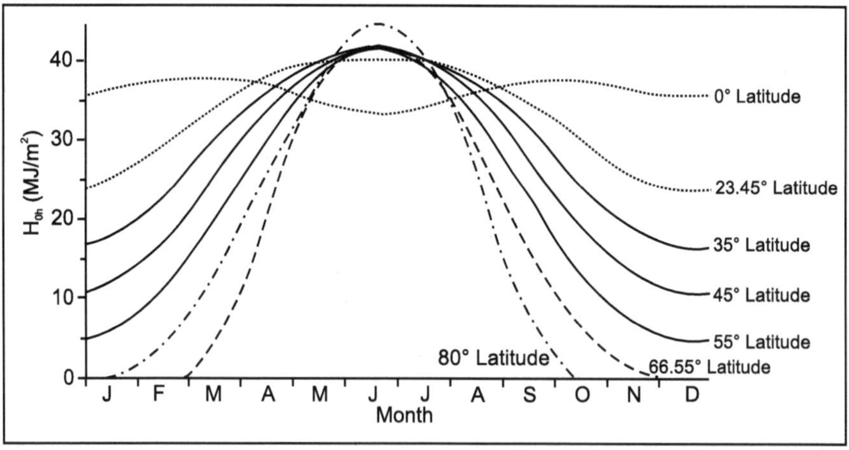

Figure 2.7 The total daily amount of extraterrestrial irradiation on a plane horizontal to the earth's surface (H_{0h}) for different latitudes

As the sun's rays pass through the atmosphere certain wavelengths are absorbed and a proportion of the total energy is scattered. Thus the solar spectrum at the earth's surface has some wavelengths missing (shown as dark portion in Fig. 2.8) and the overall intensity is reduced.

2.6.5 Atmosphere and air mass

The atmosphere scatters and absorbs some of the sun's energy that is incident on the earth's surface. Scattering of radiation by gaseous molecules (e.g. O_2, O_3, H_2O and CO_2), that are a lot smaller than the wavelengths of the radiation, is called Rayleigh scattering. Roughly half of the radiation that is scattered is lost to outer space, the remaining half is directed towards the earth's surface from all directions as diffuse radiation. Because of absorption by oxygen and ozone molecules the shortest wavelength that reaches the earth's surface is approximately 0.29 μm. Other gas molecules absorbed difference wavelengths as indicated in Fig. 2.8.

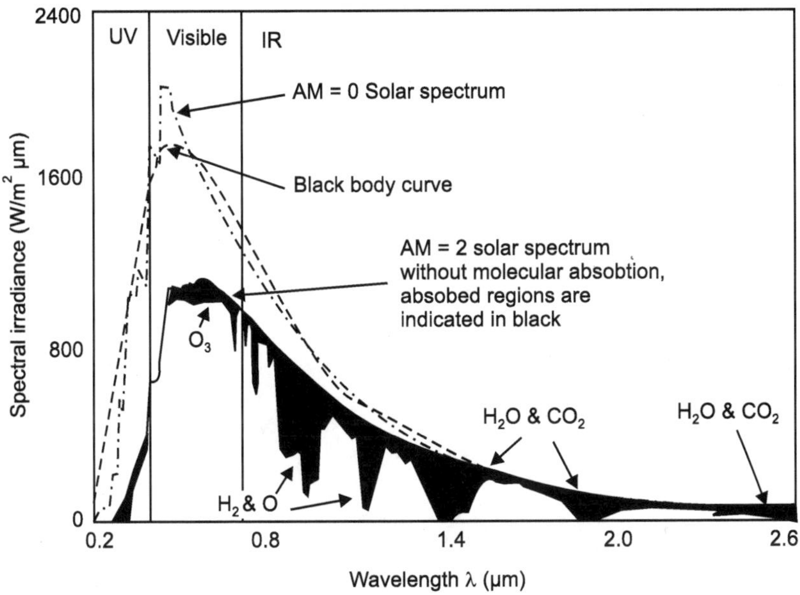

Figure 2.8 The extraterrestrial solar spectrum (AM = 0), the theoretical black body curve and the solar spectrum at the earth's surface for AM = 2 and the absorbed regions shown in black

Scattering by dust particles larger than wavelengths of light is called Mie scattering. This process includes both true scattering (where the radiation bounces of the particle) and absorption followed by emission, which heats the particles. The amount of radiation scattered by this process will vary a lot depending on location and the weather blowing particles about. A form of Mie scattering called the Tyndall effect, which preferentially scatters shorter wavelengths, is responsible for the sky being blue.

Clouds reflect a lot of radiation and also absorbed a little, the rest is transmitted through. Globally, clouds reflect a lot of radiation and help regulate the surface temperature.

The fraction of the total solar radiant energy reflected back to space from clouds, scattering and reflection from the earth's surface is called the albedo of the earth-atmosphere system and is roughly 0.3 for the earth as a whole. Figure 2.9 shows that a plane on the earth's surface receives

1. Beam (or direct) radiation—coming straight through the atmosphere to hit the plane (very directional).
2. Diffused radiation—scattered in all direction in the atmosphere and then some arrives at the plane on the earth's surface (not directional).
3. Reflected radiation—beam and diffused radiation that hits the earth's surface and is reflected onto the plane.

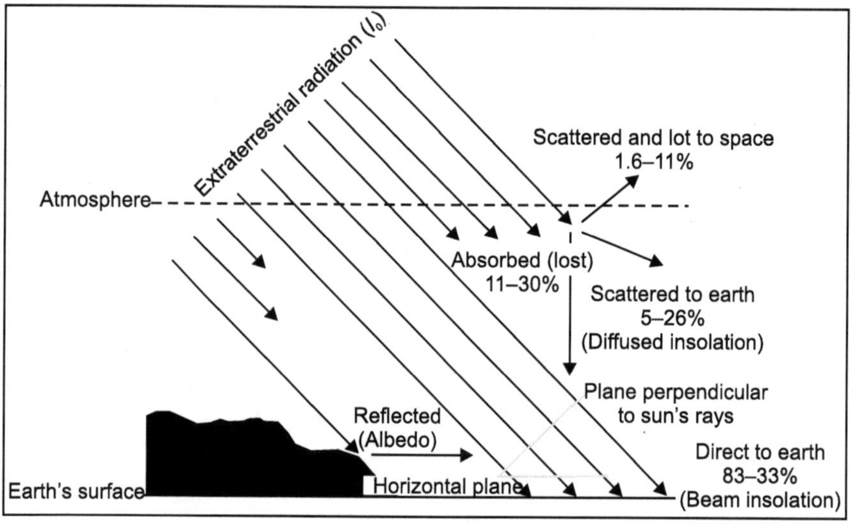

Figure 2.9 The effect of the atmosphere on the solar radiation reaching the earth's surface

The amount of energy reflected, scattered and absorbed depends on the amount of atmosphere that the incident radiation travels through as well as the levels of dust particles and water vapour present in the atmosphere. The latter is difficult to judge but the distance travelled through the atmosphere by incident radiation depends on the angle of the sun (Fig. 2.10).

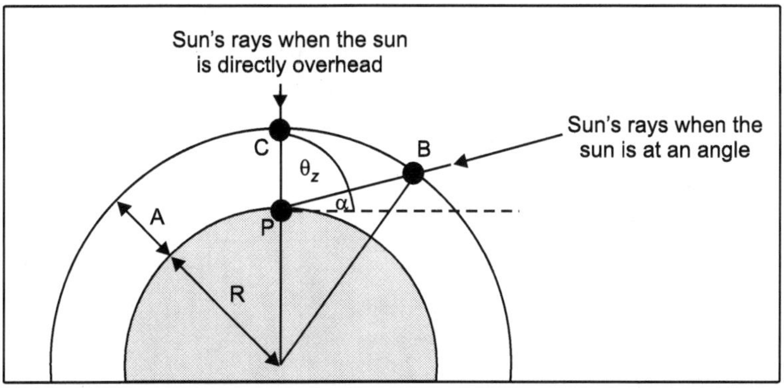

Figure 2.10 The distance travelled through the atmosphere by the sun's rays

The distance travelled through the atmosphere by the sun's rays incident on the earth is accounted for by a quantity called air mass (AM).

$$\text{Air mass} = \frac{\text{Path length travelled}}{\text{Verticle depth of the atmosphere}}$$

$$AM = \frac{BP}{CP} = \left[\left(\frac{R}{H}\cos\theta_z\right)^2 + 2\frac{R}{H} + 1\right]^{1/2} - \left(\frac{R}{H}\right)\cos\theta_z \qquad \dots (2.3)$$

where,

R = the radius of the earth, taken to be 6370 km.

H = thickness of atmosphere, taken to be 7991 km (although it is considerably thicker at the equator than the poles).

For angles of $\theta_z < 70°$.

$$AM \approx \frac{1}{\cos\theta_z} = \sec\theta_z \qquad \dots (2.4)$$

Therefore, outside the earth's atmosphere AM = 0, when the sun is directly overhead $\theta_z = 0$, AM = 1 and when the $\theta_z = 60°$ AM = 2. AM is normally taken to be an average of 1.5 for a clear sunny day and this value is used for the calibration of solar cells.

2.6 Rough estimates of the solar energy available at the earth's surface

The solar constant is the average extraterrestrial insolation at the edge of the atmosphere:

$$I_{SC} = 1367 \text{ W/m}^2$$

The earth presents a disc of area πR_2 to the sun, therefore the total amount of extraterrestrial insolation incident on the earth is $I_{SC} \times \pi R^2$. This value is then divided by half the surface areas of the earth, $4\pi R^2/2$, which gives 684 W/m², the average insolation incident on unit area of the earth facing the sun (Fig. 2.11). Note that solar panels are calibrated assuming that there is 1000 W/m² available.

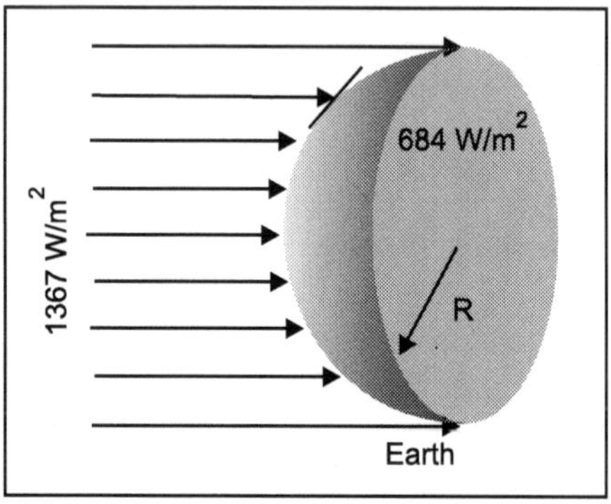

Figure 2.11 The average insolation incident on unit area of the earth facing the sun

A rough estimate of the irradiation incident per unit area (H) of the earth's surface can be made if we assume that 30 per cent of the sun's energy is lost in the atmosphere and that a day is an average of 12 hours long at any location.

$$H = 0.7 \times 684 \times 12 = 5.75 \text{ kWh/day}$$

Or if we assume that the sun is only at an appreciable strength for an average 6 hours in the day (as is likely in more northerly latitudes):

$$H = 0.7 \times 684 \times 6 = 2.88 \text{ kWh/day}$$

Figure 2.12 shows the yearly profile of mean solar radiation for different locations around the world. The solid grey line show the value of 5.75 kWh/day and the dashed grey line shows 2.88 kWh/day.

2.7 Earth–Sun geometry

We need to assess the flux density of solar radiation on flat and inclined surfaces for a variety of biometeorological problems. This information is used to compute the energy balance, evaporation and photosynthesis rates and stomatal conductance of inclined leaves, and plants. In this section we will cover issues relating to cosine laws between the incident and received rays and learn to compute angles between the sun and the normal to points on earth.

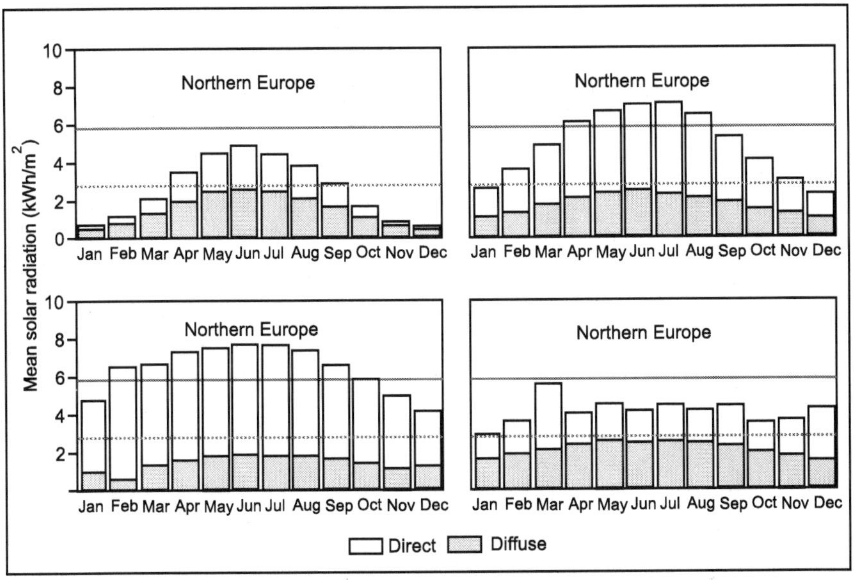

Figure 2.12 The yearly profile of mean solar radiation for different locations around the world

2.7.1 Lambert's Cosine Law

The flux density radiation on the horizontal is a function of the angle between the surface normal and the direction of the ray is defined by Lambert's Cosine Law:

$$R(i) = R_{\perp} \cos(\gamma) = R_{\perp} \sin(\beta)$$

Definition of zenith and elevation angles and the projection of area normal to incident rays on a flat surface are shown in Fig. 2.13.

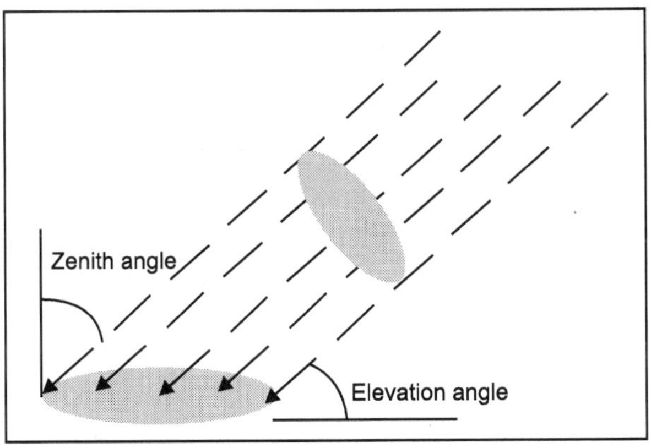

Figure 2.13 Definition of zenith and elevation angles and the projection of area normal to incident rays on a flat surface

The ratio between the flux density of radiation on the horizontal surface to the flux density normal to the incoming radiation is related to the ratios of the areas of the beam cross section and the projected horizontal area.

$$R(i) = R_{\perp} \frac{A_{beam}}{A_{horizontal}}$$

Trigonometrically, we can examine 2 complementary triangles and consider the projection of side B onto side C.

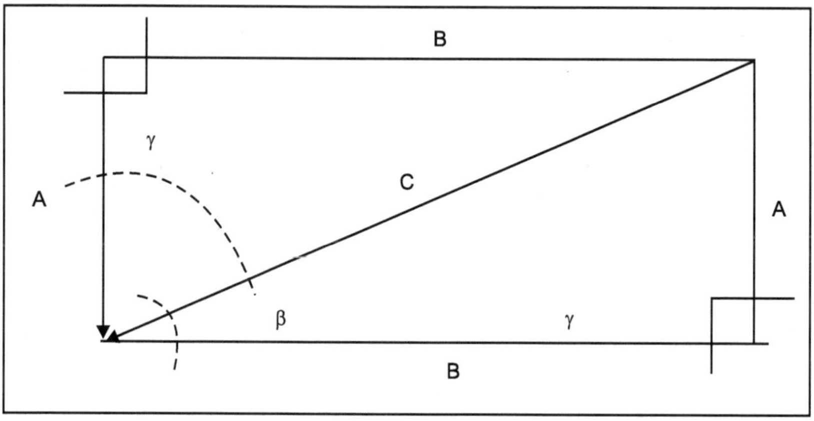

To calculate the flux density of solar radiation on the horizontal, we want to project the flux density of radiation associated with the vector C to that normal to the surface, vector A. For the lower triangle and angle β we arrive at:

$$R(i) = R_{\perp} \sin(\beta) = R_{\perp}$$

And for the upper triangle and angle γ we arrive at:

$$R_{\perp} \cos(\gamma) = R_{\perp} \frac{A}{C}$$

The lower the elevation angle, the more the unit area of beam is spread out on the ground, so its flux density on an unit area basis is reduced. As we can see in Fig. 2.14, cos(γ) is one when γ is zero and the sun is directly overhead; cos(γ) is zero when γ is 90 degrees and the sun is aligned with the horizon.

2.8 Sun–Earth geometry

The earth rotates on its axis and it revolves around the sun in an elliptical orbit. The area swept by the earth–sun radius is constant, as predicted by Kepler's Law of Planetary Motion. To perform calculations associated with the flux density of solar radiation on flat and inclined surfaces and its penetration into plant canopy, we must be able to evaluate the angle of the sun from any position on earth at any time of the day. Such a calculation involves computing the angle between a vector normal to a specified point on earth and a projection from the sun. This computation is complicated by the fact that the earth rotates on its polar axis once a day and the earth revolves around the sun once a year.

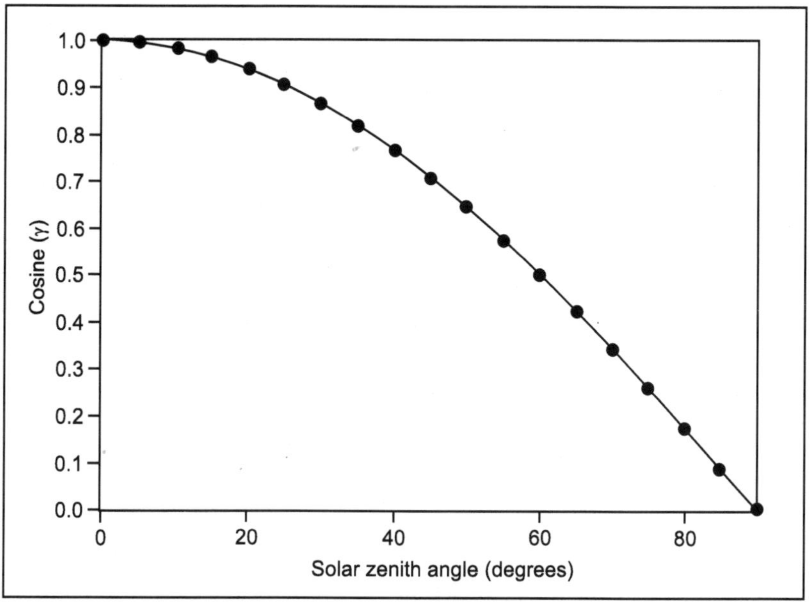

Figure 2.14 Cosine of solar zenith angle. Values vary non-linearly between 1 and zero

To compute the sun's elevation with respect to the horizon (or its complement, the angle with to the zenith) and its azimuth angle we apply concepts derived from three-dimensional geometry. The important coordinates to know include:

1. Time of day.
2. Longitude.
3. Latitude (λ).
4. Earth's declination angle (δ).

One way to think about the problem is to try and solve for the angle between two vectors. One is the normal to the earth at a given position and the other is the vector between the point on earth that is normal to the view () and the vector that points to the sun (). From vector analysis we know that the cosine of the vectors is related to their dot product:

$$\vec{a}.\vec{b} = |\vec{a}||\vec{b}| \cos \gamma$$

A schematic of the situation we are interested shows three vectors, $\vec{n}$ S , and S n - , , , and , which represent the vector equal to the radius of the earth and directed toward the surface zenith, the vector from the center of the earth to the sun, and the resultant vector.

It is the angle between the resultant vector and the vector pointing towards the zenith that gives us the angle we desire (Fig. 2.15).

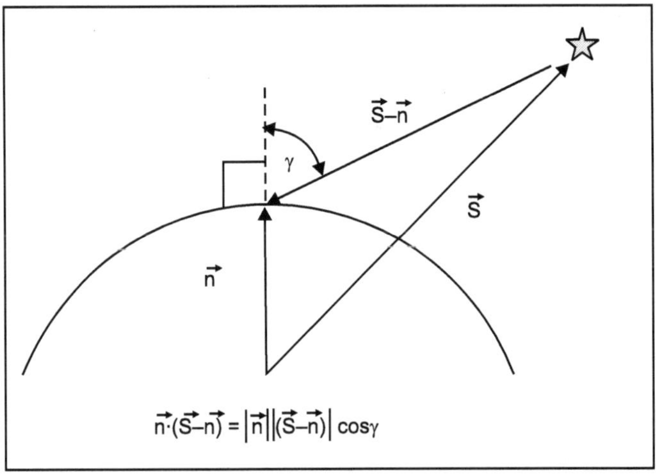

Figure 2.15 Vectors describing the angle between the sun and a point on earth

In spherical coordinates, we define three directional variables:

$$z = r \cos \gamma$$
$$x = r \sin \gamma \cos \varphi$$
$$y = r \sin \gamma \sin \varphi$$

where, γ is the zenith angle and φ is the azimuth angle. These angles are for an earth-centred perspective.

In vector notation we define:

$$\vec{S} = \begin{bmatrix} x \\ y \\ z \end{bmatrix} \begin{bmatrix} r_s \sin \gamma_s \cos \varphi_s \\ r_s \sin \gamma_s \sin \varphi_s \\ r_s \cos \gamma_s \end{bmatrix}$$

$$\vec{n} = \begin{bmatrix} r_e \sin \gamma_e \cos \varphi_e \\ r_e \sin \gamma_e \sin \varphi_e \\ r_e \cos \gamma_e \end{bmatrix}$$

Remember:

$$\vec{a}.\vec{S} = a_x S_x + a_y S_y + a_z S_z$$

$$\vec{a}.\vec{S} = \vec{S}.\vec{n}$$

$$|\vec{a}| = \sqrt{a_x^2 + a_y^2 + a_z^2}$$

and a triangle with vertices at A, B and C:

$$\overrightarrow{AB} + \overrightarrow{BC} = \overrightarrow{AC}$$

After much algebraic and trigonometric expansion and substitution we arrive at a new equation of the solar zenith or elevation angles. The cosine of the sun's angle with respect to the zenith (γ) or the sine of the solar elevation angle (β) with respect to the horizontal, can be computed as:

$$\sin(\beta) = \sin(\lambda)\sin(\delta) + \cos(\lambda)\cos(\delta)\cos(h) = \cos(\gamma)$$

λ is latitude and h is the hour angle.

We discuss the components of computing earth–sun angles in more detail.

2.8.1 Time of day

The earth is a rotating sphere. The earth rotates around its north-south axis once every 23 hours, 56 minutes and 4 seconds.

It can also be stated that the earth rotates on its axis 2π radians in one day. Over the course of an hour it rotates $\pi/12$ radians or 15 degrees. In this framework h is defined as:

Table 2.3 produces values of h and cos(h) for several reference values.

Table 2.3 The values of h and cos(h) for several reference values.

Solar time (hour)	h (radians)	Cos(h)
6	0.5π	0
12	π	–1
18	1.5π	0

In the northern hemisphere, the cosine of π equalling minus one corresponds with the sun being south of an observer at solar noon.

If we are to apply this relation using clock time, then we define h as the fraction of 2π that the earth has turned after local solar noon:

$$h = \frac{\pi}{12}(t - t_0) \text{ (radians)}$$

$$h = \frac{15\pi}{180}(t - t_0) \text{ (degrees)}$$

In this instance, t is local time and t0 is solar time.

$$t_0 = 12 - l_c - e_t$$

l_c is the longitudinal correction. Its value is 4 minutes for each degree of longitude east of the standard meridian and is –4 minutes for each degree of longitude west of the standard meridian. Standard meridians possess 15 degree increments from the prime meridian.

The longitude of the Berkeley campus is: 122° W 15′ 47″; latitude is 37° N 52′ 24″. It is about 2.25° west of the standard meridian (120° W), so solar noon occurs about 9 minutes after noon Pacific Standard Time (PST).

The variable, et, represents the equation of time (Fig. 2.16). It arises because of the eccentricity of the earth's orbit around the sun and the obliquity (tilt) of the ecliptic (the great circle that represents the earth's path around the sun). The eccentricity of the earth orbit around the sun forces the angular rotation rate to be variable; this allows equal areas to be swept with time during the earth's orbit. The equation of time (units of hours) is:

$$e_t = \left[\frac{-104.7\sin(f) + 596.2\sin(2f) + 4.3\sin(4f) - 429.3\cos(f) - 2.0\cos(2f) + 19.3\cos(3f)}{3600} \right]$$

$$f = \frac{\pi}{180}[279.5 + 0.9856d] \qquad [\text{degrees}]$$

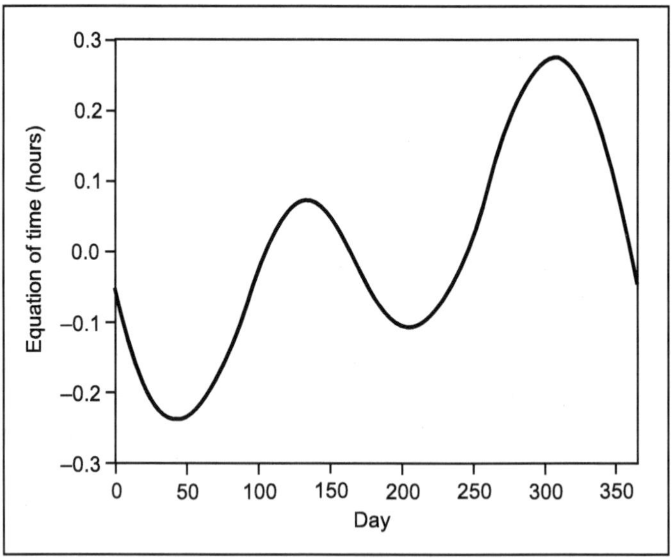

Figure 2.16 Computations based on the equation of time

d is day number (1 on January 1 and 365 on December 31). Here et and f have units of radians.

The contribution from variation in the orbital speed of the earth to the computation of the sun's zenith angle is not trivial. On February 9, the increment due to the equation of time is 14.2 minutes and on November 6 the contribution is 16.3 minutes.

Sometimes these equations are expressed in terms of Universal Time. In this case, the universal time (Greenwich) is the sum of the true solar time plus and equation for time and the longitudinal time (15 degrees per hour West of Greenwich and –15 degrees per hour East of Greenwich).

2.8.2 Longitude

Mariners, navigators and geographers have divided the sphere of the earth in the north–south plane into zones of latitude. They correspond with the degrees of a circle (0 to 360 degrees or 0 to 2π radians). The reference for zero is the prime meridian, near London, at Greenwich, England.

2.8.3 Latitude

From an imaginary point at the center of the earth's sphere angular bands of latitude are defined. At the center is the sphere is the equator, defined as zero degrees. Bands of latitude range from zero to 90 degrees for the northern and southern hemispheres.

2.8.4 Earth's declination angle

The earth's axis of rotation is tilted 66.5 degrees with respect to its orbital plane around the sun and its axis of rotation is inclined 23.5 degrees from the perpendicular, with respect to this plane. The tilt of the earth affects the angle between the sun beam and the normal over a surface. The sun angle affects the flux density of solar radiation incident on a surface.

The declination angle corresponds with the angle between the sun's rays at solar noon and the equator (Fig. 2.17). The earth's declination angle has limits that correspond with the seasons:

Summer solstice: +23.45 degrees (June 22)

During the summer solstice the sun is directly overhead (90 degrees) at the Tropic of Cancer and has an angle of 66.5 at the Equator and is at the horizon at the Antarctic Circle (66.5 S). Below the Antarctic Circle it is dark 24 hours a day. Above the Arctic Circle (66.6 N) there is sunlight 24 hours a day.

Equinox: 0 degrees (March 21 and September 22).

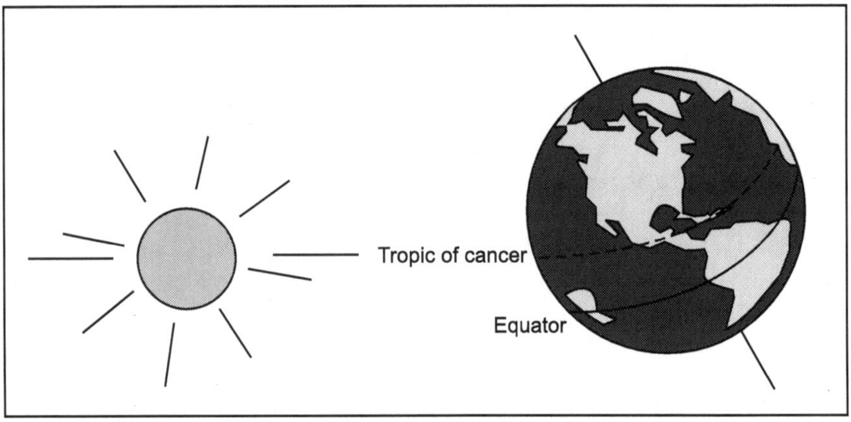

Figure 2.17 Geometrical configuration between the sun and the earth during the summer solstice

At the Spring and Autumnal Equinox, the sun is positioned over the equator and day length is 12 hours everywhere on earth (Fig. 2.18).

Winter solstice: –23.45 degrees (December 22).

At winter solstice, the sun is directly overhead the Tropic of Cancer (23.5 N) at noon. It is dark 24 hours a day above the Arctic circle (66.5° N) and it is bright below the Antarctic circle for 24 hours (Fig. 2.19). The seasonal changes in the earth–sun geometry are summarised in Fig. 2.20.

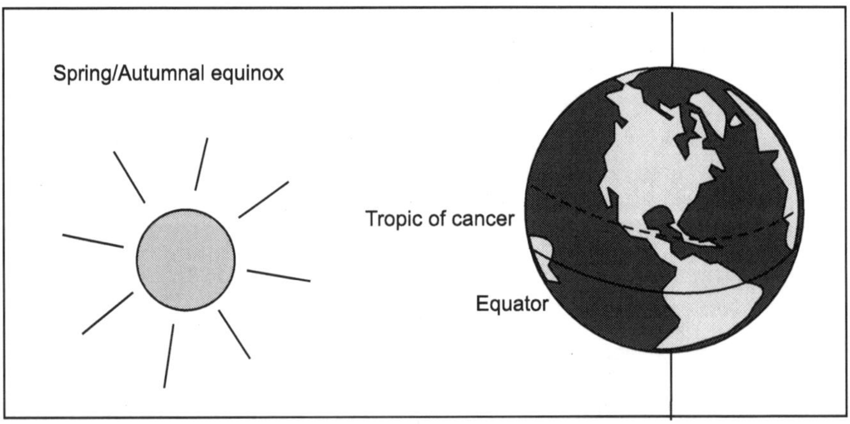

Figure 2.18 Configuration between the earth and sun during the spring and autumnal equinox

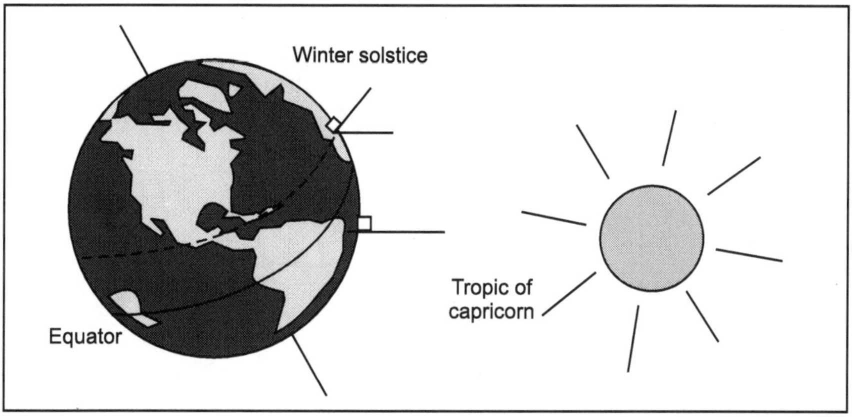

Figure 2.19 Configuration between the sun and earth during the winter solstice

A mathematical equation for computing the declination angle is:

$$\delta = -23.45.\cos\left(\frac{360(D+10)}{365}\right) \quad \text{(degrees)}$$

$$\delta = -23.45\cos\frac{\pi}{180}.\cos\left(\frac{2\pi(D+10)}{365}\right) \quad \text{(radians)}$$

where, D is day number [after Jones (1995)]. In this case the declination angle is expressed in terms of degrees. Multiplying by $\pi/180$ converts degrees to radians.

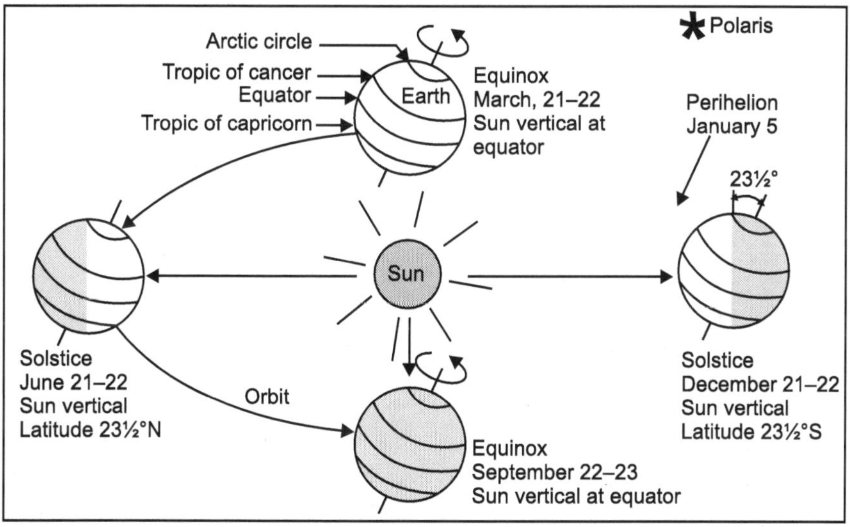

Figure 2.20 Basic earth–sun geometry

To understand how the declination angle varies with time, geometricians define a celestial sphere, of infinite radius, that rotates around the earth, the center point of the sphere. The intersection between the equatorial plane of the earth with the celestial sphere defines the celestial equator. From the point of view of the celestial sphere, the sun follows on orbit over the course of a year on the ecliptic. It is tilted 23.5 with respect to the celestial equator. Figure 2.21 shows seasonal variation in the declination angle.

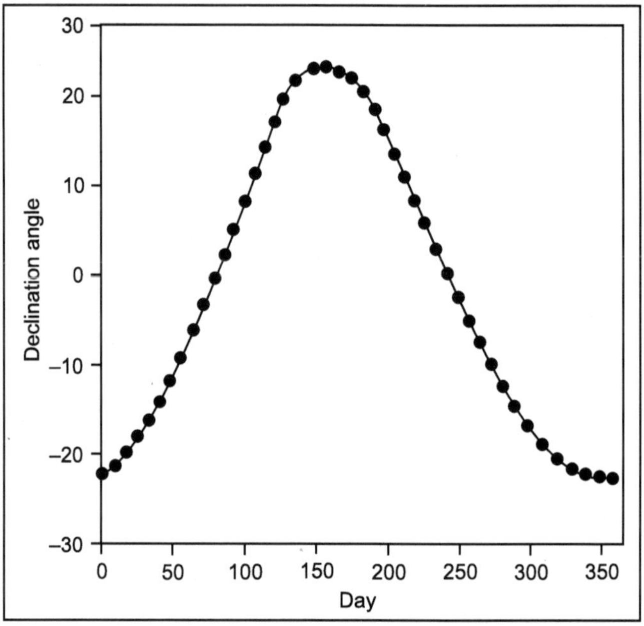

Figure 2.21 Seasonal variation in the declination angle

The sun's position on the celestial sphere can be estimated by knowing the angle of the right ascension and the declination. The right ascension is the angle between the hour circle of the vernal equinox and the hour circle of the sun (the angle between the sun and the north pole of the celestial sphere).

How azimuth angle is computed depends on the coordinate system which one uses. A mathematical perspective increments angles, starting from due south and rotating counter clockwise, from 0 to 360 degrees. A compass considers true north to be zero and increases angles in a clockwise manner from 0 to 360. An astronomical coordinate system increases azimuth angles from 0 to 180 in the clockwise direction and from 0 to –180 in the counter clockwise direction, starting with south as zero.

Campbell and Norman compute azimuth angle in the mathematical system as a function of the declination and zenith angles and latitude:

$$\cos(Az) = \frac{-\sin\delta - \cos\gamma \sin\lambda}{\cos\lambda \sin\gamma}$$

(east is 90 degrees, north is 180 and west is 270 degrees). Monteith and Unsworth compute solar azimuth, relative to south, as a function of the declination, hour and the zenith angles:

$$\sin(Az) = \frac{-\cos\delta \sin(h')}{\sin\gamma}$$

They define h′ as the hour angle which the earth has turned after local solar noon, as a fraction of 2π, $h' = 2\pi t/24$.

From Oke:

If $t < 12$,

$$\cos(Az) = \frac{\sin\delta \cos\lambda - \cos\delta \sin\lambda \cos h}{\sin\gamma}$$

If $t > 12$,

$$\cos(Az) = 2\pi \frac{\sin\delta \cos\lambda - \cos\delta \sin\lambda \cos h}{\sin\gamma}$$

An illustration of the motion of the sun at a particular location for several times during the year is presented in Fig. 2.22. Of course the sun is the highest at noon and it rises in the east and sets in the west during the equinox. For northerly climes, the sun is perceived to rise in the northeast and set in the northwest during the summer. And if we sit above the Arctic circle, the sun will revolve around the horizon during the summer.

We can manipulate this equation of the solar elevation angle to compute the times of sunrise and sunset, too.

The time of sunrise and sunset occur when the solar elevation angle is zero (or the zenith angle is $\pi/2$). It is a function of the latitude of the site and the earth's declination angle:

$$\cos(h) = -\tan\lambda \tan\delta$$

$$t_{sunrise} - t_{12} - \frac{h}{15}$$

$$t_{sunset} - t_{12} - \frac{h}{15}$$

The length of the day is $t_{sunrise}$ minus t_{sunset}, which equals 2h/15.

$$2t = \frac{24}{\pi} \cos^{-1}(\tan\lambda \tan\delta)$$

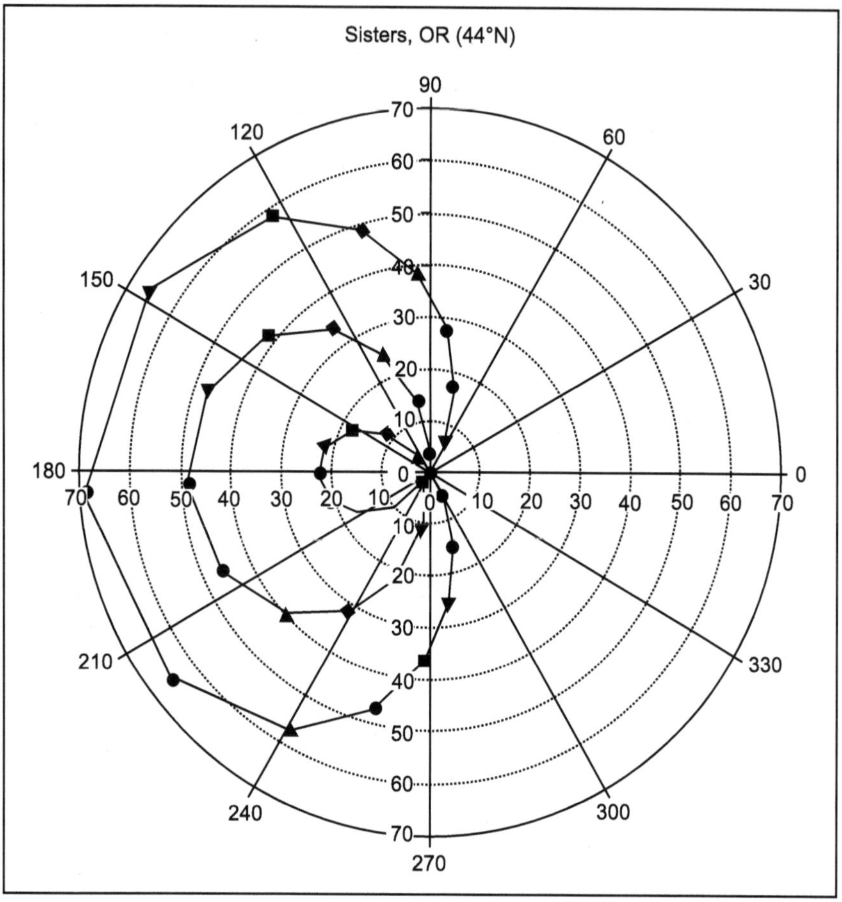

Figure 2.22 Solar elevation and azimuth angles at Sisters, OR (44° N) for summer and winter solstices and the equinoxes

Numerous algorithms are now available on the worldwide-web for computing solar elevation angles over the course of a day or year for any position on the earth. One source of information on solar elevation angles is giving at: http://susdesign.com/sunangle/

It has been updated since 2000 and is easy to use.

2.8.5 View factors

View factors are important for they give us information on the projected area of various geometry shades. The ratio of the area of shadow cast on the horizontal (Ah) and the area projected in the direction of the beam (A) yields information on the relation between the mean flux density and the horizontal flux density.

$$A_pS_p = (Ah \sin \beta)S_p = A_hS_b$$

If the area of the object is A then:

$$\bar{S}_b = \frac{A_h}{A} S_b$$

The shape factor is denoted by A_h/A. For the case of a sphere, we know that it projects a circle with an area of πr_2. The shadow projected onto the ground from a beam source is:

$$\frac{\pi r^2}{\sin \beta}$$

The surface area of a sphere is:

$$4\pi r_2$$

This yields the shape factor:

$$\frac{A_h}{A} = \frac{\pi r^2}{4\pi r^2 \sin \beta} = \frac{1}{4 \cos \gamma}$$

Geometry of a projected sphere are shown in Fig. 2.23.

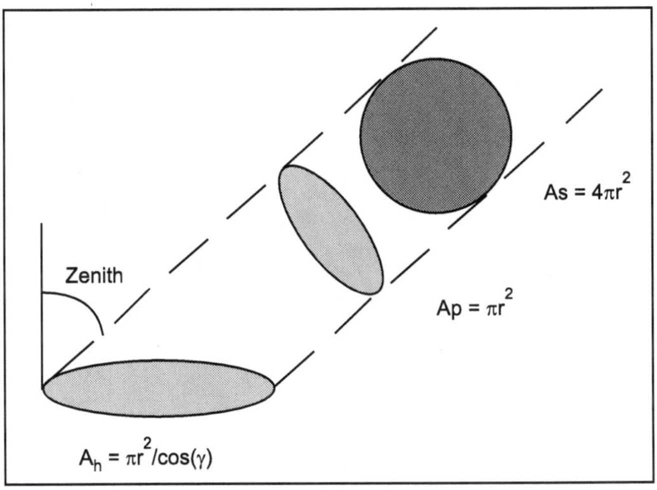

Figure 2.23 Geometry of a projected sphere

Cone:

$$\frac{A_h}{A} = \frac{(\pi - \theta_0)\cos \alpha + \sin \alpha \cot \beta \sin \theta_0}{\pi(1 + \cos \alpha)}$$

Inclined plane (Fig. 2.24):
Ah = sin α cot β – cos α

2.9 Radiation on inclined surfaces, mountain slopes

The earth is not flat. Ecosystems exist on sides of hills that may face east, west, south or north. The amount of direct sunlight incident on the side of a hill will depend on the angle between the sun's beam and the angle normal to the slope.

Direct solar radiation flux density on an inclined surface:

$$R(i) = R_{\perp} \cos (i)$$

R(i) is the radiation normal to an inclined surface.
$R_{\perp}$ is the radiation flux density normal to the sun's rays.

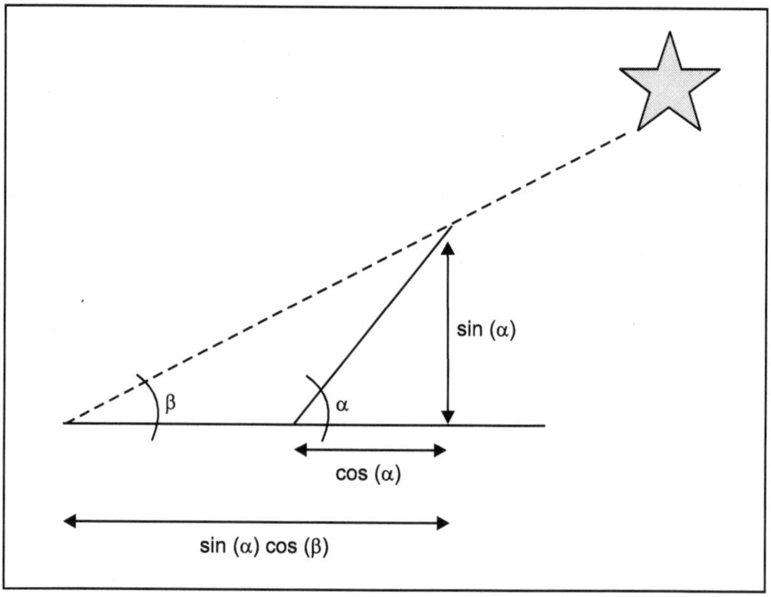

Figure 2.24 Projection of inclined plane on a surface

The angle of incidence on any given surface is computed as:

$$\cos (i) = \cos \alpha \sin \beta + \sin \alpha \cos \beta \cos \Psi$$

α is the angle of the inclination of the surface relative to a horizontal plane.

β is the solar elevation angle.

Ψ is the difference between the sun's azimuth angle and the azimuth angle of the project normal to the surface on the horizontal.

Alternatively this equation can be expressed in terms of the sun's zenith angle (γ), where $(\gamma + \beta) = \pi/2$:

$$\cos(i) = \cos\alpha \cos\gamma + \sin\alpha \sin\gamma \cos\Psi$$

Angle between any surface and the sun:

$$\cos\xi = [\sin\lambda \cos(h)(-\cos\alpha \sin\chi) - \sin(h)(\sin\alpha \sin\chi) + (\cos\lambda \cos(h)) \cos c]\cos\delta + [\cos\lambda (\cos\alpha \sin\chi) + \sin\lambda \cos c]\sin\delta$$

ξ is the angle between the sun and the surface normal.

χ is the zenith angle of the slope.

α is the azimuth angle of the slope.

3
Solar devices

3.1 Introduction

This chapter discusses solar devices such as solar photovoltaic modules, silicon solar cells, silicon wafers, solar lanterns, solar street lighting and solar home lighting systems, solar powered fans, solar air coolers and air conditioners, solar refrigerators, solar cooking systems, solar televisions, solar water pumps and solar water heater, etc. The use of solar energy has already made a mark in our lives and there is much more to come very soon.

3.2 Solar panels

Solar panels, generally comprising of arrays of photovoltaic cells, use the solar energy directly from the sun to generate electricity for our daily use. Being environment friendly in nature, solar panels collect the solar energy which is available in abundance on our planet and convert it using the advanced technology developed by human beings. This invention of humans has led to a great achievement in world's history of conserving non-renewable resources and saving the planet as well as the natural resources from depletion.

This concept is already famous in countries like Australia, United Kingdom and United States of America, etc. and is becoming very popular in solar panels India market as well. Even used solar panels are being used in these countries so that the installation cost reduces further and it is easy for the people to generate green energy. A large number of factories in the solar panels India market are now using solar panels for their daily electricity usage (Fig. 3.1).

They make optimal use of renewable energy from the sun and are very much environmental friendly in nature. Manufacturing solar panels is not an easy process. With the advancement of photovoltaic devices, we have seen a significant increase in the number of solar panels manufacturers all over India. The solar panels manufacturers manufacture the solar panels by first cutting the ingots, which are generally cylindrical in shape, into circular disc shaped silicon wafers. The silicon wafers are very carefully cut using saws or other cutting devices. These are then carefully cleaned to remove any impurities which might have arisen in the cutting process. After the cleaning process is completed, the silicon wafers are then polished and metal conductors and the products for doping the wafers for the desired electrical properties are put on each silicon wafer module. The modules are then placed on a flat sheet of solar panel and are spread in a matrix or a grid structure on the sheet. The solar panels are then installed on the roof tops or top of the buildings to get the maximum solar energy from the sun in order to run as expected.

Figure 3.1 Solar panels

We all know of global warming issues happening all over the world. This has also been proved by the rise in the earth's mean temperature and a significant increase in the natural calamities impacting the whole world. In such a critical situation, the use of alternate forms of energy like hydro energy, wind energy, solar energy, etc. are really appreciated and also encouraged by the governments in different parts of the worlds. The increased usage of solar products such as solar panels, solar water heaters, solar air coolers and the upcoming of various solar power plants has been an intelligent effort to fight global warming issues. Simple products like solar panels for generating the electricity for homes and offices, solar lighting for lighting up the homes, offices and streets, solar heaters for heating waters and swimming pools, solar cells to run small devices, etc. are really useful to prevent the global warming issues and they even do not consume any non-renewable resources of the earth.

3.3 Silicon wafers

A silicon wafer is a thin slice of crystal semiconductor, such as a material made up from silicon crystal, which is circular in shape. Silicon wafers are made up of pure and single crystalline material. They are used in the manufacture of semiconductor devices, integrated circuits and other small devices. There are multiple processes through which silicon wafers are manufactured. One such process which is widely used by most of the silicon wafers manufacturers is the Czochralski Growth process.

3.3.1 Silicon wafers manufacturing process

Ingots, made up of silicon which is very pure, are cast into cylindrical shapes. This is done by pulling the seed crystal from a molden melt. The ingots are then cut, using saws or other cutting devices, in thin circular shapes which result in formation of silicon wafers. The wafers, hence created, are thoroughly cleaned after this process to remove the impurities and are also properly shaped to increase their efficiency. Once the cleaning and shaping up of the silicon wafers is done, they can then be doped so that they acquire the electronics technicalities as desired by the silicon wafers manufacturers. This process takes a lot of time and efforts and due care must be taken in each and every stage of this process to ensure the accurate results. Silicon wafers are available in different sizes and dimensions as desired by the silicon wafers manufacturers.

Silicon wafers have amazing mechanical and electrical properties and due to the same reason, they are the most widely used wafers in the industry today. They capture a huge portion of the market and are able to satisfy almost all of the needs of their customers. One of the majorly used products in which they are used are silicon solar cells which are now very popular all over the world.

3.4 Solar cells

Solar cells are semi-conductor devices which use sunlight to produce electricity. They are manufactured and processed in a similar fashion as computer memory chips. Solar cells are primarily made up of silicon which absorbs the photons emitted by sun's rays. The process was discovered as early as 1839. Silicon wafers are doped and the electrical contacts are put in place to connect each solar cell to another. The resulting silicon disks are given an anti-reflective coating. This coating protects sunlight loss. The solar cells are then encapsulated and placed in an aluminium frame. The process requires continuous monitoring to ensure quality control over a period of time. After the manufacturing process is complete they undergo final test to check their efficiency under normal conditions (Fig. 3.2).

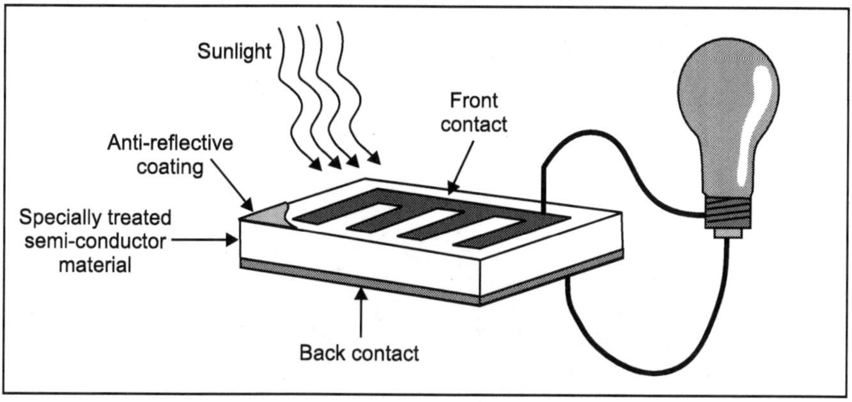

Figure 3.2 Solar cells

Solar cells provide more energy than other conventional sources with an additional advantage of being light weight and cost effective. Developing cheaper alternatives to solar cells such as amorphous silicon and polycrystalline silicon are also in the pipeline. Current research reveals that in order to increase the amount of sunlight effectively used and focusing it, prismatic lenses and layers of different materials could also be used.

3.5 Solar lanterns

Solar photovoltaic lantern is basically a portable lighting device suitable for either indoor or outdoor lighting covering a full range of 36°. The lantern can be used for 3 hours/day with autonomy of 2 days. Solar PV lantern consists of:

1. 7W CFL lamp
2. Sealed maintenance free battery
3. Electronics
4. Solar PV module
5. Module mounting structure
6. Interconnecting cable

The battery, lamp and the electronics are placed inside portable ABS plastic housing. Solar PV module generates electricity when exposed to direct sunlight, thus exploiting the unlimited source of free energy. A PV module consists of several small solar cells made of semiconductor material. Each of these solar cells generates a DC voltage and are capable of producing a finite amount of current. These solar cells are connected in series to form a PV module and are terminated to a terminal block for external connections (Fig. 3.3).

Figure 3.3 Solar lanterns

The system has been designed to meet the following requirements: The PV module has to be mounted external to the building on rooftop with maximum exposure of solar cells to the sunlight. Mounting frame provided may be used for fixing the module in position.

Compact fluorescent lamps also known as CFL are energy saving and the most efficient lamps. Due to the nature of their construction, these lamps generate higher light (lumens) output for a given amount of power or in other words these lamps consume less amount of power when compared to other lamps for a given light output. In the solar lantern, sealed lead acid maintenance free rechargeable batteries are used. The complete electronics basically consists of two functions:

1. Battery charge controller to protect deep discharging and overcharging of the battery. Suitable circuits have been provided to continuously monitor the battery voltage and based on preset voltage levels prevent either overcharging of the battery or deep discharging. Circuits for providing visual indication to indicate battery state of charge are also provided.
2. Inverter to generate the recommended AC voltage to the CFL and to ensure proper ignition and subsequent glow of the lamp. The inverter converts DC battery voltage to AC, which is fed to the lamp. Suitable filtering circuits are provided to ensure as good a sine wave as possible across the lamp.

3.6 Solar lights

Sun is the driving force behind everything. Solar lights, too, are backed up by the sun's energy. Solar lights are basically those lights that work without batteries. They get charged up by the sun's energy in the day time and then they light up in the dark. These lights consist of an LED (light emitting diode) and a photovoltaic solar panel. The photovoltaic panel directly converts the sun's energy into electricity that lights up the LED. There is a light sensitive switch inside these lights that makes them light up, the minute it begins to get dark (Fig. 3.4).

Figure 3.4 Solar lights

Solar lights are used in various areas nowadays. They are being used in the gardens to bring out the beauty of the landscape. They are used near swimming pools to light up the pathways. Solar lights are even being used to light up the streets. These lights are not expensive to install and neither do they have any maintenance cost attached to them. It uses the sun's energy for free of cost to run itself. They do not need anyone to start and stop it. They work automatically and are one of the best means to conserve power. Solar lights work on solar energy, which is a renewable source of energy. In this day and age, when the fossil fuels are getting exhausted at a very high rate for power generation purposes, solar lights come as the saving grace for humans. These lights help in saving power for the future generations and they are very eco-friendly. Solar lights are quite convenient and hassle free to install. They do not require wires in their installation and can light up to 8–10 hours.

Solar lights are being improved upon with each passing day. The solar technology being utilised in them is the technology of tomorrow. When nothing will be left on earth to create power from, humans will turn to sun only. Solar lights work on cloudy days as well if there's enough sun's energy stored in them. Researchers and scientists are working to improve this technology only, so that people can get the best out of these lights on cloudy and rainy days as well. Solar lights are very versatile in nature. Since they do not require any cords and extensions, they can be placed about anywhere, be it on the ground, on the wall, patios or fences.

These lights are available in many stylish designs and sizes. They are quite an attractive item to decorate the house, besides being very useful. The light emitted by these lights depends on the number of light emitting diodes placed in them. People can choose the amount of lighting they want, depending on the design of the solar lights they opt for. Solar lights are a good option to go with. They help people cut down their electricity bills and also help in making their houses more eye-catching with their elegant looks. These lights are available in different designs to suit the household needs and various colours like brown, copper, white, black and antique finish, etc.

3.7 Solar lighting systems

With the passage of each single day, the importance of saving power for the future generations is being realised by every individual. Measures are being taken up to save the fossil fuels from getting depleted, as they are non-renewable source of energy. Solar lighting systems are one of the measures to utilise the unbound sun's energy. Solar lighting systems work on simple basics, i.e. they trap the solar energy and convert it into electricity that lights up the LED present inside them, when there is darkness. Solar lighting systems have photovoltaic modules integrated in them that convert the solar energy directly into electricity (Fig. 3.5).

These solar lighting systems are very handy and easy to install. There are no extensions or wires required to install them. They can be placed about anywhere, from wall, fences, grounds, patios, anywhere. All it needs is to be placed in direct sunlight, so that it gets recharged in the day time and lights up during the night. There's a variety of solar lighting systems available in the market today. It all depends on the customer's requirements, as to what kind of solar lighting systems they want to set up in their house.

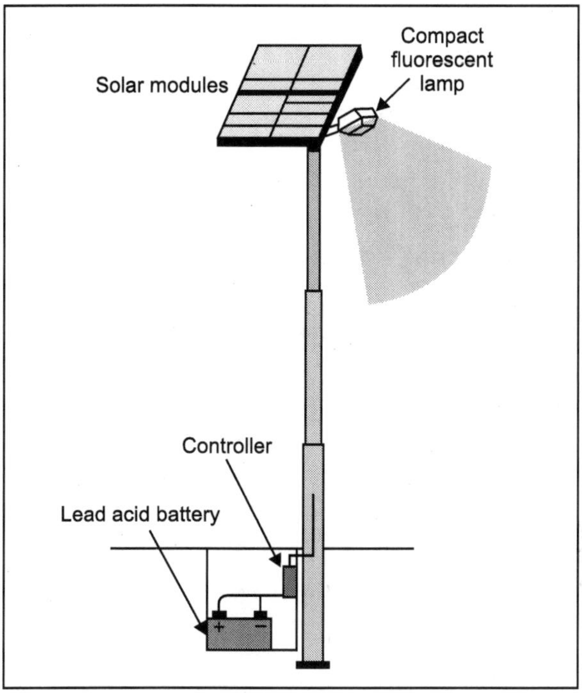

Figure 3.5 Solar lighting systems

Solar lighting systems differ in design, colour and the amount of lighting they provide. They also vary in their working systems. There are some solar lighting systems that light up automatically, after a certain point, when the light gets diminished. There are other types which have motion sensors positioned in them, that makes them light up the minute when there is some motion detected in a certain amount of radius. Then, there is another type of solar lighting systems that have a manual switch attached to it. Finally, there are some that are time monitored, i.e. they light up for a set time period according to the user's needs. Solar lighting systems are being used in different areas because of its versatility. Motion sensor solar lights are mounted in the households for security reasons. Light sensitive solar lights are placed in the gardens to add to its beauty and light up the pathways during the nights, without any extra cost. Solar lighting systems are being used in the shop signs, which get charged up during the day and light up during the night. Traffic signal lights are one of the best places that are utilising the solar energy. They are better than those lights that run on unreliable batteries. It helps in saving a huge amount of government's money as they are powered by the free source, sun. Solar lighting systems are also being applied in the bus stands, parking lots and streets lights. All these places are always under the sun and to make the best use of sun's energy here is a wise decision.

Solar lights are cost effective, hassle free and do not have any maintenance cost. Solar panels are the answer to the energy problems of the present as well as the future. The solar lighting systems help the people in going green, who want to save the nature for future.

3.8 Solar street lights

In this day and age, when the energy resources are being wasted every single second, it has become important for the people to go green. Solar street lights are a step taken up in that direction only. These lights do not run on conventional power, but are run by the solar energy. Solar street lights are the invention of the latest technology. They help in reducing the electricity consumption of any country, dramatically. They are worth every single penny that goes in its installation. Installation cost is the only cost, which the government has to bear, rest it does not call for any maintenance (Fig. 3.6).

Figure 3.6 Solar street systems

Solar street lights are economically viable and are being put to use in most of the developing countries. Change has begun in the right direction with the increase in use of these lights. Solar street lights do not require any excavation or any huge amount of labour work in their installation, as there are no cables attached to them. They have solar panels affixed to them that get charged by the sunlight and that makes them light up during the night. These panels must be properly aligned in the right direction, so that they can collect as much sunlight as possible. The operation of solar street lights, from dusk to dawn, is controlled by the photocell integrated within them.

Solar street lights are getting popular by day because they make the roads safe at night at a very effective cost. Basically, the general public pays the taxes for such services, and when there will be no cost attached to the street lights working, automatically the taxes will lower down. It is a win-win situation for both the public and the environment, as it does not work on the power of fossil fuels. Another major advantage of solar street lights is that they work in any condition whether there's power or not. For example, after any natural calamity, like a storm, hurricane or a blizzard, many lights get damaged and they do not work because the power lines fail. But these lights work, because of sun's natural power and they help in safeguarding the streets at night in a horrendous moment like the one aforesaid.

Solar street lights have lower power consumption and higher power efficacy. These lights have a longer lifetime as well when compared to conventional street lights. Solar street lights can work up to 50,000 hours, if there's no major damage done to them. These lights do not require a regular maintenance; all they need is a regular check up to see if everything is working optimally.

Solar street lights are no doubt the lights of the future. They are getting accepted by the day, by all the countries because of their numerous advantages for the earth. Street lights are a necessity of every country and going solar with these lights, will definitely have a huge impact in the conservation of power. That day is not far away, when every street in the world will be lit up by the solar street lights and every country will go green with its usage.

3.9 Solar ingots

Solar ingots are materials that can be cast into different shapes so that they can further be processed to make solar products like silicon solar cells, silicon wafers, etc. In the manufacturing of silicon wafers, solar ingots, which are formed of silicon which is very pure in nature, are cast into cylindrical shapes for further processing. The solar ingots are then cut, using saws or other cutting devices, in thin circular shapes which result in formation of silicon wafers. Using the solar ingots, the silicon wafers, which are prepared so far, are then thoroughly cleaned to remove the impurities and are also properly shaped to increase their efficiency.

There are many reasons why solar products should be used as a cleaner and more efficient source of energy. It becomes imperative to save on the manufacturing costs of solar ingots, silicon wafers and silicon solar cells so that their market price is very cheap and a lot of people use them in their daily lives.

There has also been an increase in the use of solar ingots and other solar products like solar water heaters, solar panels, silicon wafers, etc. in the last decade. To increase the growth rate of such products and hence their daily use, various new furnaces as well as new crystallisation processes have been developed in due course of time.

We hope that there will be a significant increase in the number of solar products which are being sold in each and every country.

3.10 Solar attic fans

These fans work on solar energy through solar panels and do not require any conventional power. Solar attic fans have photovoltaic modules incorporated in them, which convert the sun's light directly into electricity that powers the motor of the fan. The main function of solar attic fans is to reduce the temperature of the attics that can rise up to 160°F.

In the summers, when the sun shines in its full glory, the temperature of the attics rises up and that heat gets transferred to the entire house. That time an air conditioner is put to best use and the electricity bills touches the ceilings. Solar fans helps in venting out the hot air from the attics and thus reducing the workload on the air conditioners. These fans are not just about the summers; they are very beneficial during the winters as well. When the snow get accumulated on the roof of the house, the heat trapped in the attic makes that snow melt and hence, creates a puddle surrounding the entire house. Solar fans exhaust the trapped heat and prevent such happening.

Solar attic fans run the fastest when the sun shines the strongest, thereby keeping the overall temperature of the house neutral all year round, be it hot or cold. These fans even prevent the growth of mildew and moulds in the attic due to the moisture that gathers up in the attic, owing to the household activities like cooking, washing, showering, etc. Solar fans aerate this moisture and protect the wooden attic structure. Solar fans have no maintenance or operating cost attached to it. They do not even require any installation cost. You don't need any electrician to install these fans in your attic. You can do it yourself, very easily. You just need to put a hole in the drywall and install it in there. Solar fans are cost-effective and they have a very long lifespan.

Solar fans are available in three styles: The gable fan, the curb mounted fan and the self flashing roof fan. The first one exhausts the air horizontally through the gable end of the roof. The second type gets attached over a curb that is appended to the roof and the third type is affixed directly to the roof because the fan base is flat.

Solar fans are godsend to humans in this day and age, when global warming is happening at an unbelievable rate. Not only they help in lowering down the house temperature, they make use of the renewable source of energy. Utilising the solar energy to its full potential is one step forward towards a green and healthy future. Saving the electricity and in turn the power, with these fans, will surely help the future generations to see some amount of beautiful earth that is present today.

3.11 Solar water pumps

Water is the lifeline of every human being; without it, there's nothing. But there are places where there's a huge problem of water. The unavailability of water in remote places is causing the death of millions. To solve this problem, solar water pumps have been invented. Poor people do not have the money to run their households, so there's no way that they can afford a conventional grid-tied water pump to suffice their water needs. Even if they are able to afford the initial installation cost of the pump, it's impossible to match up with its running cost. Solar water pumps come as a saving grace for these poor souls, who desperately need help to meet their water requirements.

Solar water pumps run on solar energy. They are more eco-friendly and economical in their operation than the pumps powered by an internal combustion engine. The conventional water pumps operate on gas and diesel for their powering, which is a very expensive affair. Water pumps are generally used for irrigation purposes and farmers cannot afford such expensive pumps for their needs. Solar water pumps are an ideal solution for them. They only cost the installation charges and after that they work free on the solar energy. Solar water pumps are very easy to install and use. They do not even require extensive chords to power them. All they need is just a solar panel, which has photovoltaic modules included in it that convert the sun's energy into electricity. Solar water pumps are very eco-friendly, as there is no pollution caused by them. They help in preventing the global warming and cut down the electricity bills to bare minimum.

Solar water pumps helps in bringing water to the remote places, specially the desert areas where there is scarcity of water and abundance of solar heat. It is the ideal place to install solar water pumps. Besides that, solar water pumps are becoming popular among the home owners as well. They install the solar water pumps to irrigate their gardens and nurseries. Having a huge garden and maintaining it is not an easy task. It requires regular watering and for that, one needs a constant supply of water. To get that constant supply of water would obviously cost a lot.

Hence, solar water pumps come as a boon for such people. Solar energy is the best renewable source of energy that will help in prevention of the exhaustion of the fossil fuels. Solar water pumps are one of the various means to harness the sun's energy in most efficient manner. These pumps are an ideal solution in this age, to save the earth from this over increasing pollution. They are one of the best ways to help out the masses, who can't afford high electricity bills just to quench their thirst. Various countries, around the globe, are opting for these pumps because of their numerous benefits for the people as well as the environment. Solar pumps are used principally for three applications:

1. Village water supply
2. Livestock watering
3. Irrigation

A solar pump for village water supply is shown schematically in Fig. 3.7. The village will have a constant water demand although there is need to store water for periods of low insolation (low solar radiation). In environments where rainy seasons occur some of this demand can be met by rainwater harvesting during the rainy season.

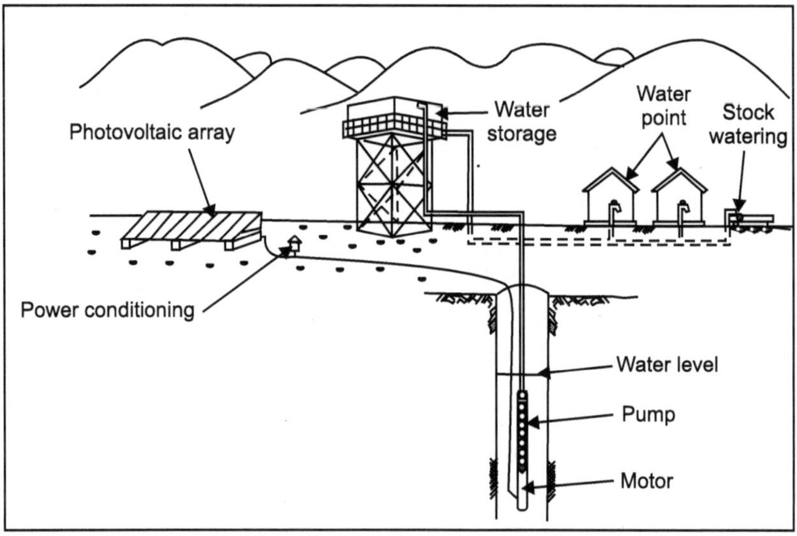

Figure 3.7 Village water supply

Ideally in Sahelian Africa the storage would be 3–5 days of water demand. In practice some installed tanks do not have sufficient capacity and are smaller than a days demand leaving the tank empty at the end of the day. This is due to a mismatch between the sizing, pump capacity and the demand profile during the day. The main applications for solar water pumping are for livestock watering in the USA and Australia. In Africa the systems are used for village water systems and livestock watering. While applications of solar water pumping for irrigation are on the increase especially in India and China. A solar irrigation system (Fig. 3.8) needs to take account of the fact that demand for irrigation water will vary throughout the year. Peak demand during the irrigation seasons is often more than twice the average demand. This means that solar pumps for irrigation can be under-utilised for most of the year although there can be a reduction in strength of the sun during these times reducing supply side of the equation. Attention should be paid to the system of water distribution and application to the crops. The system should minimise water losses, without imposing significant additional head on the pumping system and be of low cost.

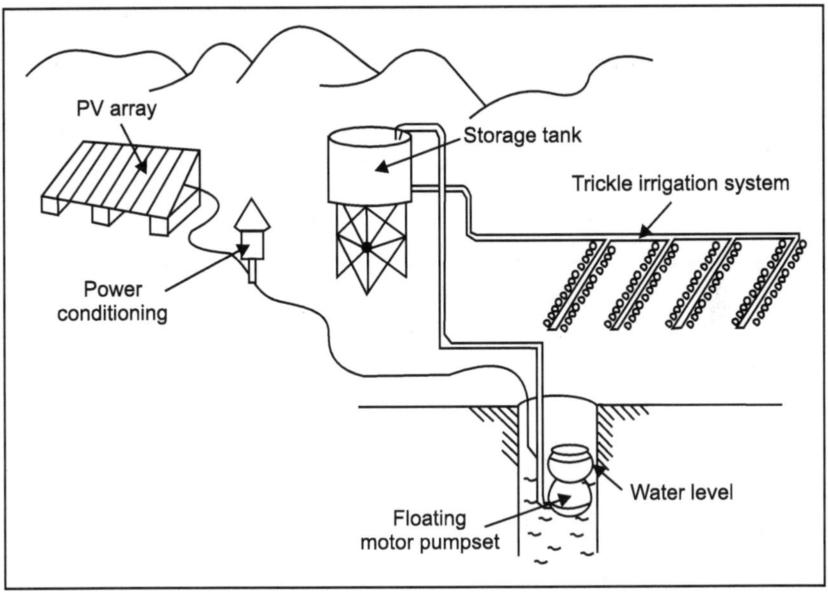

Figure 3.8 Solar irrigation system

Photovoltaic pumps are made up of a number of components. There is a photovoltaic array which converts solar energy directly into electricity as DC. The pump will have an electric motor to drive it. The characteristics of these components need to be matched to get the best performance. The pump motor unit will have its own optimum speed and load depending on the type and size of the pump.

3.12 Solar water heater

We can use sun's energy in a lot of ways for our benefit. Solar powered products like solar water heaters help us achieve this potential at no cost at all.

Process steam generation: The systems can be used to support almost any kind of thermal load for industrial or commercial applications, including:

1. Steam generation
2. Process hot water production
3. Desalination
4. Industrial and domestic hot water heating
5. Food processing (washing, cleaning, cooking, etc.)

Let us understand the make up of a solar water heater. The solar energy gets accumulated with the help of a collector. It consists of a solar panel with a selected coating that absorbs the heat and transfers it to the riser pipes underneath it. The temperature of the water being circulated in the pipes rises up and gets stored up in a tank. The storage tank is insulated to maintain the temperature of the water (Fig. 3.9).

The climate of a region and location are the deciding factors in choosing the solar water heater suitable for your use. Warmer areas benefit from active systems with direct circulation whereas colder climates require indirect circulation which makes use of a heat transfer fluid. For rooftops you can install passive systems where city water pushes the heated water to the outlet and storage is the collector itself. One can also go for the thermosiphon system where the storage tank is placed above the collector and gravity pushes the heated water.

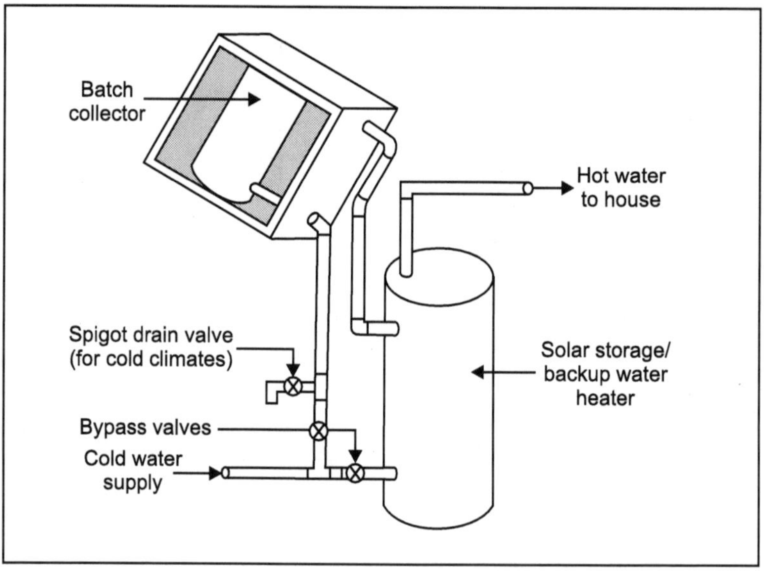

Figure 3.9 Solar water heater

Changing weather condition, limitation of space, amount of sunlight available or the capacity are some other factors that can influence your choice for buying a solar water heater.

3.12.1 Types of solar water heaters

Other types of a solar water heater are the flat plate collectors and evacuated tube collectors.

(a) *Flat plate collectors:* Flat plate collectors have glass on top and have blackened metallic absorbers inside and riser tubes to carry water.

(b) *Evacuated tube collectors:* Evacuated tube collectors are made of a glass tube where the outer covering absorbs the solar radiation and the heat generated is transferred to the flowing water.

After taking a look at the types of a solar water heater available in the market, one can decide which one would be best suitable for their homes. Similarly, during summers, one can choose from solar air conditioners and solar coolers. It is an excellent option for saving your electricity bills by 50–80 per cent and approximately 1500 units of electricity annually over the other conventional water heating alternatives like geysers of a capacity of 100 litres. Not only domestic but solar water heater is also used for commercial purposes such as hotels, hospitals, industries, dairies and enjoys a wider international market now. One of the major advantages of solar water heaters is that they are non-polluting and prevent the emission of carbon dioxide to a large extent. Once you have got your solar water heater installed, you can stop worrying about the return on investment.

3.13 Solar refrigerators

Solar refrigerators are not something that will come into existence in the future and will be invented in the developed countries, for that matter. Solar refrigerators are already in existence and were developed in the developing countries where there's always a shortage of power supply and excess of poverty. When one thinks of the word 'solar', photovoltaic modules and panels automatically come to mind, that converts the sunlight into electricity but that is not the case here. Solar refrigerators work on the evaporative cooling system. They just need sunlight and some water for their functioning. Solar panels in India are now being manufactured on a large scale. This has helped a lot in the popularity of solar refrigerators in India.

Solar refrigerators are an invention of necessity. In poverty-stricken developing countries, where there is no power and no money, usage of conventional refrigerators is out of the question. To start with, people cannot afford to buy such refrigerators and secondly, they cannot maintain such grid-power based refrigerators because of no regular supply of power. Thus, for such people an alternative was desperately required and that led to solar refrigerators. Solar refrigerators operate on the solar energy that cost not a single penny to the user. The main reason to build such refrigerators is to protect the perishable foods like meat and dairy items from getting spoiled as well as prevent the spoilage of the vaccines. These refrigerators are widely popular in many countries of Africa like Namibia and Zambia.

Solar refrigerators are very easy to make. They basically work on the principle of heat conduction and convection. There are two cylinders in solar refrigerators; the inner cylinder can be made of any recycled metal and it is placed inside the outside cylinder, which can be made of anything available in the household, like cardboard, plastic or wood. The outer cylinder must have holes on it, so that the sunlight can reach inside for its powering. The gap between the two cylinders is packed with organic material, either sand or wool and it is then saturated with water. The perishable items are kept inside the inner cylinder and then the whole set up is placed in direct sunlight. The heat of the sun evaporates the water in the organic material. This heat, in effect pulls the heat from the inner chamber and that leads to the cooling and reduction in temperature up to 6°C.

In the present age, conventional refrigerators prove to be very harmful for the environment. The chlorofluorocarbons present in the refrigerators contribute to global warming and ozone depletion. Using solar energy to power the solar refrigerators, aids in minimising the destructive impacts of conventional refrigerators on the environment. Hence, solar refrigerators are not only an economical solution to the problems of billions of people but it is also, one of the ways to safeguard the mother earth from complete destruction. Solar refrigerators can help us to protect the nature, so that even the future generations are able to see what we can see right now. It is in our hands to create a better future and the first step towards it can be the usage of solar refrigerators.

3.14 Solar televisions (TV)

Television has been called names like idiot box, since forever. The reason behind this is the addictiveness, which it impinges on the human brain. People get addicted to it because of its alluring programs and cannot seem to switch it off. Televisions around the globe consume the most power supply as they are turned on almost 24×7. This leads to sky-scraping electricity bills and wastage of the non-renewable energy. Solar televisions come as a solution to all the problems, which a traditional television puts up for the populace.

Solar televisions are not just a myth anymore; they have been invented in countries like Japan. Solar televisions are not run by grid-power and this is its biggest advantage. There are still, many places around the globe where power supply has not been made possible. Solar televisions are the answer to the prayers of those people living there. Living in the third world countries with no power supply is not a happy picture. Without the television, it is impossible for those people to know about what is happening around the world. Solar televisions run on solar power and thus can be accessed from anywhere. People will only gain from solar televisions; there is no doubt about that.

Solar televisions operate due to the thin-film photovoltaic panel that is attached to the television set. This solar panel traps the sun's energy and turns it into electricity. This electricity, in effect then powers the television. Solar televisions have very low power consumption. These televisions utilise less than one-third of the total energy required by traditional CRT-televisions. The area of the solar panel attached to them, is about the same as of the television screen.

Solar televisions are the creation of a futurist mind. The person who has invented these amazing televisions has surely thought about the future. Everyone knows that future is not far off, when all the non-renewable sources of energy will get completely used up. There will be no power to run the numerous inventions, which man has invented so arduously. There will be no more refrigerators, televisions, air conditioners, blenders, etc. because there will be no electricity to power them.

Hence, we need to start thinking about the alternatives and solar power is the most effective and beneficial alternative to power the various appliances. Solar televisions, thus, are the alternatives to help the future generation. These televisions are not popular right now but the scientists are working day and night to improve these solar televisions, so that they are used ubiquitously. They will not only help in saving the energy for the future but they will also help the present generation to lower their electricity bills. Thus, choose the right way. Go solar with solar televisions.

3.15 Solar coolers

During the summers, the sun shines in its full glory over our heads and makes us sweat like pigs. This results in the extra consumption of electricity to power our air conditioners and our coolers. Electricity bills touch the ceilings during the summers because of 24 hours of usage. But now, this can be warded off by making use of solar coolers. Solar coolers are just like conventional water coolers. The only difference between the two is that, the former operates on the DC power produced by the solar panels and the latter one operates on AC power.

The best quality about the solar coolers is that they can be made out of any conventional water cooler and still costs only 1/4 to 1/3 of power consumption of conventional units. It is an able decision to make use of solar power to run the water coolers because it is due to the sun only that their need arose in the first place. Making the optimum use of solar power during the daytime to power up the solar coolers would surely help in saving of a lot of electricity thereby putting off the burden from the electricity bills. During the night time, the solar coolers can be run by the battery backup that stores the solar power during the daytime. This makes the solar coolers function 24/7 and serves as a feasible alternative for the humans (Fig. 3.10).

Solar coolers are very easy to make from any conventional water cooler. To make a solar cooler, attach a 12 V automobile radiator fan and a bilge pump to the traditional cooler. The original fan and motor need not be removed for the purpose. The radiator fan can be mounted in front of the squirrel cage fan present in the water cooler. The bilge pump, too, can be placed next to the original pump inside the cooler. Conventional water coolers run on 110 V of AC power, whereas the solar cooler runs on a 12 V solar panel attached to it. The solar electric panel can be mounted on the home made tracker, which helps in extending the time period of solar cooler's working. It has been noticed that almost 2 to 3 hours of extra usage is possible with the tracker.

Solar coolers are very economical and to get the best out of them. One must switch it on since early morning when the sun comes out. Ideally, an AC or a cooler is turned on when it begins to get really hot. But with solar coolers, it would not cost a penny to start them since early morning. It will be put to maximum utilisation and will keep the house temperature at about 15 degrees.

Solar coolers work best in dry climate areas. They are eco-friendly and do not add to the unlimited pollution already surmounting the earth. Solar coolers help in preserving the priceless non-renewable sources of energy, which goes to waste in an unrestricted manner during the summers. Going solar with solar coolers is the coolest decision one can take for themselves as well as for the environment.

Figure 3.10 Solar coolers

3.16 Solar air conditioners (AC)

In this day and age, when there is a rapid decease in the fossil fuels and a swift increase in pollution happening, going Green is a wise decision. Solar air conditioners are one of the best ways of contributing to the nature's preservation. During the summers, when the heat of the sun surpasses all the boundaries and creates inhospitable conditions for the humans, an air conditioner is turned on to seek relief. Conventional air conditioners use excessive power and add to the electricity bills tremendously. During the summers, an air conditioner is a necessity of every household. In today's world, one can't stop using them but one can always go solar with them.

Solar air conditioners work on the power of sunlight. The more the sunlight, the more effective the solar air conditioner would be. Solar air conditioners are a practical appliance and are getting popular by the day. People, now value the power of solar energy and they appreciate the solar products being launched in the market. Solar air conditioners help the populace to get relieved of the summer temperature in a guilt free manner. Conventional air conditioners have harmful elements in them like chlorofluorocarbons, which contribute excessively to global warming and the expansion of the ozone hole. Solar air conditioners, on the other hand are totally free of such hazardous elements. They run on electricity, which is produced by the photovoltaic integrated on the solar panel attached to the solar air conditioners. This power is absolutely free of cost and does not add to any electricity bill, either. Solar air conditioners work on renewable energy resource and thus aid in the conservation of the non-renewable resources.

Solar air conditioners have many other benefits as well. If you install solar air conditioners in your home, you will get great incentives by the government on its installation cost. The savings, which one generates by using solar air conditioners, would even enhance their property rates. Solar air conditioners are very silent, have no compressor appended to it and hence have no costly maintenance. It is just the solar panels, which require a timely check up to ensure optimum sunlight absorption during the daytime.

Solar air conditioners are used as an alternative to conventional air conditioners in countries like Japan, China and India. There the electricity price is very high and to avoid heavy electricity bills, going solar is the safest bet. These air conditioners are proving to be, one of the finest invention by man kind as they cut back the electric bills to 1/3 of the original. They provide the same amount of cooling, which a traditional air conditioner does. The chief difference between the two conditioners is that, the air produced by the solar air conditioners is 100 per cent eco-friendly. It helps the humans as well as the environment to breathe healthily. Solar power is the power of the future. Getting access to solar power, as fast as one can, is the smart thing to do because future is not far-off. The time has come to make the better judgement yourself.

3.17 Solar cooking systems

Cooking food is a necessity of every household. You cannot stop cooking, no matter how much concern you have for the preservation of the resources. You cannot stop using gas, until and unless you have a better substitute. Solar cooking systems are those alternatives, which will help you to cook food efficiently and conserve the energy resources for the future. Employment of LPG for cooking, leads to the wastage of non-renewable resources. It is a very dangerous gas as well, since it ignites immediately. If left unopened, gas would spread very rapidly and would set fire to the entire house, in matter of seconds. It just needs a spark to get ignited (Fig. 3.11).

Figure 3.11 Solar cooking system

Solar cooking systems are altogether different from conventional gas systems in all its aspects. Solar cooking systems do not work on any kind of gas, so there is no chance of wastage of any resources. They do not operate on the burning of wood either, so there is no chance of deforestation. Solar cooking systems work on solar energy. They just require sunlight to cook food and nothing else. It is hence clear, that these cooking systems do not cost a single penny to the pocket as they function using the free energy source. Solar cooking systems are very eco-friendly and do not assist in any kind of pollution, global warming or ozone layer depletion, for that matter.

Solar cooking systems are basically solar cookers, solar kettles, solar bowls and hybrid solar grills, that are used to cook food. Solar cooking systems essentially work on a simple rule of converting the light energy into heat energy. Reflective mirrors are used in the solar cookers to converge the sunrays to a single focus point, where they produce heat. At this point the food is kept and hence gets cooked.

There is a variety of solar cookers available in the market nowadays. Some of the most popular solar cookers are parabolic cookers, panel cookers and box cookers. Solar kettles are mainly used to boil water using the solar glass tube technology that stores solar energy, and powers the kettle. Hybrid solar grills have moveable grill surface, which has an adjustable parabolic reflector balanced on a tripod. This parabolic reflector reflects the sunrays and converge them to grill food.

All around the globe, solar cooking systems are being promoted because of their various advantages for the environment. Third world countries are endorsing these solar cooking systems because they do not require any gas or electricity for their functioning. Solar cooking systems are an answer to the woes of poverty-stricken people living in remote areas, who depend on wood for cooking their food. These people won't have to go to the forests anymore to gather wood and they won't even have to get their throats choked due to the fumes coming out of wood's combustion. Solar cooking systems and solar panels are gaining acceptability among the masses due to their utilitarian aspect as well as their contribution to the conservation of the natural resources.

3.18 Solar steam generators

In present time, the excessive use of electricity for every small thing has led to a shortage all around the globe. Resources are getting exhausted swiftly and population is increasing on the double. This situation is heading towards a disastrous end and to avert that end solar steam generators have been invented. Solar steam generators utilise the solar energy to produce electricity rather than the conventional non-renewable resources. Traditionally, electricity is produced in power plants by the turning of heavy turbines, which are powered by steam produced by the burning of coal, gas or oil. Hence, these power plants emits out unsafe gases in the atmosphere that is causing the global warming.

Solar steam generators are extremely eco-friendly and helps in keeping the air clean and produce maximum electricity. These generators' popularity is increasing by the day around the sphere, as they work on a renewable source of energy, i.e. solar thermal energy. Another factor for increasing acceptability of solar steam generators is that solar energy is absolutely free whereas the price of coal and oil is escalating by the minute.

Solar steam generators consist of a boiler, heliostats, heat transfer fluid and steam turbine. Boiler is the place where the heat is collected from the sun. The spherical mirrors (heliostats) used in the solar steam generators can either be cylindrical parabola or the circular parabola. The former one focuses the sunlight to a line whereas the latter one focuses it to a point. Heliostats, the moveable parabolic mirrors, face automatically towards the sun to absorb the heat of the optimally. The heat transfer fluid can either be water or liquid sodium. When the fluid placed inside the boiler gets heated up by the sun's heat, it is transferred to the heat exchanger where it gets transformed into steam. Water's temperature can be raised up to 220°C using the solar steam generators. After 100°C, water converts into steam under very high pressure and this steam is then pushed on to the steam turbine. The pressure of the steam makes it to move in circular motions and thus creates electricity (Fig. 3.12).

Solar steam generators' utility is helping people to get access to electricity where there is extreme shortage of it. In distant future, when there will be no more fossil fuels to burn that can power the turbine, solar steam generators will be the only solution. Solar steam generators help in protecting the environment as well as the human species from harmful diseases, which are caused due to the deadly fumes and gases released by the conventional power plants. Using solar steam generators from today itself is a step taken towards a secure future. Safeguarding oneself from utter despair, when there will be no more resources to burn, is the only priority in everyone's mind right now. Solar steam generators are being adopted by responsible countries across the globe, which understand the value of non-renewable resources and the harmful effects of their sheer wastage.

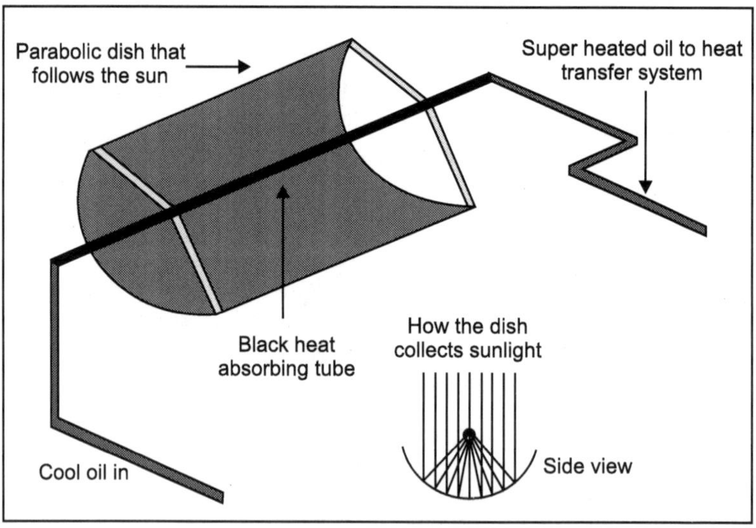

Figure 3.12 Solar steam generators

3.19 Solar flat-plate collectors

Solar energy is the most viable source of power, which is gaining popularity by the day. Heating the water using the solar energy is one of the most efficient utilisation of solar power. Basically, in solar heating systems there are solar flat-plate collectors and storage tanks to store the thermal water. Solar flat-plate collectors are one of the types of solar collectors that are used to absorb and intercept the sun's radiations for water heating. Thermal water obtained using the solar flat-plate collectors is generally used in the hotels, businesses, homes, swimming pools, under floor heating, etc.

Solar flat-plate collectors are the solar thermal collectors that consist of an insulated and weatherproof box, which has a black metal absorber sheet and built in pipes where the water is present. The water inside the pipes gets heated up by the solar energy and by natural convection the water circulates through the system. There are chiefly two types of solar heating—active and passive. In passive solar water heating, the hot water from the solar collectors gets transferred to the storage tanks through means of gravity. The motion of water is slow compared to the demand of the water supply. But when there are some electrical pumps attached in the system to push the thermal water from the collector pipes to the storage tanks then it becomes an active solar heating system. Solar flat-plate collectors are varied in design but mainly all of them contain a flat-plate collector, a transparent cover and a heat transport fluid (generally water or an antifreeze fluid). Flat-plate collector absorbs the sunrays and the transparent cover lets the solar energy to pass through reducing the heat loss from the absorber. The heat transport fluid that circulates in the pipes affixed to the collector removes the heat from the absorber and heats the water, which is then transferred to the storage tanks. Solar flat plate collectors, best suit the demand of places where the required water temperature ranges 30° to 70°C. Solar flat-plate collectors are the best alternative to the conventional geysers used in the winters to heat the water. The electricity consumption that reaches its peak using the power-grid geysers can be cut back to 20 per cent consumption with the installation of solar flat-plate collectors in any household. The solar flat plate collectors are now designed very diligently for areas where freezing is a possibility. They are plumbed in such a way that the water gets drained down completely using gravity, so that their cracking does not happen due to freezing.

Solar flat-plate collectors are being installed everywhere nowadays because of their utilitarian aspect as well as their economical aspect. They do not use grid-tied power, and hence, do not cause any wastage of the natural resources. Solar-flat plate collectors work on free solar power and are thus aiding the prevention of global warming. Solar-flat plate collectors utilise the solar power optimally to heat the water and hence meet all the needs of humans all year round.

A flat-plate collector is shown in Fig. 3.13. The basic parts noted are a full-aperture absorber, transparent or translucent cover sheets, and an insulated box. The absorber is usually a sheet of high-thermal-conductivity metal with tubes or ducts either integral or attached. Its surface is painted or coated to maximise radiant energy absorption and in some cases to minimise radiant emission. The cover sheets, called glazing, let sunlight pass through to the absorber but insulate the space above the absorber to prohibit cool air from flowing into this space. The insulated box provides structure and sealing and reduces heat loss from the back or sides of the collector.

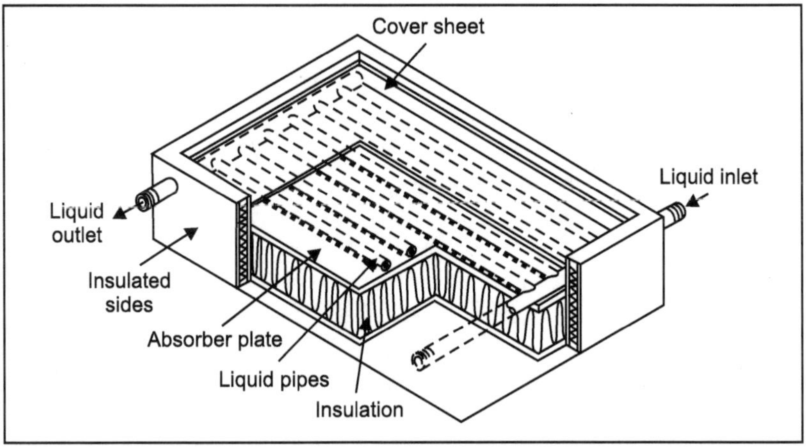

Figure 3.13 A typical liquid flat-plate collector

3.20 Solar power plants

Sun is the ultimate source of energy for each and every living thing on earth. Harnessing sun's unbound energy to the best of its ability is done in solar power plants. There are many ways to get the power from the sun through solar panels in solar power plants. There are direct as well as indirect ways to convert the sun's heat into power. Directly, the heat from sun is converted into electricity using photovoltaic. Indirectly, the sun's heat is converted into power by concentrating the heat of the sun to boil the water and then the steam from it is used for electricity generation (Fig. 3.14). The construction of solar power plants is the best thing that is happening right now for the future generations. With each passing day, the consumption of power, which is produced by burning of the fossil fuels, is increasing at a very high rate. Fossil fuels are getting depleted at a much faster rate than they are being produced. If things keep on going the way they are right now, that time is not far away when there will be no energy or power to consume. To prevent this dreadful day from coming into being, solar power plants are being set up around the globe. These solar power plants have proved to be and still are a boon to mankind.

Figure 3.14 Solar power plants

Sun's radiations are intermittent and to get a continuous supply of power various methods are adopted at the solar power plants. One of such ways taken on at the solar power plants is storage. The solar power is produced by concentrating the sun's radiations, with the help of mirrors and lenses, into a single beam and that beam is used to heat the water. The steam from the hot water is then used in a conventional power plant to generate electricity. This produced electricity is then stored so that it can be utilised when the sun is not there outside, shining in its full glory. Although the solar power plants are expensive to install, in spite of that, countries around the globe are opting for solar power plants instead of other power plants because solar energy is a renewable source of energy. This solar energy when combined with the other sources of energy such as the wind, biomass and hydro power, results in 100 per cent renewable energy. This has been accomplished in a large combined power plant, which is bigger than any of the solar power plants, set up in Germany. Solar power plants are the best source of power supply not only for the humans but for the environment as well. Solar energy is very eco-friendly and it does not contribute towards global warming at all as the other fossil fuel power plants do. Solar power plants make the ideal utilisation of sun's energy and are helping the future generations from being powerless. Solar power plants are a like a ray of hope for the human race to survive when all the fossil fuels will get exhausted and there will be nothing left to produce power from anymore.

4
Solar thermal energy

4.1 Introduction

Solar thermal energy (STE) is a technology for harnessing solar energy for thermal energy (heat). Solar thermal collectors are classified as Low-, medium- or High-temperature collectors. Low temperature collectors are flat-plates generally used to heat swimming pools. Medium-temperature collectors are also usually flat-plates but are used for heating water or air for residential and commercial use. High-temperature collectors concentrate sunlight using mirrors or lenses and are generally used for electric power production. STE is different from (and indeed much more efficient than) photovoltaics, which converts solar energy directly into electricity.

4.2 Low-temperature collectors

Low-temperature collectors are generally installed to heat swimming pools, although they can also be used for space heating. Collectors can use air or water as the medium to transfer the heat to their destination.

4.2.1 Heating, cooling and ventilation

In the United States, heating, ventilation, and air conditioning (HVAC) systems account for over 25 per cent (4.75 EJ) of the energy used in commercial buildings and nearly half (10.1 EJ) of the energy used in residential buildings. Solar heating, cooling, and ventilation technologies can be used to offset a portion of this energy.

Thermal mass materials store solar energy during the day and release this energy during cooler periods. Common thermal mass materials include stone, concrete, and water. The proportion and placement of thermal mass should consider several factors such as climate, day lighting and shading conditions. When properly incorporated, thermal mass can passively maintain comfortable temperatures while reducing energy consumption. A solar chimney (or thermal chimney) is a passive solar ventilation system composed of a hollow thermal mass connecting the interior and exterior of a building. As the chimney warms, the air inside is heated causing an updraft that pulls air through the building. These systems have been in use since Roman times and remain common in the Middle East.

Solar space heating with solar air heat collectors is more popular in the USA and Canada than heating with solar liquid collectors since most buildings already have a ventilation system for heating and cooling. The two main types of solar air panels are glazed and unglazed. Glazed solar collectors are designed primarily for space heating and they re-circulate building air through a solar air panel where the air is heated and then directed back into the building. These solar space heating systems require at least two penetrations into the building and only perform when the air in the solar collector is warmer than the building room temperature. Most glazed collectors are used in the residential sector.

Unglazed solar collectors are primarily used to pre-heat make-up ventilation air in commercial, industrial and institutional buildings with a high ventilation load. They turn building walls or sections of walls into low cost, high performance, unglazed solar collectors. Also called, 'transpired solar panels', they employ a painted perforated metal solar heat absorber that also serves as the exterior wall surface of the building. Heat conducts from the absorber surface to the thermal boundary layer of air 1 mm thick on the outside of the absorber and to air that passes behind the absorber. The boundary layer of air is drawn into a nearby perforation before the heat can escape by convection to the outside air. The heated air is then drawn from behind the absorber plate into the building's ventilation system.

A Trombe wall is a passive solar heating and ventilation system consisting of an air channel sandwiched between a window and a sun-facing thermal mass. During the ventilation cycle, sunlight stores heat in the thermal mass and warms the air channel causing circulation through vents at the top and bottom of the wall. During the heating cycle the Trombe wall radiates stored heat.

Solar roof ponds are unique solar heating and cooling systems developed by Harold Hay in the 1960s. A basic system consists of a roof-mounted water bladder with a moveable insulating cover. This system can control heat exchange between interior and exterior environments by covering and uncovering the bladder between night and day. When heating is a concern the bladder is uncovered during the day allowing sunlight to warm the water bladder and store heat for evening use. When cooling is a concern the covered bladder draws heat from the building's interior during the day and is uncovered at night to radiate heat to the cooler atmosphere. The Skytherm house in Atascadero, California, uses a prototype roof pond for heating and cooling.

Active solar cooling can be achieved via absorption refrigeration cycles, desiccant cycles, and solar mechanical processes. In 1878, Auguste Mouchout pioneered solar cooling by making ice using a solar steam engine attached to a refrigeration device. Thermal mass, smart windows and shading methods can also be used to provide cooling. The leaves of deciduous trees provide natural shade during the summer while the bare limbs allow light and warmth into a building during the winter. The water content of trees will also help moderate local temperatures.

Process heat

Solar process heating systems are designed to provide large quantities of hot water or space heating for non-residential buildings. Evaporation ponds are shallow ponds that concentrate dissolved solids through evaporation. The use of evaporation ponds to obtain salt from sea water is one of the oldest applications of solar energy. Modern uses include concentrating brine solutions used in leach mining and removing dissolved solids from waste streams. Altogether, evaporation ponds represent one of the largest commercial applications of solar energy in use today.

Unglazed transpired collectors (UTC) are perforated sun-facing walls used for preheating ventilation air. UTCs can raise the incoming air temperature up to 22°C and deliver outlet temperatures of 45–60°C. The short payback period of transpired collectors (3 to 12 years) make them a more cost-effective alternative to glazed collection systems. As of 2009, over 1500 systems with a combined collector area of 3,00,000 m^2 had been installed worldwide. Representatives include an 860 m^2 collector in Costa Rica used for drying coffee beans and a 1300 m^2 collector in Coimbatore, India used for drying marigolds.

A food processing facility in Modesto, California uses parabolic troughs to produce steam used in the manufacturing process. The 5000 m^2 collector area is expected to provide 4.3 GJ per year.

4.3 Medium-temperature collectors

These collectors could be used to produce approximately 50 per cent and more of the hot water needed for residential and commercial use in the United States. In the United States 30 per cent of the system qualifies for a federal tax credit + additional state credit exists in about half of the states. Labour for a simple open loop system in southern climates can take 3–5 hours for the installation and 4–6 hours in Northern areas. Northern system requires more collector area and more complex plumbing to protect the collector from freezing. With this incentive, the payback time for a typical household is four to nine years, depending on the state. Similar subsidies exist in parts of Europe. A crew of one solar plumber and two assistants with minimal training can install a system per day. Thermosiphon installation have negligible maintenance costs (costs rise if antifreeze and mains power are used for circulation) and in the US reduces a household's operating costs by $6 per person per month. Solar water heating can reduce CO_2 emissions of a family of four by 1 ton/year (if replacing natural gas) or 3 ton/year (if replacing electricity). Medium-temperature installations can use any of several designs: Common designs are pressurised glycol, drain back, batch systems and newer low pressure freeze tolerant systems using polymer pipes containing water with photovoltaic pumping. European and International standards are being reviewed to accommodate innovations in design and operation of medium temperature collectors. Operational innovations include 'permanently wetted collector' operation. This innovation reduces or even eliminates the occurrence of no-flow high-temperature stresses called stagnation which would otherwise reduce the life expectancy of collectors.

4.3.1 Solar drying

Solar thermal energy can be useful for drying wood for construction and wood fuels such as wood chips for combustion. Solar is also used for food products such as fruits, grains, and fish. Crop drying by solar means is environmentally friendly as well as cost effective while improving the quality. The less money it takes to make a product, the less it can be sold for, pleasing both the buyers and the sellers. Technologies in solar drying include ultra low cost pumped transpired plate air collectors based on black fabrics. Solar thermal energy is helpful in the process of drying products such as wood chips and other forms of biomass by raising the heat while allowing air to pass through and get rid of the moisture.

4.3.2 Cooking

Solar cookers use sunlight for cooking, drying and pasteurisation. Solar cooking offsets fuel costs, reduces demand for fuel or firewood, and improves air quality by reducing or removing a source of smoke. The simplest type of solar cooker is the box cooker first built by Horace de Saussure in 1767. A basic box cooker consists of an insulated container with a transparent lid. These cookers can be used effectively with partially overcast skies and will typically reach temperatures of 50–100°C.

Concentrating solar cookers use reflectors to concentrate light on a cooking container. The most common reflector geometries are flat-plate,

disc and parabolic trough type. These designs cook faster and at higher temperatures (up to 350°C) but require direct light to function properly.
The Solar Kitchen in Auroville, India, uses a unique concentrating technology known as the solar bowl. Contrary to conventional tracking reflector/fixed receiver systems, the solar bowl uses a fixed spherical reflector with a receiver which tracks the focus of light as the sun moves across the sky. The solar bowl's receiver reaches temperature of 150°C that is used to produce steam that helps cook 2000 daily meals.

Many other solar kitchens in India use another unique concentrating technology known as the Scheffler reflector. This technology was first developed by Wolfgang Scheffler in 1986. A Scheffler reflector is a parabolic dish that uses single axis tracking to follow the sun's daily course. These reflectors have a flexible reflective surface that is able to change its curvature to adjust to seasonal variations in the incident angle of sunlight. Scheffler reflectors have the advantage of having a fixed focal point which improves the ease of cooking and are able to reach temperatures of 450°–650°C. Built in 1999, the world's largest Scheffler reflector system in Abu Road, Rajasthan India is capable of cooking up to 35,000 meals a day. By early 2008, over 2000 large cookers of the Scheffler design had been built worldwide.

4.3.3 Distillation

Solar stills can be used to make drinking water in areas where clean water is not common. Solar distillation is necessary in these situations to provide people with purified water. Solar energy heats up the water in the still. The water then evaporates and condenses on the bottom of the covering glass.

4.4 High-temperature collectors

Where temperatures below about 95°C are sufficient, as for space heating, flat-plate collectors of the non-concentrating type are generally used. Because of the relatively high heat losses through the glazing, flat plate collectors will not reach temperatures much above 200°C even when the heat transfer fluid is stagnant. Such temperatures are too low for efficient conversion to electricity.

The efficiency of heat engines increases with the temperature of the heat source. To achieve this in solar thermal energy plants, solar radiation is concentrated by mirrors or lenses to obtain higher temperatures—a technique called concentrated solar power (CSP). The practical effect of high efficiencies is to reduce the plant's collector size and total land use per unit power generated, reducing the environmental impacts of a power plant as well as its expense.

As the temperature increases, different forms of conversion become practical. Up to 600°C, steam turbines, standard technology, have efficiency up to 41 per cent. Above 600°C, gas turbines can be more efficient. Higher temperatures are problematic because different materials and techniques are needed. One proposal for very high temperatures is to use liquid fluoride salts operating between 700°C and 800°C, using multi-stage turbine systems to achieve 50 per cent or more thermal efficiencies. The higher operating temperatures permit the plant to use higher-temperature dry heat exchangers for its thermal exhaust, reducing the plant's water use–critical in the deserts where large solar plants are practical. High temperatures also make heat storage more efficient, because more watt-hours are stored per unit of fluid.

Since the CSP plant generates heat first of all, it can store the heat before conversion to electricity. With current technology, storage of heat is much cheaper and more efficient than storage of electricity. In this way, the CSP plant can produce electricity day and night. If the CSP site has predictable solar radiation, then the CSP plant becomes a reliable power plant. Reliability can further be improved by installing a back-up system that uses fossil energy. The back-up system can reuse most of the CSP plant, which decreases the cost of the back-up system.

With reliability, unused desert, no pollution, and no fuel costs, the obstacles for large deployment for CSP are cost, aesthetics, land use and similar factors for the necessary connecting high tension lines. Although only a small percentage of the desert is necessary to meet global electricity demand, still a large area must be covered with mirrors or lenses to obtain a significant amount of energy. An important way to decrease cost is the use of a simple design.

4.4.1 System designs

During the day the sun has different positions. For low concentration systems (and low temperatures) tracking can be avoided (or limited to a few positions per year) if non-imaging optics are used. For higher concentrations, however, if the mirrors or lenses do not move, then the focus of the mirrors or lenses changes (but also in these cases non-imaging optics provides the widest acceptance angles for a given concentration). Therefore it seems unavoidable that there needs to be a tracking system that follows the position of the sun (for solar photovoltaic a solar tracker is only optional). The tracking system increases the cost and complexity. With this in mind, different designs can be distinguished in how they concentrate the light and track the position of the sun.

Parabolic trough designs

Parabolic trough power plants use a curved, mirrored trough which reflects the direct solar radiation onto a glass tube containing a fluid (also called a receiver, absorber or collector) running the length of the trough, positioned at the focal point of the reflectors. The trough is parabolic along one axis and linear in the orthogonal axis. For change of the daily position of the sun perpendicular to the receiver, the trough tilts east to west so that the direct radiation remains focused on the receiver. However, seasonal changes in the in angle of sunlight parallel to the trough does not require adjustment of the mirrors, since the light is simply concentrated elsewhere on the receiver. Thus the trough design does not require tracking on a second axis (Fig. 4.1).

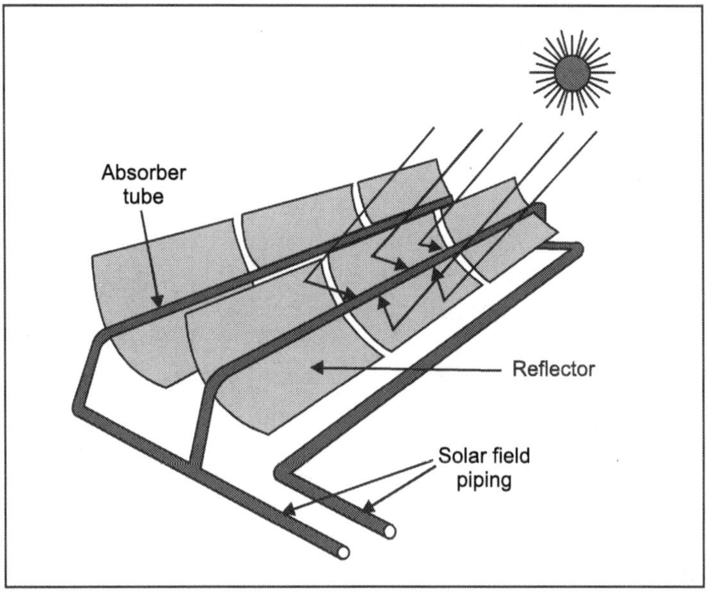

Figure 4.1 A parabolic trough design. A change in position of the sun parallel to the receiver does not require adjustment of the mirrors.

The receiver may be enclosed in a glass vacuum chamber. The vacuum significantly reduces convective heat loss. A fluid (also called heat transfer fluid) passes through the receiver and becomes very hot. Common fluids are synthetic oil, molten salt and pressurised steam. The fluid containing the heat is transported to a heat engine where about a third of the heat is converted to electricity.

Andasol 1 in Gaudix, Spain, uses the Parabolic Trough design which consists of long parallel rows of modular solar collectors. Tracking the sun from East to West by rotation on one axis, the high precision reflector panels concentrate the solar radiation coming directly from the sun onto an absorber pipe located along the focal line of the collector. A heat transfer medium, such as synthetic oil in car engines, is circulated through the absorber pipes at temperatures up to 400°C and generates live steam to drive the steam turbine generator of a conventional power block.

Full-scale parabolic trough systems consist of many such troughs laid out in parallel over a large area of land. Since 1985 a solar thermal system using this principle has been in full operation in California in the United States. It is called the SEGS system. Other CSP designs lack this kind of long experience and therefore it can currently be said that the parabolic trough design is the most thoroughly proven CSP technology. The solar energy generating system (SEGS) is a collection of nine plants with a total capacity of 350 MW. It is currently the largest operational solar system (both thermal and non-thermal). A newer plant is Nevada Solar One plant with a capacity of 64 MW. Under construction are Andasol 1 and Andasol 2 in Spain with each site having a capacity of 50 MW.

Note however, that those plants have heat storage which requires a larger field of solar collectors relative to the size of the steam turbine-generator to store heat and send heat to the steam turbine at the same time. Heat storage enables better utilisation of the steam turbine. With day and some night time operation of the steam-turbine Andasol 1 at 50 MW peak capacity produces more energy than Nevada Solar One at 64 MW peak capacity, due to the former plant's thermal energy storage system and larger solar field.

553 MW new capacity is proposed in Mojave Solar Park, California. Furthermore, 59 MW hybrid plant with heat storage is proposed near Barstow, California. Near Kuraymat in Egypt, some 40 MW steam is used as input for a gas powered plant. Finally, 25 MW steam input is used for a gas power plant in Hassi R'mel, Algeria.

Power tower designs

Power towers (also known as 'central tower' power plants or 'heliostat' power plants) capture and focus the sun's thermal energy with thousands of tracking mirrors (called heliostats) in roughly a two square mile field. A tower resides in the centre of the heliostat field. The heliostats focus concentrated sunlight on a receiver which sits on top of the tower. Within the receiver the concentrated sunlight heats molten salt to over 1000°F (538°C). The heated molten salt then flows into a thermal storage tank where it is stored, maintaining 98 per cent thermal efficiency, and eventually pumped to a steam generator. The steam drives a standard turbine to generate electricity. This process, also known as the 'Rankine cycle' is similar to a standard coal-fired power plant, except it is fuelled by clean and free solar energy.

The advantage of this design above the parabolic trough design is the higher temperature. Thermal energy at higher temperatures can be converted to electricity more efficiently and can be more cheaply stored for later use. Furthermore, there is less need to flatten the ground area. In principle a power tower can be built on a hillside. Mirrors can be flat and plumbing is concentrated in the tower. The disadvantage is that each mirror must have its own dual-axis control, while in the parabolic trough design one axis can be shared for a large array of mirrors. In June 2008, eSolar, a California-based company founded by Idealab CEO Bill Gross with funding from Google, announced a power purchase agreement (PPA) with the utility Southern California Edison to produce 245 megawatts of power. Also, in February 2009, eSolar announced it had licensed its technology to two development partners, the Princeton, N.J.-based NRG Energy, Inc., and the India-based ACME Group. In the deal with NRG, the companies announced plans to jointly build 500 MW of concentrating solar thermal plants throughout the United States. The target goal for the ACME Group was nearly double; ACME plans to start construction on its first eSolar power plant in the year 2012, and will build a total of 1 gigawatt over the next 10 years.

eSolar's proprietary sun-tracking software coordinates the movement of 24,000 1 m^2 mirrors per 1 tower using optical sensors to adjust and calibrate the mirrors in real time. This allows for a high density of reflective material which enables the development of modular CSP power plants in 46 megawatt (MW) units on approximately π square mile parcels of land, resulting in a land-to-power ratio of 4 acres (16,000 m^2) per 1 MW.

Bright Source Energy entered into a series of power purchase agreements with Pacific Gas and Electric Company in March 2008 for up to 900 MW of electricity, the largest solar power commitment ever made by a utility. Bright Source is currently developing a number of solar power plants in Southern California, with construction of the first plant planned to start in 2009.

In June 2008, Bright Source Energy dedicated its 4–6 MW Solar Energy Development Center (SEDC) in Israel's Negev Desert. The site, located in the Rotem Industrial Park, features more than 1600 heliostats that track the sun and reflect light onto a 60-m high tower. The concentrated energy is then used to heat a boiler atop the tower to 550 degrees Celsius, generating superheated steam. A working tower power plant is PS10 in Spain with a capacity of 11 MW. The 15 MW Solar Tres plant with heat storage is under construction in Spain. In South Africa, a 100 MW solar power plant is planned with 4000–5000 heliostat mirrors, each having an area of 140 m^2.

Morocco is building five solar thermal power plants around Ouasarzate. The sites will produce about 2000 MW by 2012. Over ten thousand hectors of land will be needed to sustain all of the sites.

Out of commission are the 10 MW Solar One (later redeveloped and made into Solar Two) and the 2 MW Themis plants.

A cost/performance comparison between power tower and parabolic trough concentrators was made by the NREL which estimated that by 2020 electricity could be produced from power towers for 5.47 ë/kWh and for 6.21 ë/kWh from parabolic troughs. The capacity factor for power towers was estimated to be 72.9 per cent and 56.2 per cent for parabolic troughs. There is some hope that the development of cheap, durable, mass producible heliostat power plant components could bring this cost down.

Dish designs

A dish stirling system uses a large, reflective, parabolic dish (similar in shape to satellite television dish). It focuses all the sunlight that strikes the dish up onto to a single point above the dish, where a receiver captures the heat and transforms it into a useful form. Typically the dish is coupled with a stirling engine in a Dish-Stirling System, but also sometimes a steam engine is used. These create rotational kinetic energy that can be converted to electricity using an electric generator.

The advantage of a dish system is that it can achieve much higher temperatures due to the higher concentration of light (as in tower designs). Higher temperatures lead to better conversion to electricity and the dish system is very efficient on this point. However, there are also some disadvantages. Heat to electricity conversion requires moving parts and that result in maintenance. In general, a centralised approach for this conversion is better than the decentralised concept in the dish design. Second, the (heavy) engine is part of the moving structure, which requires a rigid frame and strong tracking system. Furthermore, parabolic mirrors are used instead of flat mirrors and tracking must be dual-axis.

In 2005 Southern California Edison announced an agreement to purchase solar powered Stirling engines from Stirling Energy Systems over a 20 year period and in quantities (20,000 units) sufficient to generate 500 MW of electricity. Stirling Energy Systems announced another agreement with San Diego Gas and Electric to provide between 300 and 900 megawatts of electricity. In January 2010, Stirling Energy Systems and Tessera Solar commissioned the first demonstration 1.5 MW power plant (Maricopa solar) using Stirling technology in Peoria, Arizona. In July 2011, Iran inaugurated Iran's biggest solar power plant in Mashhad which produce 72,000 kW/h electricity per year.

Fresnel reflectors

A linear Fresnel reflector power plant uses a series of long, narrow, shallow-curvature (or even flat) mirrors to focus light onto one or more linear receivers positioned above the mirrors. On top of the receiver, a small parabolic mirror can be attached for further focusing the light. These systems aim to offer lower overall costs by sharing a receiver between several mirrors (as compared with trough and dish concepts), while still using the simple line-focus geometry with one axis for tracking. This is similar to the trough design (and different from central towers and dishes with dual-axis). The receiver is stationary and so fluid couplings are not required (as in troughs and dishes). The mirrors also do not need to support the receiver, so they are structurally simpler. When suitable aiming strategies are used (mirrors aimed at different receivers at different times of day), this can allow a denser packing of mirrors on available land area.

Recent prototypes of these types of systems have been built in Australia (CLFR) and by Solarmundo in Belgium. The Solarmundo research and development project, with its pilot plant at Liège, was closed down after successful proof of concept of the Linear Fresnel technology. Subsequently, Solar Power Group GmbH (SPG), based in Munich, Germany, was founded by some Solarmundo team members. A Fresnel-based prototype with direct steam generation was built by SPG in conjunction with the German Aerospace Center (DLR).

In May 2008, the German Solar Power Group GmbH and the Spanish Laer S.L. agreed the joint execution of a solar thermal power plant in central Spain. This will be the first commercial solar thermal power plant in Spain based on the Fresnel collector technology of the Solar Power Group. The planned size of the power plant will be 10 MW a solar thermal collector field with a fossil co-firing unit as backup system.

The start of constructions is planned for 2012. The project is located in Gotarrendura, a small renewable energy pioneering village, about 100 km northwest of Madrid, Spain.

A Multi-Tower Solar Array (MTSA) concept, that uses a point-focus Fresnel reflector idea, has also been developed, but has not yet been prototyped. Since March 2009, the Fresnel solar power plant PE 1 of the German company Novatec Biosol is in commercial operation in southern Spain. The solar thermal power plant is based on linear Fresnel collector technology and has an electrical capacity of 1.4 MW. Beside a conventional power block, PE 1 comprises a solar boiler with mirror surface of around 18,000 m^2. The steam is generated by concentrating direct solar irradiation onto a linear receiver which is 7.40 m above the ground. An absorber tube is positioned in the focal line of the mirror field in which water is evaporated directly into saturated steam at 270°C and at a pressure of 55 bar by the concentrated solar energy.

Linear Fresnel reflector technologies

Rival single axis tracking technologies include the relatively new linear Fresnel reflector (LFR) and compact-LFR (CLFR) technologies. The LFR differs from that of the parabolic trough in that the absorber is fixed in space above the mirror field. Also, the reflector is composed of many low row segments, which focus collectively on an elevated long tower receiver running parallel to the reflector rotational axis.

This system offers a lower cost solution as the absorber row is shared among several rows of mirrors. However, one fundamental difficulty with the LFR technology is the avoidance of shading of incoming solar radiation and blocking of reflected solar radiation by adjacent reflectors. Blocking and shading can be reduced by using absorber towers elevated higher or by increasing the absorber size, which allows increased spacing between reflectors remote from the absorber. Both these solutions increase costs, as larger ground usage is required.

The CLFR offers an alternate solution to the LFR problem. The classic LFR has only one linear absorber on a single linear tower. This prohibits any option of the direction of orientation of a given reflector. Since this technology would be introduced in a large field, one can assume that there will be many linear absorbers in the system. Therefore, if the linear absorbers are close enough, individual reflectors will have the option of directing reflected solar radiation to at least two absorbers. This additional factor gives potential for more densely packed arrays, since patterns of alternative reflector inclination can be set up such that closely packed reflectors can be positioned without shading and blocking.

CLFR power plants offer reduced costs in all elements of the solar array. These reduced costs encourage the advancement of this technology. Features that enhance the cost effectiveness of this system compared to that of the parabolic trough technology include minimised structural costs, minimised parasitic pumping losses, and low maintenance. Minimised structural costs are attributed to the use of flat or elastically curved glass reflectors instead of costly sagged glass reflectors are mounted close to the ground. Also, the heat transfer loop is separated from the reflector field, avoiding the cost of flexible high pressure lines required in trough systems. Minimised parasitic pumping losses are due to the use of water for the heat transfer fluid with passive direct boiling. The use of glass-evacuated tubes ensures low-radiative losses and is inexpensive. Studies of existing CLFR plants have been shown to deliver tracked beam to electricity efficiency of 19 per cent on an annual basis as a preheater.

Fresnel lenses

Prototypes of Fresnel lens concentrators have been produced for the collection of thermal energy by International Automated Systems. No full-scale thermal systems using Fresnel lenses are known to be in operation, although products incorporating Fresnel lenses in conjunction with photovoltaic cells are already available. The advantage of this design is that lenses are cheaper than mirrors. Furthermore, if a material is chosen that has some flexibility, then a less rigid frame is required to withstand wind load. A new concept of a lightweight, 'non-disruptive' solar concentrator technology using asymmetric Fresnel lenses that occupies minimal ground surface area and allows for large amounts of concentrated solar energy per concentrator is seen in the 'Desert Blooms' project, though a prototype has yet to be made.

MicroCSP

'MicroCSP' references solar thermal technologies in which concentrating solar power (CSP) collectors are based on the designs used in traditional Concentrating Solar Power systems found in the Mojave Desert but are smaller in collector size, lighter and operate at lower thermal temperatures usually below 315°C (600°F). These systems are designed for modular field or rooftop installation where they are easy to protect from high winds, snow and humid deployments. Solar manufacturer Sopogy completed construction on a 1 MW CSP plant at the Natural Energy Laboratory of Hawaii. MicroCSP is used for community-sized power plants (1 MW to 50 MW), for industrial, agricultural and manufacturing 'process heat' applications, and when large amounts of hot water are needed, such as resort swimming pools, water parks, large laundry facilities, sterilisation, distillation and other such uses.

4.4.2 Heat exchange

Heat in a solar thermal system is guided by five basic principles: heat gain, heat transfer, heat storage, heat transport, and heat insulation. Here, heat is the measure of the amount of thermal energy an object contains and is determined by the temperature, mass and specific heat of the object. Solar thermal power plants use heat exchangers that are designed for constant working conditions, to provide heat exchange.

Heat gain is the heat accumulated from the sun in the system. Solar thermal heat is trapped using the greenhouse effect; the greenhouse effect in this case is the ability of a reflective surface to transmit short wave radiation and reflect long wave radiation. Heat and infrared radiation (IR) are produced when short wave radiation light hits the absorber plate, which is then trapped inside the collector. Fluid, usually water, in the absorber tubes collect the trapped heat and transfer it to a heat storage vault.

Heat is transferred either by conduction or convection. When water is heated, kinetic energy is transferred by conduction to water molecules throughout the medium. These molecules spread their thermal energy by conduction and occupy more space than the cold slow moving molecules above them. The distribution of energy from the rising hot water to the sinking cold water contributes to the convection process. Heat is transferred from the absorber plates of the collector in the fluid by conduction. The collector fluid is circulated through the carrier pipes to the heat transfer vault. Inside the vault, heat is transferred throughout the medium through convection. Heat storage enables solar thermal plants to produce electricity during hours without sunlight. Heat is transferred to a thermal storage medium in an insulated reservoir during hours with sunlight, and is withdrawn for power generation during hours lacking sunlight. Thermal storage mediums will be discussed in a heat storage section. Rate of heat transfer is related to the conductive and convection medium as well as the temperature differences. Bodies with large temperature differences transfer heat faster than bodies with lower temperature differences. Heat transport refers to the activity in which heat from a solar collector is transported to the heat storage vault. Heat insulation is vital in both heat transport tubing as well as the storage vault. It prevents heat loss, which in turn relates to energy loss or decrease in the efficiency of the system.

4.4.3 Heat storage

Heat storage allows a solar thermal plant to produce electricity at night and on overcast days. This allows the use of solar power for base-load generation as well as peak power generation, with the potential of displacing both coal and natural gas fired power plants. Additionally, the utilisation of the generator is higher which reduces cost. Heat is transferred to a thermal storage medium in an insulated reservoir during the day, and withdrawn for power generation at night. Thermal storage media include pressurised steam, concrete, a variety of phase change materials, and molten salts such as sodium and potassium nitrate.

Steam accumulator

The PS10 solar power tower stores heat in tanks as pressurised steam at 50 bar and 285°C. The steam condenses and flashes back to steam, when pressure is lowered. Storage is for one hour. It is suggested that longer storage is possible, but that has not been proven yet in an existing power plant.

Molten salt storage

A variety of fluids have been tested to transport the sun's heat, including water, air, oil, and sodium, but molten salt was selected as best. Molten salt is used in solar power tower systems because it is liquid at atmosphere pressure, it provides an efficient, low-cost medium in which to store thermal energy, its operating temperatures are compatible with today's high-pressure and high-temperature steam turbines, and it is non-flammable and nontoxic. In addition, molten salt is used in the chemical and metals industries as a heat-transport fluid, so experience with molten-salt systems exists in non-solar settings.

The molten salt is a mixture of 60 per cent sodium nitrate and 40 per cent potassium nitrate, commonly called saltpetre. New studies show that calcium nitrate could be included in the salts mixture to reduce costs and with technical benefits. The salt melts at 220°C (430°F) and is kept liquid at 290°C (550°F) in an insulated storage tank. The uniqueness of this solar system is in de-coupling the collection of solar energy from producing power, electricity can be generated in periods of inclement weather or even at night using the stored thermal energy in the hot salt tank. Normally tanks are well insulated and can store thermal energy for up to a week. As an example of their size, tanks that provide enough thermal storage to power a 100-megawatt turbine for four hours would be about 9 m (30 ft) tall and 24 m (80 ft) in diameter. The Andasol power plant in Spain is the first commercial solar thermal power plant to utilise molten salt for heat storage and night-time generation. It came online March 2009. On July 4, 2011, a company in Spain celebrated an historic moment for the solar industry: Torresol's 19.9 MW concentrating solar power plant became the first ever to generate uninterrupted electricity for 24 hours straight. It achieved this using a molten salt heat storage design.

Graphite heat storage

Direct: The proposed power plant in Cloncurry Australia will store heat in purified graphite. The plant has a power tower design. The graphite is located on top of the tower. Heat from the heliostats goes directly to the storage. Heat for energy production is drawn from the graphite. This simplifies the design.

Indirect: Molten salt coolants are used to transfer heat from the reflectors to heat storage vaults. The heat from the salts is transferred to a secondary heat transfer fluid via a heat exchanger and then to the storage media or alternatively, the salts can be used to directly heat graphite. Graphite is used as it has relatively low costs and compatibility with liquid fluoride salts. The high mass and volumetric heat capacity of graphite provide an efficient storage medium.

Phase-change materials for storage

Phase-change material (PCMs) offers an alternate solution in energy storage. Using a similar heat transfer infrastructure, PCMs have the potential of providing a more efficient means of storage. PCMs can be either organic or inorganic materials. Advantages of organic PCMs include no corrosives, low or no undercooling, and chemical and thermal stability. Disadvantages include low phase-change enthalpy, low thermal conductivity, and flammability. Inorganics are advantageous with greater phase-change enthalpy, but exhibit disadvantages with undercooling, corrosion, phase separation, and lack of thermal stability. The greater phase-change enthalpy in inorganic PCMs make hydrate salts a strong candidate in the solar energy storage field.

4.4.4 Use of water

A design which requires water for condensation or cooling may conflict with location of solar thermal plants in desert areas with good solar radiation but limited water resources. The conflict is illustrated by plans of Solar Millennium, a German company, to build a plant in the Amargosa Valley of Nevada which would require 20 per cent of the water available in the area. Some other projected plants by the same and other companies in the Mojave Desert of California may also be affected by difficulty in obtaining adequate and appropriate water rights. California water law currently prohibits use of potable water for cooling.

Other designs require less water. The proposed Ivanpah Solar Power Facility in south-eastern California will conserve scarce desert water by using air-cooling to convert the steam back into water. Compared to conventional wet-cooling, this results in a 90 per cent reduction in water usage at the cost of some loss of efficiency. The water is then returned to the boiler in a closed process which is environmentally friendly.

4.5 Conversion rates from solar energy to electrical energy

Of all of these technologies, the solar dish/stirling engine has the highest energy efficiency. A single solar dish-Stirling engine installed at Sandia National Laboratories National Solar Thermal Test Facility produces as much as 25 kW of electricity, with a conversion efficiency of 31.25 per cent. Solar parabolic trough plants have been built with efficiencies of about 20 per cent. Fresnel reflectors have an efficiency that is slightly lower (but this is compensated by the denser packing).

The gross conversion efficiencies (taking into account that the solar dishes or troughs occupy only a fraction of the total area of the power plant) are determined by net generating capacity over the solar energy that falls on the total area of the solar plant. The 500-MW SCE/SES plant would extract about 2.75 per cent of the radiation (1 kW/m^2) that falls on its 4500 acres (18.2 km^2). For the 50 MW AndaSol Power Plant that is being built in Spain (total area of 1300 × 1500 m = 1.95 km^2) gross conversion efficiency comes out at 2.6 per cent.

Furthermore, efficiency does not directly relate to cost: On calculating total cost, both efficiency and the cost of construction and maintenance should be taken into account.

4.5.1 Levelised cost

Since a solar power plant does not use any fuel, the cost consists mostly of capital cost with minor operational and maintenance cost. If the lifetime of the plant and the interest rate is known, then the cost per kWh can be calculated. This is called the levelised energy cost.

The first step in the calculation is to determine the investment for the production of 1 kWh in a year. Example, the fact sheet of the Andasol 1 project shows a total investment of 310 million Euros for a production of 179 GWh a year. Since 179 GWh is 179 million kWh, the investment per kWh a year production is 310/179 = 1.73 euro. Another example is Cloncurry solar power station in Australia. It is planned to produce 30 million kWh a year for an investment of 31 million Australian dollars. So, if this is achieved in reality, the cost would be 1.03 Australian dollar for the production of 1 kWh in a year. This would be significantly cheaper than Andasol 1, which can partially be explained by the higher radiation in Cloncurry over Spain. The investment per kWh cost for one year should not be confused with the cost per kWh over the complete lifetime of such a plant.

In most cases the capacity is specified for a power plant (for instance Andasol 1 has a capacity of 50 MW). This number is not suitable for comparison, because the capacity factor can differ. If a solar power plant has heat storage, then it can also produce output after sunset, but that will not change the capacity factor, it simply displaces the output. The average capacity factor for a solar power plant, which is a function of tracking, shading and location, is about 20 per cent, meaning that a 50 MW capacity power plant will typically provide a yearly output of 50 MW × 24 hours × 365 days × 20 per cent = 87,600 MWh/year or 87.6 GWh/year.

Although the investment for 1 kWh year production is suitable for comparing the price of different solar power plants, it does not give the price per kWh yet. The way of financing has a great influence on the final price. If the technology is proven, an interest rate of 7 per cent should be possible. However, for a new technology investors want a much higher rate to compensate for the higher risk. This has a significant negative effect on the price per kWh. Independent of the way of financing, there is always a linear relation between the investment per kWh production in a year and the price for 1 kWh (before adding operational and maintenance cost). In other words, if by enhancements of the technology the investments drop by 20 per cent, then the price per kWh also drops by 20 per cent.

If a way of financing is assumed where the money is borrowed and repaid every year, in such way that the debt and interest decreases. For a lifetime of 25 years and an interest rate of 7 per cent, the division factor is 11.65. For example, the investment of Andasol 1 was 1.73 euro per kWh, divided by 11.65 results in a price of 0.15 euro per kWh. If 1 per cent operation and maintenance cost is added, then the levelised cost is 0.16 euro per kWh. Other ways of financing, different way of debt repayment, different lifetime expectation, different interest rate, may lead to a significantly different number.

If the cost per kWh may follow the inflation, then the inflation rate can be added to the interest rate. If an investor puts his money on the bank for 7 per cent, then he is not compensated for inflation. However, if the cost per kWh is raised with inflation, then he is compensated and he can add 2 per cent (a normal inflation rate) to his return. The Andasol 1 plant has a guaranteed feed-in tariff of 0.21 euro for 25 years. If this number is fixed, after 25 years with 2 per cent inflation, 0.21 euro will have a value comparable with 0.13 euro now. Finally, there is some gap between the first investment and the first production of electricity. This increases the investment with the interest over the period that the plant is not active yet. The modular solar dish (but also solar photovoltaic and wind power) have the advantage that electricity production starts after first construction. Given the fact that solar thermal power is reliable, can deliver peak load and does not cause pollution, a price of US$0.10 per kWh starts to become competitive. Although a price of US$0.06 has been claimed with some operational cost a simple target is 1 dollar (or lower) investment for 1 kWh production in a year.

4.6 Solar pond

A solar pond is a pool of saltwater which acts as a large-scale solar thermal energy collector with integral heat storage for supplying thermal energy. A solar pond can be used for various applications, such as process heating, desalination, refrigeration, drying and solar power generation.

A solar pond is simply a pool of saltwater which collects and stores solar thermal energy. The saltwater naturally forms a vertical salinity gradient also known as a 'halocline', in which low-salinity water floats on top of high-salinity water. The layers of salt solutions increase in concentration (and therefore density) with depth. Below a certain depth, the solution has a uniformly high salt concentration.

There are three distinct layers of water in the pond:

1. The top layer, which has a low salt content
2. An intermediate insulating layer with a salt gradient, which establishes a density gradient that prevents heat exchange by natural convection
3. The bottom layer, which has a high salt content

If the water is relatively translucent, and the pond's bottom has high optical absorption, then nearly all of the incident solar radiation (sunlight) will go into heating the bottom layer.

When solar energy is absorbed in the water, its temperature increases, causing thermal expansion and reduced density. If the water were fresh, the low-density warm water would float to the surface, causing a convection current. The temperature gradient alone causes a density gradient that decreases with depth. However the salinity gradient forms a density gradient that increases with depth, and this counteracts the temperature gradient, thus preventing heat in the lower layers from moving upwards by convection and leaving the pond. This means that the temperature at the bottom of the pond will rise to over 90°C while the temperature at the top of the pond is usually around 30°C. A natural example of these effects in a saline water body is Solar Lake in the Sinai Peninsula of Egypt.

The heat trapped in the salty bottom layer can be used for many different purposes, such as the heating of buildings or industrial hot water or to drive an organic Rankine cycle turbine or Stirling engine for generating electricity.

Advantages and disadvantages

1. The approach is particularly attractive for rural areas in developing countries. Very large area collectors can be set up for just the cost of the clay or plastic pond liner.
2. The evaporated surface water needs to be constantly replenished.
3. The accumulating salt crystals have to be removed and can be both a valuable by-product and a maintenance expense.
4. No need of a separate collector for this thermal storage system.

4.6.1 Efficiency

The energy obtained is in the form of low-grade heat of 70–80°C compared to an assumed 20°C ambient temperature. According to the second law of thermodynamics (Carnot-cycle), the maximum theoretical efficiency of a solar concentrator system with molten salt is: 1 – (273 + 20) / (273 + 80) = 17 per cent. By comparison, a power plant's heat engine delivering high-grade heat at 800°C would have a maximum theoretical limit of 73 per cent for converting heat into useful work (and thus would be forced to divest as little as 27 per cent in waste heat to the cold temperature reservoir at 20°C). The low efficiency of solar ponds is usually justified with the argument that the 'collector', being just a plastic-lined pond, might potentially result in a large-scale system that is of lower overall levelised energy cost than a solar concentrating system.

Examples: The largest operating solar pond for electricity generation was the Beit HaArava pond built in Israel and operated up until 1988. It had an area of 2,10,000 m^2 and gave an electrical output of 5 MW.

The first solar pond in India (6000 sq. metres) was built at Bhuj. The project was sanctioned under the National Solar Pond Program by the Ministry of Non-conventional Energy Sources in 1987 and completed in 1993 after a sustained collaborative effort by TERI, the Gujarat Energy Development Agency, and the GDDC. The solar pond successfully demonstrated the expediency of the technology by supplying 80,000 litres of hot water daily to the plant. The Energy and Resources Institute provided all technical inputs and took up the complete execution of research, development, and demonstration. TERI operated and maintained this facility until 1996 before handing it over to the GDDC. The solar pond functioned effortlessly till the year 2000 when severe financial losses crippled GDDC. Subsequently, the Bhuj earthquake left the Kutch Dairy non-functional.

The 0.8-acre (3,200 m^2) solar pond powering 20 per cent of Bruce Foods Corporation's operations El Paso, Texas is the second largest in the US. It is also the first ever salt-gradient solar pond in the US.

4.7 Solar water heating

Solar water heating or solar hot water is water heated by the use of solar energy. Solar water heating systems are generally composed of solar thermal collectors, a water storage tank or another point of usage, interconnecting pipes and a fluid system to move the heat from the collector to the tank. This thermodynamic approach is distinct from semiconductor photovoltaic (PV) cells that generate electricity from light; solar water heating deals with the direct heating of liquids by the sun where no electricity is directly generated. A solar water heating system may use electricity for pumping the fluid, and have a reservoir or tank for heat storage and subsequent use. The water can be heated for a wide variety of uses, including home, business and industrial uses. Heating swimming pools, underfloor heating or energy input for space heating or cooling are common examples of solar water heating.

A solar water heating system can form part of a solar thermal cooling system, promoting efficient temperature control of buildings or parts thereof. During cool conditions, the same system can provide hot water. Solar heating of buildings in temperate climates has a season-problem: In winter, when most heating is needed, least is available from the sun. This can often be solved by storing solar heat in the ground or in groundwater (Seasonal thermal store). Hot water heated by the sun is used in many ways. While perhaps best known in a residential setting to provide domestic hot water, solar hot water also has industrial applications, e.g. to generate electricity. Designs suitable for hot climates can be much simpler and cheaper, and can be considered an appropriate technology for these places. The global solar thermal market is dominated by China, Europe, Japan and India.

In order to heat water using solar energy, a collector, often fastened to a roof or a wall facing the sun, heats working fluid that is either pumped (active system) or driven by natural convection (passive system) through it. The collector could be made of a simple glass topped insulated box with a flat solar absorber made of sheet metal attached to copper pipes and painted black or a set of metal tubes surrounded by an evacuated (near vacuum) glass cylinder. In industrial cases a parabolic mirror can concentrate sunlight on the tube. Heat is stored in a hot water storage tank.

The volume of this tank needs to be larger with solar heating systems in order to allow for bad weather, and because the optimum final temperature for the solar collector is lower than a typical immersion or combustion heater. The heat transfer fluid (HTF) for the absorber may be the hot water from the tank, but more commonly (at least in active systems) is a separate loop of fluid containing anti-freeze and a corrosion inhibitor which delivers heat to the tank through a heat exchanger (commonly a coil of copper tubing within the tank). Another lower-maintenance concept is the 'drain-back': no anti-freeze is required; instead all the piping is sloped to cause water to drain back to the tank. The tank is not pressurised and is open to atmospheric pressure. As soon as the pump shuts off, flow reverses and the pipes are empty before freezing could occur.

Residential solar thermal installations fall into two groups: passive (sometimes called compact) and active (sometimes called pumped) systems. Both typically include an auxiliary energy source (electric heating element or connection to a gas or fuel oil central heating system) that is activated when the water in the tank falls below a minimum temperature setting such as 55°C. Hence, hot water is always available. The combination of solar water heating and using the back-up heat from a wood stove chimney to heat water can enable a hot water system to work all year round in cooler climates, without the supplemental heat requirement of a solar water heating system being met with fossil fuels or electricity.

When a solar water heating and hot-water central heating system are used in conjunction, solar heat will either be concentrated in a pre-heating tank that feeds into the tank heated by the central heating or the solar heat exchanger will replace the lower heating element and the upper element will remain in place to provide for any heating that solar cannot provide. However, the primary need for central heating is at night and in winter when solar gain is lower. Therefore, solar water heating for washing and bathing is often a better application than central heating because supply and demand are better matched. In many climates, a solar hot water system can provide up to 85 per cent of domestic hot water energy. This can include domestic non-electric concentrating solar thermal systems. In many northern European countries, combined hot water and space heating systems (solar combisystems) are used to provide 15–25 per cent of home heating energy.

4.7.1 System design requirements

The type, complexity and size of a solar water heating system are mostly determined by:

1. The temperature and amount of the water required from the system
2. The changes in ambient temperature and solar radiation between summer and winter
3. The changes in ambient temperature during the day–night cycle
4. The possibility of the potable water or collector fluid overheating
5. The possibility of the potable water or collector fluid freezing

The minimum requirements of the system are typically determined by the amount or temperature of hot water required during winter, when a system's output and incoming water temperature are typically at their lowest. The maximum output of the system is determined by the need to prevent the water in the system from becoming too hot or, in the systems that overheating is avoided, the desire to waste money on unneeded components.

4.7.2 Freeze protection

Freeze protection measures prevent damage to the system due to the expansion of freezing transfer fluid. Drainback systems drain the transfer fluid from the system when the pump stops. Many indirect systems use antifreeze (e.g. propylene glycol) in the heat transfer fluid.

In some direct systems, the collectors can be manually drained when freezing is expected. This approach is common in climates where freezing temperatures do not occur often, but is somewhat unreliable since the operator can forget to drain the system. Other direct systems use freeze-tolerant collectors made with flexible polymers such as silicone rubber.

4.7.3 Overheat protection

When no hot water has been used for a day or two, the fluid in the collectors and storage can reach very high temperatures in all systems except for those of the drain-back variety. When the storage tank in a drain-back system reaches its desired temperature, the pumps are shut off, putting an end to the heating process and thus preventing the storage tank from overheating. One method of providing over heat protection is to dump the heat into a hot tub. Some active systems deliberately cool the water in the storage tank by circulating hot water through the collector at times when there is little sunlight or at night, causing increased heat loss. This is particularly ineffective in systems that use evacuated tube collectors, due to their superior insulation. No matter the collector type, however, they can still overheat and ultimately rely on the operation of temperature and pressure relief valves.

4.7.4 Types of solar water heating systems

Solar water heaters can be either active or passive. An active system uses an electric pump to circulate the heat-transfer fluid; a passive system has no pump. The amount of hot water a solar water heater produces depends on the type and size of the system, the amount of sun available at the site, proper installation, and the tilt angle and orientation of the collectors.

Solar water heaters are also characterised as open loop (also called direct) or closed loop (also called indirect). An open-loop system circulates household (potable) water through the collector. A closed-loop system uses a heat-transfer fluid (water or diluted antifreeze, for example) to collect heat and a heat exchanger to transfer the heat to household water.

Direct and indirect systems

Direct or open-loop systems circulate potable water through the collectors. They are cheaper than indirect systems and offer superior heat transfer from the collectors to the storage tank, but have many drawbacks:

1. They offer little or no overheat protection.
2. They offer little or no freeze protection.
3. The collectors will accumulate scale in hard water areas.

They are often not considered suitable for cold climates since, in the event of the collector being damaged by a freeze, pressurised water lines will force water to gush from the freeze-damaged collector until the problem is noticed and rectified (Fig. 4.2). Indirect or closed loop systems use a heat exchanger that separates the potable water from the fluid, known as the 'heat-transfer fluid' (HTF), which circulates through the collector. The two most common HTFs are water and an antifreeze/water mix that typically uses non-toxic propylene glycol. After being heated in the panels, the HTF travels to the heat exchanger, where its heat is transferred to the potable water. Though slightly more expensive, indirect systems offer freeze protection and typically offer overheat protection as well.

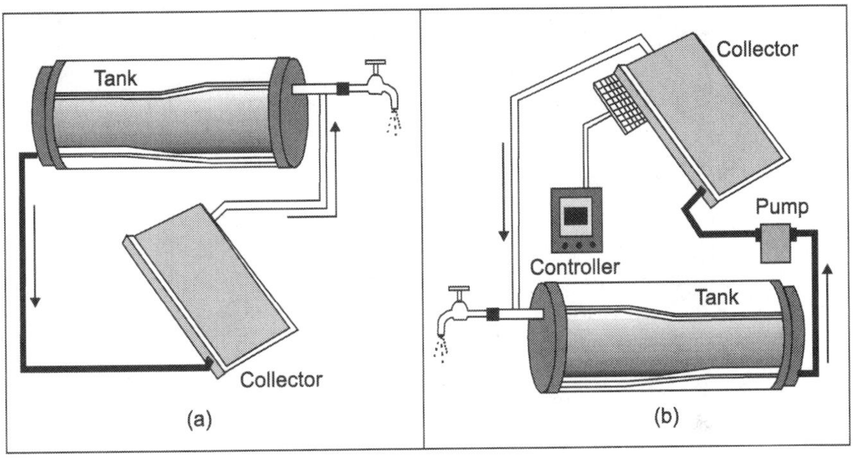

Figure 4.2 Direct systems: (a) passive CHS system with tank above collector, and (b) active system with pump and controller driven by a photovoltaic panel.

Passive and active systems

Passive systems rely on heat-driven convection or heat pipes to circulate water or heating fluid in the system. Passive solar water heating systems cost less and have extremely low or no maintenance, but the efficiency of a passive system is significantly lower than that of an active system, and overheating and freezing are major concerns. Active systems use one or more pumps to circulate water and/or heating fluid in the system (Fig. 4.3).

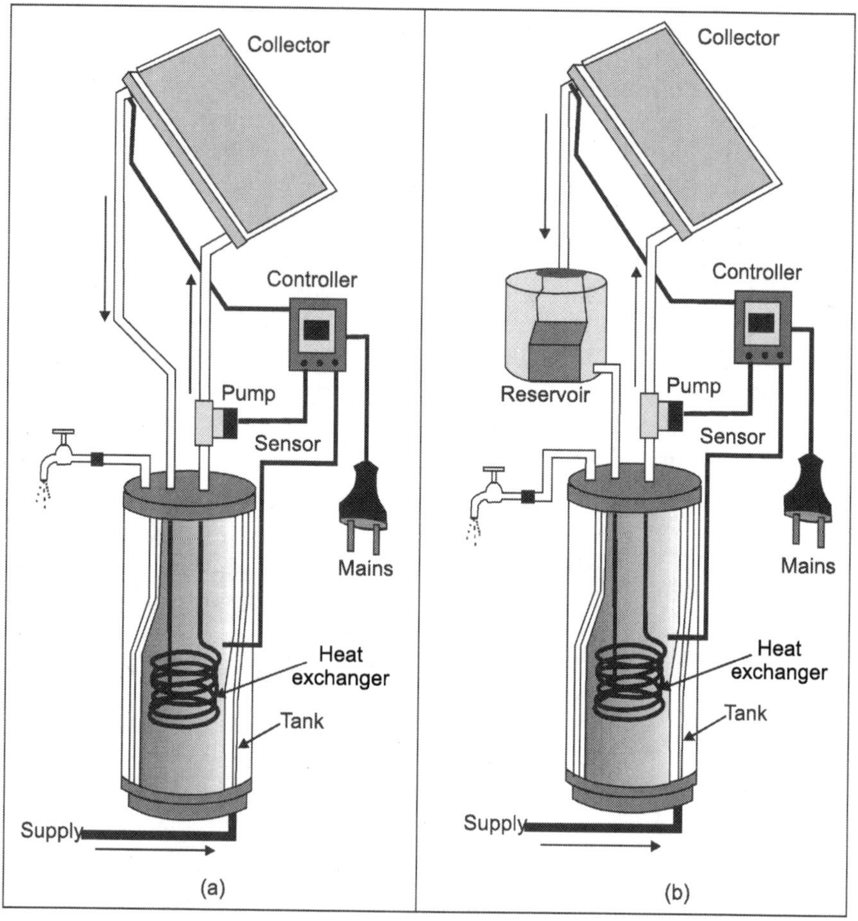

Figure 4.3 Indirect active systems: (a) indirect system with heat exchanger in tank, (b) drainback system with drainback reservoir. In these schematics the controller and pump are driven by mains electricity.

Though slightly more expensive, active systems offer several advantages:

1. The storage tank can be situated lower than the collectors, allowing increased freedom in system design and allowing pre-existing storage tanks to be used.
2. The storage tank can always be hidden from view.
3. The storage tank can be placed in conditioned or semi-conditioned space, reducing heat loss.
4. Drainback tanks can be used.
5. Superior efficiency
6. Increased control over the system

Modern active solar water systems have electronic controllers that offer a wide-range of functionality, such as the modification of settings that control the system, interaction with a backup electric or gas-driven water heater, calculation and logging of the energy saved by a SWH system, safety functions, remote access, and various informative displays, such as temperature readings.

The most popular pump controller is a differential controller that senses temperature differences between water leaving the solar collector and the water in the storage tank near the heat exchanger. In a typical active system, the controller turns the pump on when the water in the collector is about 8–10°C warmer than the water in the tank, and it turns the pump off when the temperature difference approaches 3–5°C. This ensures the water always gains heat from the collector when the pump operates and prevents the pump from cycling on and off too often. (In direct systems this 'on differential' can be reduced to around 4C because there is no heat exchanger impediment.)

Some active SWH systems use energy obtained by a small photovoltaic (PV) panel to power one or more variable-speed DC pump(s). In order to ensure proper performance and longevity of the pump(s), the DC-pump and PV panel must be suitably matched. These systems are almost always of the antifreeze variety and often do not use controllers, as the collectors will almost always be hot when the pump(s) are operating (i.e. when the sun is bright). Sometimes, however, a differential controller (that can also be powered by the DC output of a PV panel) is used to prevent the operation of the pumps when there is sunlight to power the pump but the collectors are still cooler than the water in storage. One advantage of a PV-driven system is that solar hot water can still be collected during a power outage if the sun is shining.

An active solar water heating system can also be equipped with a bubble pump (also known as geyser pump) instead of an electric pump. A bubble pump circulates the heat transfer fluid (HTF) between collector and storage tank using solar power and without any external energy source and is suitable for flat panel as well as vacuum tube systems. In a bubble pump system, the closed HTF circuit is under reduced pressure, which causes the liquid to boil at low temperature as it is heated by the sun. The steam bubbles form a geyser pump, causing an upward flow. The system is designed such that the bubbles are separated from the hot fluid and condensed at the highest point in the circuit, after which the fluid flows downward towards the heat exchanger caused by the difference in fluid levels. The HTF typically arrives at the heat exchanger at 70°C and returns to the circulating pump at 50°C. In frost prone climates the HTF is water with propylene glycol antifreeze added, usually in the ratio of 60 to 40. Pumping typically starts at about 50°C and increases as the sun rises until equilibrium is reached depending on the efficiency of the heat exchanger, the temperature of the water being heated and the strength of the sun.

Passive direct systems

An integrated collector storage (ICS or Batch Heater) system uses a tank that acts as both storage and solar collector. Batch heaters are basically thin rectilinear tanks with a glass side, facing south. They are simple and less costly than plate and tube collectors, but they sometimes require extra bracing if installed on a roof (since they are heavy when filled with water [400–700 lbs],) suffer from significant heat loss at night since the side facing the sun is largely uninsulated, and are only suitable in moderate climates. A convection heat storage unit (CHS) system is similar to an ICS system, except the storage tank and collector are physically separated and transfer between the two is driven by convection. CHS systems typically use standard flat-plate type or evacuated tube collectors, and the storage tank must be located above the collectors for convection to work properly. The main benefit of a CHS systems over an ICS system is that heat loss is largely avoided since: (i) the storage tank can be better insulated, and (ii) since the panels are located below the storage tank, heat loss in the panels will not cause convection, as the cold water will prefer to stay at the lowest part of the system (Fig. 4.4).

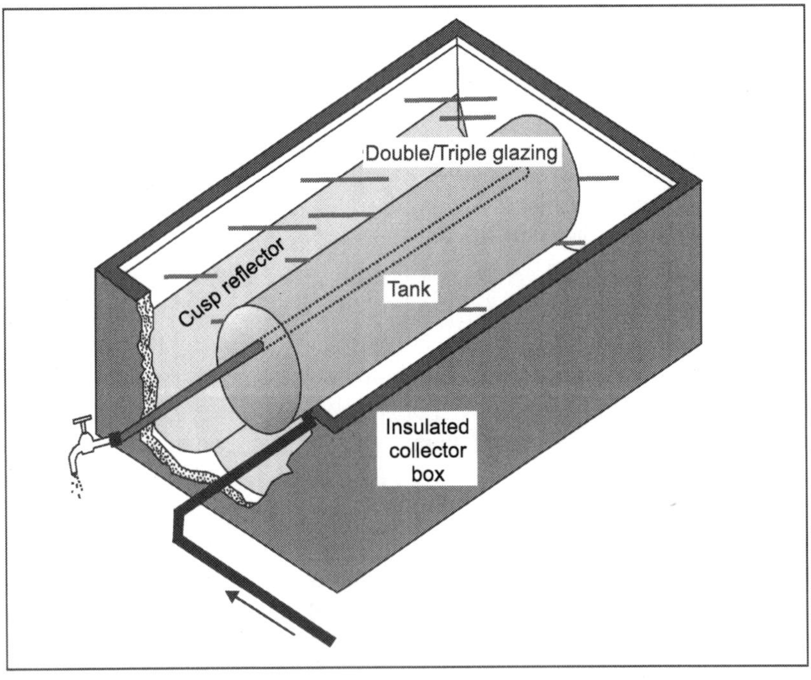

Figure 4.4 An integrated collector storage (ICS) system

Active indirect systems: Drainback and antifreeze

Pressurised antifreeze or pressurised glycol systems use a mix of antifreeze (almost always non-toxic propylene glycol) and water mix for HTF in order to prevent freeze damage.

Though effective at preventing freeze damage, antifreeze systems have many drawbacks:

1. If the HTF gets too hot (for example, when the homeowner is on vacation), the glycol degrades into acid. After degradation, the glycol not only fails to provide freeze protection, but also begins to eat away at the solar loop's components: The collectors, the pipes, the pump, etc. Due to the acid and excessive heat, the longevity of parts within the solar loop is greatly reduced.
2. Most do not feature drainback tanks, so the system must circulate the HTF—regardless of the temperature of the storage tank—in order to prevent the HTF from degrading. Excessive temperatures in the tank cause increased scale and sediment build-up, possible severe burns if a tempering valve is not installed, and, if a water heater is being used for storage, possible failure of the water heater's thermostat.
3. The glycol/water HTF must be replaced every 3–8 years, depending on the temperatures it has experienced.
4. Some jurisdictions require double-walled heat exchangers even though propylene glycol is non-toxic.
5. Even though the HTF contains glycol to prevent freezing, it will still circulate hot water from the storage tank into the collectors at low temperatures (e.g. below 40 degrees Fahrenheit), causing substantial heat loss.

A drainback system is an indirect active system where the HTF (almost always pure water) circulates through the collector, being driven by a pump. The collector piping is not pressurised and includes an open drainback reservoir that is contained in conditioned or semi-conditioned space. If the pump is switched off, the HTF drains into the drainback reservoir and none remains in the collector. Since the system relies upon being able to drain properly, all piping above the drainback tank, including the collectors, must slope downward in the direction of the drainback tank. Installed properly, the collector cannot be damaged by freezing or overheating. Drainback systems require no maintenance other than the replacement of failed system components.

4.7.5 Collectors used in modern domestic SWH systems

Solar thermal collectors capture and retain heat from the sun and transfer this heat to a liquid. Two important physical principles govern the technology of solar thermal collectors:

1. Any hot object ultimately returns to thermal equilibrium with its environment, due to heat loss from the hot object. The processes that result in this heat loss are conduction, convection and radiation. The efficiency of a solar thermal collector is directly related to heat losses from the collector surface (efficiency being defined as the proportion of heat energy that can be retained for a predefined period of time). Within the context of a solar collector, convection and radiation are the most important sources of heat loss. Thermal insulation is used to slow down heat loss from a hot object to its environment. This is actually a direct manifestation of the second law of thermodynamics but we may term this the 'equilibrium effect'.
2. Heat is lost more rapidly if the temperature difference between a hot object and its environment is larger. Heat loss is predominantly governed by the thermal gradient between the temperature of the collector surface and the ambient temperature. Conduction, convection as well as radiation occur more rapidly over large thermal gradients. We may term this the 'delta-t effect'.

The simplest approach to solar heating of water is to simply mount a metal tank filled with water in a sunny place. The heat from the sun would then heat the metal tank and the water inside. Indeed, this was how the very first SWH systems worked more than a century ago. However, this setup would be inefficient due to an oversight of the equilibrium effect, above: as soon as heating of the tank and water begins, the heat gained starts to be lost back into the environment, and this continues until the water in the tank reaches the ambient temperature. The challenge is therefore to limit the heat loss from the tank, thus delaying the time when thermal equilibrium is regained.

ICS or batch collectors

ICS or batch collectors reduce heat loss by placing the water tank in a thermally insulated box. This is achieved by encasing the water tank in a glass-topped box that allows heat from the sun to reach the water tank. However, the other walls of the box are thermally insulated, reducing convection as well as radiation to the environment. In addition, the box can also have a reflective surface on the inside. This reflects heat lost from the tank back towards the tank. In a simple way one could consider an ICS solar water heater as a water tank that has been enclosed in a type of 'oven' that retains heat from the sun as well as heat of the water in the tank. Using a box does not eliminate heat loss from the tank to the environment, but it largely reduces this loss.

Standard ICS collectors

Standard ICS collectors have a characteristic that strongly limits the efficiency of the collector: A small surface-to-volume ratio. Since the amount of heat that a tank can absorb from the sun is largely dependent on the surface of the tank directly exposed to the sun, it follows that a small surface would limit the degree to which the water can be heated by the sun. Cylindrical objects such as the tank in an ICS collector inherently have a small surface-to-volume ratio and most modern collectors attempt to increase this ratio for efficient warming of the water in the tank. There are many variations on this basic design, with some ICS collectors comprising several smaller water containers and even including evacuated glass tube technology, a type of ICS system known as an Evacuated Tube Batch (ETB) collector.

Flat-plate collectors

Flat-plate collectors are an extension of the basic idea to place a collector in an 'oven'-like box with glass in the direction of the sun. Most flat-plate collectors have two horizontal pipes at the top and bottom, called headers, and many smaller vertical pipes connecting them, called risers. The risers are welded (or similarly connected) to thin absorber fins. Heat-transfer fluid (water or water/antifreeze mix) is pumped from the hot water storage tank (direct system) or heat exchanger (indirect system) into the collectors' bottom header, and it travels up the risers, collecting heat from the absorber fins, and then exits the collector out of the top header. Serpentine flat-plate collectors differ slightly from this 'harp' design, and instead use a single pipe that travels up and down the collector. However, since they cannot be properly drained of water, serpentine flat-plate collectors cannot be used in drainback systems. The type of glass used in flat-plate collectors is almost always low-iron, tempered glass. Being tempered, the glass can withstand significant hail without breaking, which is one of the reasons that flat-plate collectors are considered the most durable collector type.

Unglazed or formed collectors

Unglazed or formed collectors are similar to flat-plate collectors, except they are not thermally insulated nor physically protected by a glass panel. Consequently these types of collectors are much less efficient for domestic water heating. For pool heating applications, however, the water being heated is often colder than the ambient roof temperature, at which point the lack of thermal insulation allows additional heat to be drawn from the surrounding environment.

Evacuated tube collectors (ETC)

Evacuated tube collectors are a way in which heat loss to the environment, inherent in flat-plates, has been reduced. Since heat loss due to convection cannot cross a vacuum, it forms an efficient isolation mechanism to keep heat inside the collector pipes. Since two flat sheets of glass are normally not strong enough to withstand a vacuum, the vacuum is rather created between two concentric tubes. Typically, the water piping in an ETC is therefore surrounded by two concentric tubes of glass with a vacuum in between that admits heat from the sun (to heat the pipe) but which limits heat loss back to the environment. The inner tube is coated with a thermal absorbent. Life of the vacuum varies from collector to collector, anywhere from 5 years to 15 years. Flat-plate collectors are generally more efficient than ETC in full sunshine conditions. However, the energy output of flat-plate collectors is reduced slightly more than evacuated tube collectors in cloudy or extremely cold conditions. Most ETCs are made out of annealed glass, which is susceptible to hail, breaking in roughly golf ball-sized hail. ETCs made from 'coke glass', which has a green tint, are stronger and less likely to lose their vacuum, but efficiency is slightly reduced due to reduced transparency.

4.8 Heating of swimming pools

Both pool covering systems floating atop the water and separate solar thermal collectors may be used for pool heating. Pool covering systems, whether solid sheets or floating disks, act as solar collectors and provide pool heating benefits which, depending on climate, may either supplement the solar thermal collectors discussed below or make them unnecessary.

Solar thermal collectors for nonpotable pool water use are often made of plastic. Pool water, mildly corrosive due to chlorine, is circulated through the panels using the existing pool filter or supplemental pump. In mild environments, unglazed plastic collectors are more efficient as a direct system. In cold or windy environments evacuated tubes or flat-plates in an indirect configuration do not have pool water pumped through them, they are used in conjunction with a heat exchanger that transfers the heat to pool water. This causes less corrosion. A fairly simple differential temperature controller is used to direct the water to the panels or heat exchanger either by turning a valve or operating the pump. Once the pool water has reached the required temperature, a diverter valve is used to return pool water directly to the pool without heating. Many systems are configured as drainback systems where the water drains into the pool when the water pump is switched off. The collector panels are usually mounted on a nearby roof or ground-mounted on a tilted rack. Due to the low temperature difference between the air and the water, the panels are often formed collectors or unglazed flat-plate collectors. A simple rule-of-thumb for the required panel area needed is 50 per cent of the pool's surface area. This is for areas where pools are used in the summer season only, not year round. Adding solar collectors to a conventional outdoor pool, in a cold climate, can typically extend the pool's comfortable usage by some months or more if an insulating pool cover is also used. An active solar energy system analysis program may be used to optimise the solar pool heating system before it is built.

4.9 Economics, energy, environment and system costs

4.9.1 Energy production

The amount of heat delivered by a solar water heating system depends primarily on the amount of heat delivered by the sun at a particular place (the insolation). In tropical places the insolation can be relatively high, e.g. 7 kWh/m^2 per day, whereas the insolation can be much lower in temperate areas where the days are shorter in winter, e.g. 3.2 kWh/m^2 per day. Even at the same latitude the average insolation can vary a great deal from location to location due to differences in local weather patterns and the amount of overcast. Useful calculators for estimating insolation at a site can be found with the Joint Research Laboratory of the European Commission and the American National Renewable Energy Laboratory.

Table 4.1 gives a rough indication of the specifications and energy that could be expected from a solar water heating system involving some 2 m^2 of absorber area of the collector, demonstrating two evacuated tube and three flat-plate solar water heating systems. Certification information or figures calculated from those data are used. The bottom two rows give estimates for daily energy production (kWh/day) for a tropical and a temperate scenario. These estimates are for heating water to 50°C above ambient temperature. With most solar water heating systems, the energy output scales linearly with the surface area of the absorbers. Therefore, when comparing figures, take into account the absorber area of the collector because collectors with less absorber area yield less heat, even within the 2 m^2 range. Specifications for many complete solar water heating systems and separate solar collectors can be found at Internet site of the SRCC.

Table 4.1 Daily energy production (kW/thh) of five solar thermal systems. Both the evac tube systems used have 20 tubes.

Technology Configuration	Flat-plate Direct active	Flat-plate Thermosiphon	Flat-plate Indirect active	Evac tube Indirect active	Evac tube Direct active
Overall size (m^2)	2.49	1.98	1.87	2.85	2.97
Absorber size (m^2)	2.21	1.98	1.72	2.85	2.96
Maximum efficiency	0.68	0.74	0.61	0.57	0.46
Energy production (kWh/day) Insolation 3.2 kWh/m^2/day (temperate) e.g. Zurich, Switzerland	5.3	3.9	3.3	4.8	4.0
Insolation 6.5 kWh/m^2/day (tropical) e.g. Phoenix, USA	11.2	8.8	7.1	9.9	8.4

The figures are fairly similar between the above collectors, yielding some 4 kWh/day in a temperate climate and some 8 kWh/day in a more tropical climate when using a collector with an absorber area of about 2 m^2 in size. In the temperate scenario this is sufficient to heat 200 litres of water by some 17°C. In the tropical scenario the equivalent heating would be by some 33°C. Many thermosiphon systems are quite efficient and have comparable energy output to equivalent active systems. The efficiency of evacuated tube collectors is somewhat lower than for flat plate collectors because the absorbers are narrower than the tubes and the tubes have space between them, resulting in a significantly larger percentage of inactive overall collector area. Some methods of comparison calculate the efficiency of evacuated tube collectors based on the actual absorber area and not on the 'roof area' of the system as has been done in the above table. The efficiency of the collectors becomes lower if one demands water with a very high temperature.

4.9.2 System cost

In sunny, warm locations, where freeze protection is not necessary, an ICS (batch type) solar water heater can be extremely cost effective. In higher latitudes, there are often additional design requirements for cold weather, which add to system complexity. This has the effect of increasing the initial cost (but not the life-cycle cost) of a solar water heating system, to a level much higher than a comparable hot water heater of the conventional type. The biggest single consideration is therefore the large initial financial outlay of solar water heating systems. Offsetting this expense can take several years and the payback period is longer in temperate environments where the insolation is less intense. When calculating the total cost to own and operate, a proper analysis will consider that solar energy is free, thus greatly reducing the operating costs, whereas other energy sources, such as gas and electricity, can be quite expensive over time. Thus, when the initial costs of a solar system are properly financed and compared with energy costs, then in many cases the total monthly cost of solar heat can be less than other more conventional types of hot water heaters (also in conjunction with an existing hot water heater). At higher latitudes, solar heaters may be less effective due to lower solar energy, possibly requiring larger and/or dual-heating systems.

The calculation of long term-cost and payback period for a household SWH system depends on a number of factors. Some of these are:

1. Price of purchasing solar water heater (more complex systems are more expensive)
2. Efficiency of SWH system purchased
3. Installation cost
4. State or government subsidy for installation of a solar water heater
5. Price of electricity per kWh
6. Number of kWh of electricity used per month by a household
7. Annual tax rebates or subsidy for using renewable energy
8. Annual maintenance cost of SWH system
9. Savings in annual maintenance of conventional (electric/gas/oil) water heating system

Operational carbon/energy footprint and life-cycle assessment

Terminology

1. Operational energy footprint (OEF) is also called energy parasitics ratio (EPR) or coefficient of performance (CoP).
2. Operational carbon footprint (OCF) is also called carbon clawback ratio (CCR).
3. Life-cycle assessment is usually referred to as LCA.

4.9.3 Carbon/energy footprint

The source of electricity in an active SWH system determines the extent to which a system contributes to atmospheric carbon during operation. Active solar thermal systems that use mains electricity to pump the fluid through the panels are called 'low carbon solar'. In most systems the pumping cancels the energy savings by about 8 per cent and the carbon savings of the solar by about 20 per cent. However, some new low power pumps will start operation with 1 W and use a maximum of 20 W. Assuming a solar collector panel delivering 4 kWh/day and a pump running intermittently from mains electricity for a total of 6 hours during a 12-hour sunny day, the potentially negative effect of such a pump can be reduced to about 3 per cent of the total power produced.

The carbon footprint of such household systems varies substantially, depending on whether electricity or other fuels such as natural gas are being displaced by the use of solar. Except where a high proportion of electricity is already generated by non-fossil fuel means, natural gas, a common water heating fuel, in many countries, has typically only about 40 per cent of the carbon intensity of mains electricity per unit of energy delivered. Therefore the 3 per cent or 8 per cent energy clawback in a gas home referred to above could therefore be considered 8 per cent to 20 per cent carbon clawback, a very low figure compared to technologies such as heat pumps. However, PV-powered active solar thermal systems typically use a 5–30 W PV panel which faces in the same direction as the main solar heating panel and a small, low power diaphragm pump or centrifugal pump to circulate the water. This reduces the operational carbon and energy footprint: a growing design goal for solar thermal systems.

Work is also taking place in a number of parts of the world on developing alternative non-electrical pumping systems. These are generally based on thermal expansion and phase changes of liquids and gases, a variety of which are under development.

Life-cycle carbon/energy assessment

Now looking at a wider picture than just the operational environmental impacts, recognised standards can be used to deliver robust and quantitative life-cycle assessment (LCA). LCA takes into account the total environmental cost of acquisition of raw materials, manufacturing, transport, using, servicing and disposing of the equipment. There are several aspects to such an assessment, including:

1. The financial costs and gains incurred during the life of the equipment.
2. The energy used during each of the above stages.
3. The CO_2 emissions due to each of the above stages.

Each of these aspects may present different trends with respect to a specific SWH device.

Financial assessment

Table 4.1 in the previous section as well as several other studies suggest that the cost of production is gained during the first 5–12 years of use of the equipment, depending on the insolation, with cost efficiency increasing as the insolation does.

In terms of energy, some 60 per cent of the materials of a SWH system go into the tank, with some 30 per cent towards the collector (thermosiphon flat plate in this case). In Italy, some 11 GJ of electricity are used in producing the equipment, with about 35 per cent of the energy going towards the manufacturing the tank, with another 35 per cent towards the collector and the main energy-related impact being emissions. The energy used in manufacturing is recovered within the first two to three years of use of the SWH system through heat captured by the equipment.

Moving further north into colder, less sunny climates, the energy payback time of a solar water heating system in a UK climate is reported as only 2 years. This figure was derived from the studied solar water heating system being: direct, retrofitted to an existing water store, PV pumped, freeze tolerant and of 2.8 sq m aperture. For comparison, a solar electric (PV) installation took around 5 years to reach energy payback, according to the same comparative study.

In terms of CO_2 emissions, a large degree of the emissions-saving traits of a SWH system is dependent on the degree to which water heating by gas or electricity is used to supplement solar heating of water. Using the Eco-indicator 99 points system as a yardstick (i.e. the yearly environmental load of an average European inhabitant) in Greece, a purely gas-driven system may be cheaper in terms of emissions than a solar system. This calculation assumes that the solar system produces about half of the hot water requirements of a household. The production of a test SWH system in Italy produced about 700 kg of CO_2, with all the components of manufacture, use and disposal contributing small parts towards this. Maintenance was identified as an emissions-costly activity when the heat transfer fluid (glycol-based) was periodically replaced. However, the emissions cost was recovered within about two years of use of the equipment through the emissions saved by solar water heating. In Australia, the life-cycle emissions of a SWH system are also recovered fairly rapidly, where a SWH system has about 20 per cent of the impact of an electrical water heater and half of the emissions impact of a gas water heater.

Analysing their lower impact retrofit solar water heating system, Allen (qv) report a production of CO_2 impact of 337 kg, which is around half the environmental impact reported in the Ardente (qv) study.

Where information based on established standards are available, the environmental transparency afforded by life-cycle analysis allows consumers (of all products) to make increasingly well-informed product selection decisions. As for identifying sectors where this information is likely to appear first, environmental technology suppliers in the microgeneration and renewable energy technology arena are increasingly being pressed by consumers to report typical CoP and LCA figures for their products.

In summary, the energy and emissions cost of a SWH system forms a small part of the life-cycle cost and can be recovered fairly rapidly during use of the equipment. Their environmental impacts can be reduced further by sustainable materials sourcing, using non-mains circulation, by reusing existing hot water stores and, in cold climates, by eliminating antifreeze replacement visits.

4.10 Do-It-Yourself (DIY) systems

People have begun building their own (small-scale) solar water heating systems from scratch or buying kits. Plans for solar water heating systems are available on the Internet, and people have set about building them for their own domestic requirements. DIY SWH systems are usually cheaper than commercial ones, and they are used both in the developed and developing world.

4.10.1 System specification and installation

1. Except in rare instances, it will be insufficient to install a SWH system with no electrical or gas or other fuel backup. Many SWH systems have a back-up electric heating element in the integrated tank, the operation of which may be necessary on cloudy days to ensure a reliable supply of hot water.
2. The temperature stability of a system is dependent on the ratio of the volume of warm water used per day as a fraction of the size of the water reservoir/tank that stores the hot water. If a large proportion of hot water in the reservoir is used each day, a large fraction of the water in the reservoir needs to be heated. This brings about significant fluctuations in water temperature every day, with possible risks of overheating or underheating, depending on the design of the system. Since the amount of heating that needs to take place every day is proportional to hot water usage and not to the size of the reservoir, it is desirable to have a fairly large reservoir (i.e. equal to or greater than daily usage), which will help prevent fluctuations in water temperature.

3. If ample storage is pre-existing or can otherwise be reasonably acquired, a large SWH system is more efficient economically than a small system. This is because the price of a system is not linearly proportional to the size of the collector array, so the price per square metre of collector is cheaper in a larger system. If this is the case, it pays to use a system that covers nearly all of the domestic hot water needs, and not only a small fraction of the needs. This facilitates more rapid cost recovery.
4. Not all installations require new replacement solar hot water stores. Existing stores may be large enough and in suitable condition. Direct systems can be retrofitted to existing stores, while indirect systems can be also sometimes be retrofitted using internal and external heat exchangers.
5. The installation of a SWH system needs to be complemented with efficient insulation of all the water pipes connecting the collector and the water storage tank, as well as the storage tank (or 'geyser') and the most important warm water outlets. The installation of efficient lagging significantly reduces the heat loss from the hot water system. The installation of lagging on at least two metres of pipe on the cold water inlet of the storage tank reduces heat loss, as does the installation of a 'geyser blanket' around the storage tank (if inside a roof). In cold climates the installation of lagging and insulation is often performed even in the absence of a SWH system.
6. The most efficient PV pumps are designed to start very slowly in very low light levels, so if connected uncontrolled, they may cause a small amount of unwanted circulation early in the morning—for example when there is enough light to drive the pump but while the collector is still cold. To eliminate the risk of hot water in the storage tank from being cooled that way this is very important. Solar controller may be required.
7. The modularity of an evacuated tube collector array allows the adjustment of the collector size by removing some tubes or their heat pipes. Budgeting for a larger than required array of tubes therefore allows for the customisation of collector size to the needs of a particular application, especially in warmer climates.
8. Particularly in locations further towards the poles than 45 degrees from the equator, roof mounted sun facing collectors tend to outperform wall mounted collectors in terms of total energy output. However it is total useful energy output which usually matters most to consumers. So arrays of sunny wall mounted steep collectors can sometimes produce more useful energy because there can be a small increase in winter gain at the expense of a large unused summer surplus.

4.11 Solar tracker

A solar tracker is a generic term used to describe devices that orient various payloads toward the sun. Payloads can be photovoltaic panels, reflectors, lenses or other optical devices. In flat-panel photovoltaic (PV) applications trackers are used to minimise the angle of incidence between the incoming light and a photovoltaic panel. This increases the amount of energy produced from a fixed amount of installed power generating capacity. In standard photovoltaic applications, it is estimated that trackers are used in at least 85 per cent of commercial installations greater than 1 MW from 2009 to 2012.

In concentrated photovoltaic (CPV) and CSP applications trackers are used to enable the optical components in the CPV and CSP systems. The optics in concentrated solar applications accepts the direct component of sunlight and therefore must be oriented appropriately to collect energy. Tracking systems are found in all concentrator applications because such systems do not produce energy unless oriented closely toward the sun.

4.11.1 Types of solar collector

Different types of solar collector and their location (latitude) require different types of tracking mechanism. Solar collectors may be:

1. Non-concentrating flat-panels, usually photovoltaic or hot-water
2. Concentrating systems, of a variety of types

Solar collector mounting systems may be fixed (manually aligned) or tracking. Tracking systems may be configured as:

1. Fixed collector/moving mirror, i.e. Heliostat
2. Moving collector

Fixed mount

Domestic and small-scale commercial photovoltaic and hot-water panels are usually fixed, often flush-mounted on an appropriately facing pitched roof. Advantages of fixed mount systems (i.e. factors tending to indicate against trackers) include the following:

1. Mechanical simplicity, and hence lower installation and ongoing maintenance costs.
2. Wind-loading: It is easier and cheaper to provision a sturdy mount; all mounts other than fixed flush-mounted panels must be carefully designed having regard to their wind loading due to their greater exposure.
3. Indirect light: Approximately 10 per cent of the incident solar radiation is diffuse light, available at any angle of misalignment with the direct sun.

4. Tolerance to misalignment: Effective collection area for a flat-panel is relatively insensitive to quite high levels of misalignment with the sun (Table 4.2 and Fig. 4.5) at Accuracy Requirements section below—for example even a 25° misalignment reduces the direct solar energy collected by less than 10 per cent.

Fixed mounts are usually used in conjunction with non-concentrating systems, however an important class of non-tracking concentrating collectors, of particular value in the 3rd world, are portable solar cookers. These utilise relatively low levels of concentration, typically around 2 to 8 suns and are manually aligned.

Table 4.2 Direct power lost (per cent) due to misalignment (angle i).

i	Lost	i	Hours	Lost
0°	0%	15°	1	3.4%
1°	0.015%	30°	2	13.4%
3°	0.14%	45°	3	30%
8°	1%	60°	4	>50%
23.4°	8.3%	75°	5	>75%

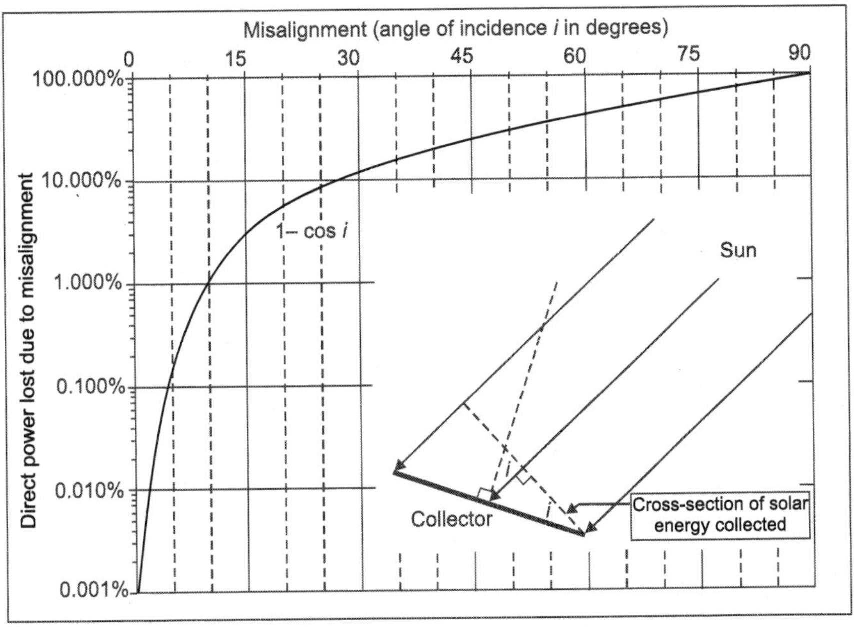

Figure 4.5 The effective collection area of a flat-panel solar collector varies with the cosine of the misalignment of the panel with the sun.

Trackers

Even though a fixed flat-panel can be set to collect a high proportion of available noon-time energy, significant power is also available in the early mornings and late afternoons when the misalignment with a fixed panel becomes excessive to collect a reasonable proportion of the available energy. For example, even when the sun is only 10° above the horizon the available energy can already be around half the noon-time energy levels (or even greater depending on latitude, season, and atmospheric conditions).

Thus the primary benefit of a tracking system is to collect solar energy for the longest period of the day, and with the most accurate alignment as the sun's position shifts with the seasons.

In addition, the greater the level of concentration employed the more important accurate tracking becomes, because the proportion of energy derived from direct radiation is higher, and the region where that concentrated energy is focused becomes smaller.

Fixed collector/moving mirror

Many collectors cannot be moved, for example high-temperature collectors where the energy is recovered as hot liquid or gas (e.g. steam). Other examples include direct heating and lighting of buildings and fixed in-built solar cookers, such as Scheffler reflectors. In such cases it is necessary to employ a moving mirror so that, regardless of where the sun is positioned in the sky, the sun's rays are redirected onto the collector. Due to the complicated motion of the sun across the sky, and the level of precision required to correctly aim the sun's rays onto the target, a heliostat mirror generally employs a dual axis tracking system, with at least one axis mechanised. In different applications, mirrors may be flat or concave.

Moving collector

Trackers can be grouped into classes by the number and orientation of the tracker's axes. Compared to a fixed mount, a single axis tracker increases annual output by approximately 30 per cent, and a dual axis tracker an additional 6 per cent. Photovoltaic trackers can be classified into two types: Standard Photovoltaic (PV) Trackers and Concentrated Photovoltaic (CPV) Trackers. Each of these tracker types can be further categorised by the number and orientation of their axes, their actuation architecture and drive type, their intended applications, their vertical supports and foundation type. Solar tracking systems are mostly used in Bangladesh.

4.11.2 Non-concentrating photovoltaic (PV) trackers

Photovoltaic panels accept both direct and diffuse light from the sky. The panels on a standard photovoltaic trackers always gather the available direct light. The tracking functionality in standard photovoltaic trackers is used to minimise the angle of incidence between incoming light and the photovoltaic panel. This increases the amount of energy gathered from the direct component of the incoming light.

Accuracy requirements

In non-concentrating flat-panel systems, the energy contributed by the direct beam drops off with the cosine of the angle between the incoming light and the panel. In addition, the reflectance (averaged across all polarisations) is approximately constant for angles of incidence up to around 50°, beyond which reflectance degrades rapidly. For example trackers that have accuracies of ±5° can deliver greater than 99.6 per cent of the energy delivered by the direct beam plus 100 per cent of the diffuse light. As a result, high accuracy tracking is not typically used in non-concentrating PV applications.

Technologies supported

The physics behind standard photovoltaic (PV) trackers works with all standard photovoltaic module technologies. These include all types of crystalline silicon panels (monocrystalline, multicrystalline, polycrystalline) and all types of thin film panels (amorphous silicon, CdTe, CIGS, microcrystalline).

4.11.3 Concentrated photovoltaic (CPV) trackers

The optics in CPV modules accept the direct component of the incoming light and therefore must be oriented appropriately to maximise the energy collected. In low concentration applications a portion of the diffuse light from the sky can also be captured. The tracking functionality in CPV modules is used to orient the optics such that the incoming light is focused to a photovoltaic collector.

CPV modules that concentrate in one dimension must be tracked normal to the sun in one axis. CPV modules that concentrate in two dimensions must be tracked normal to the sun in two axes.

Accuracy requirements

The physics behind CPV optics requires that tracking accuracy increase as the systems concentration ratio increases. However, for a given concentration, non-imaging optics provides the widest possible acceptance angles, which may be used to reduce tracking accuracy.

In typical high concentration systems tracking accuracy must be in the $\pm 0.1°$ range to deliver approximately 90 per cent of the rated power output. In low concentration systems, tracking accuracy must be in the $\pm 2.0°$ range to deliver 90 per cent of the rated power output. As a result, high accuracy tracking systems are typically used.

Technologies supported

Concentrated photovoltaic trackers are used with refractive and reflective based concentrator systems. There is a range of emerging photovoltaic cell technologies used in these systems. These range from crystalline silicon are based on photovoltaic receivers to germanium based triple junction receivers.

4.11.4 Single axis trackers

Single axis trackers have one degree of freedom that acts as an axis of rotation. The axis of rotation of single axis trackers is typically aligned along a true North meridian. It is possible to align them in any cardinal direction with advanced tracking algorithms.

There are several common implementations of single axis trackers. These include horizontal single axis trackers (HSAT), vertical single axis trackers (VSAT), tilted single axis trackers (TSAT) and polar aligned single axis trackers (PSAT). The orientation of the module with respect to the tracker axis is important when modelling performance.

Horizontal single axis tracker (HSAT)

The axis of rotation for horizontal single axis tracker is horizontal with respect to the ground. The posts at either end of the axis of rotation of a horizontal single axis tracker can be shared between trackers to lower the installation cost. Field layouts with horizontal single axis trackers are very flexible. The simple geometry means that keeping all the axis of rotation parallel to one another is all that is required for appropriately positioning the trackers with respect to one another.

Appropriate spacing can maximise the ratio of energy production to cost, this being dependent upon local terrain and shading conditions and the time-of-day value of the energy produced. Backtracking is one means of computing the disposition of panels. Horizontal trackers typically have the face of the module oriented parallel to the axis of rotation. As a module tracks, it sweeps a cylinder that is rotationally symmetric around the axis of rotation.

Several manufacturers can deliver single axis horizontal trackers. In these, a long horizontal tube is supported on bearings mounted upon pylons or frames. The axis of the tube is on a north-south line. Panels are mounted upon the tube, and the tube will rotate on its axis to track the apparent motion of the sun through the day.

Vertical single axis tracker (VSAT)

The axis of rotation for vertical single axis trackers is vertical with respect to the ground. These trackers rotate from East to West over the course of the day. Such trackers are more effective at high latitudes than are horizontal axis trackers.

Field layouts must consider shading to avoid unnecessary energy losses and to optimise land utilisation. Also optimisation for dense packing is limited due to the nature of the shading over the course of a year.

Vertical single axis trackers typically have the face of the module oriented at an angle with respect to the axis of rotation. As a module tracks, it sweeps a cone that is rotationally symmetric around the axis of rotation.

Tilted single axis tracker (TSAT)

All trackers with axes of rotation between horizontal and vertical are considered tilted single axis trackers. Tracker tilt angles are often limited to reduce the wind profile and decrease the elevated end's height off the ground. Field layouts must consider shading to avoid unnecessary losses and to optimise land utilisation. With backtracking, they can be packed without shading perpendicular to their axis of rotation at any density. However, the packing parallel to their axis of rotation is limited by the tilt angle and the latitude. Tilted single axis trackers typically have the face of the module oriented parallel to the axis of rotation. As a module tracks, it sweeps a cylinder that is rotationally symmetric around the axis of rotation.

Polar-aligned single axis trackers (PASAT)

One scientifically interesting variation of a tilted single axis tracker is a polar-aligned single axis trackers (PASAT). In this particular implementation of a tilted single axis tracker the tilt angle is equal to the latitude of the installation. This aligns the tracker axis of rotation with the earth's axis of rotation. These are rarely deployed because of their high wind profile.

4.11.5 Dual axis trackers

Dual axis trackers have two degrees of freedom that act as axes of rotation. These axes are typically normal to one another. The axis that is fixed with respect to the ground can be considered a primary axis. The axis that is referenced to the primary axis can be considered a secondary axis.

There are several common implementations of dual axis trackers. They are classified by the orientation of their primary axes with respect to the ground. Two common implementations are tip–tilt dual axis trackers (TTDAT) and azimuth-altitude dual axis trackers (AADAT). The orientation of the module with respect to the tracker axis is important when modelling performance. Dual axis trackers typically have modules oriented parallel to the secondary axis of rotation. Dual axis trackers allow for optimum solar energy levels due to their ability to follow the sun vertically and horizontally. No matter where the sun is in the sky, dual axis trackers are able to angle themselves to be in direct contact with the sun.

Tip–tilt dual axis tracker (TTDAT)

A tip–tilt dual axis tracker has its primary axis horizontal to the ground. The secondary axis is then typically normal to the primary axis. The posts at either end of the primary axis of rotation of a tip–tilt dual axis tracker can be shared between trackers to lower installation costs.

Field layouts with tip–tilt dual axis trackers are very flexible. The simple geometry means that keeping the axes of rotation parallel to one another is all that is required for appropriately positioning the trackers with respect to one another. In addition, with backtracking, they can be packed without shading at any density. The axes of rotation of tip–tilt dual axis trackers are typically aligned either along a true North meridian or an east-west line of latitude. It is possible to align them in any cardinal direction with advanced tracking algorithms.

Azimuth-altitude dual axis tracker (AADAT)

An azimuth–altitude dual axis tracker has its primary axis vertical to the ground. The secondary axis is then typically normal to the primary axis. Field layouts must consider shading to avoid unnecessary energy losses and to optimise land utilisation. Also optimisation for dense packing is limited due to the nature of the shading over the course of a year. This mount is used as a large telescope mount owing to its structure and dimensions. One axis is a vertical pivot shaft or horizontal ring mount that allows the device to be swung to a compass point. The second axis is a horizontal elevation pivot mounted upon the azimuth platform. By using combinations of the two axis, any location in the upward hemisphere may be pointed. Such systems may be operated under computer control according to the expected solar orientation or may use a tracking sensor to control motor drives that orient the panels toward the sun. This type of mount is also used to orient parabolic reflectors that mount a Stirling engine to produce electricity at the device.

4.11.6 Tracker type selection

The selection of tracker type is dependent on many factors including installation size, electric rates, government incentives, land constraints, latitude, and local weather. Horizontal single axis trackers are typically used for large distributed generation projects and utility scale projects. The combination of energy improvement and lower product cost and lower installation complexity results in compelling economics in large deployments. In addition the strong afternoon performance is particularly desirable for large grid-tied photovoltaic systems so that production will match the peak demand time. Horizontal single axis trackers also add a substantial amount of productivity during the spring and summer seasons when the sun is high in the sky. The inherent robustness of their supporting structure and the simplicity of the mechanism also result in high reliability which keeps maintenance costs low. Since the panels are horizontal, they can be compactly placed on the axle tube without danger of self-shading and are also readily accessible for cleaning. A vertical axis trackers pivots only about a vertical axle, with the panels either vertical, at a fixed, adjustable or tracked elevation angle. Such trackers with fixed or (seasonably) adjustable angles are suitable for high latitudes, where the apparent solar path is not especially high, but which leads to long days in summer, with the sun travelling through a long arc. Dual axis trackers are typically used in smaller residential installations and locations with very high government feed-in tariffs.

4.11.7 Multi-mirror concentrating PV

This device uses multiple mirrors in a horizontal plane to reflect sunlight upward to a high temperature photovoltaic or other system requiring concentrated solar power. Structural problems and expense are greatly reduced since the mirrors are not significantly exposed to wind loads. Through the employment of a patented mechanism, only two drive systems are required for each device. Because of the configuration of the device it is especially suited for use on flat roofs and at lower latitudes. The units illustrated each produce approximately 200 peak DC watts. A multiple mirror reflective system combined with a central power tower is employed at the Sierra Sun Tower, located in Lancaster, California. This system, which uses multiple heliostats in a north–south alignment, uses pre-fabricated parts and construction as a way of decreasing start-up and operating costs.

4.11.8 Drive types

Active tracker

Active trackers use motors and gear trains to direct the tracker as commanded by a controller responding to the solar direction. In order to control and manage the movement of these massive structures special slewing drives are designed and rigorously tested. Active two-axis trackers are also used to orient heliostats-moveable mirrors that reflect sunlight toward the absorber of a central power station. As each mirror in a large field will have an individual orientation these are controlled programmatically through a central computer system, which also allows the system to be shut down when necessary. Light-sensing trackers typically have two photosensors, such as photodiodes, configured differentially so that they output a null when receiving the same light flux. Mechanically, they should be omnidirectional (i.e. flat) and are aimed 90° apart.

This will cause the steepest part of their cosine transfer functions to balance at the steepest part, which translates into maximum sensitivity. Since the motors consume energy, one wants to use them only as necessary. So instead of a continuous motion, the heliostat is moved in discrete steps. Also, if the light is below some threshold there would not be enough power generated to warrant reorientation. This is also true when there is not enough difference in light level from one direction to another, such as when clouds are passing overhead. Consideration must be made to keep the tracker from wasting energy during cloudy periods.

Passive tracker

Passive trackers use a low boiling point compressed gas fluid that is driven to one side or the other (by solar heat creating gas pressure) to cause the tracker to move in response to an imbalance. As this is a non-precision orientation it is unsuitable for certain types of concentrating photovoltaic collectors but works fine for common PV panel types. These will have viscous dampers to prevent excessive motion in response to wind gusts. Shader/reflectors are used to reflect early morning sunlight to 'wake up' the panel and tilt it toward the sun, which can take nearly an hour. The time to do this can be greatly reduced by adding a self-releasing tie down that positions the panel slightly past the zenith (so that the fluid does not have to overcome gravity) and using the tie down in the evening. (A slack-pulling spring will prevent release in windy overnight conditions.)

The term 'passive tracker' is also used for photovoltaic modules that include a hologram behind stripes of photovoltaic cells. That way, sunlight passes through the transparent part of the module and reflects on the hologram. This allows sunlight to hit the cell from behind, thereby increasing the module's efficiency. Also, the module does not have to move since the hologram always reflects sunlight from the correct angle towards the cells.

Chronological tracker

A chronological tracker counteracts the earth's rotation by turning at an equal rate as the earth, but in the opposite direction. Actually the rates are not quite equal, because as the earth goes around the sun, the position of the sun changes with respect to the earth by 360° every year or 365.24 days. A chronological tracker is a very simple yet potentially a very accurate solar tracker specifically for use with a polar mount. The drive method may be as simple as a gear motor that rotates at a very slow average rate of one revolution per day (15° per hour). In theory the tracker may rotate completely, assuming there is enough clearance for a complete rotation, and assuming that twisting wires are not an issue.

4.12 Photovoltaic thermal hybrid solar collector

Photovoltaic thermal hybrid solar collectors, sometimes known as hybrid PV/T systems or PVT, are systems that convert solar radiation into thermal and electrical energy. These systems combine a photovoltaic cell, which converts electromagnetic radiation (photons) into electricity, with a solar thermal collector, which captures the remaining energy and removes waste heat from the PV module. Photovoltaic (PV) cells suffer from a drop in efficiency with the rise in temperature due to increased resistance. Such systems can be engineered to carry heat away from the PV cells thereby cooling the cells and thus improving their efficiency by lowering resistance.

4.12.1 System types

A number of PV/T collectors in different categories are commercially available and can be divided into the following categories:

1. PV/T liquid collector
2. PV/T air collector
3. PV/T concentrator (CPVT)

PV/T liquid collector

The basic water-cooled design uses conductive-metal piping or plates attached to the back of a PV module. A working fluid, typically water, glycol or mineral oil is then piped through these pipes. The heat from the PV cells are conducted through the metal and absorbed by the working fluid (presuming that the working fluid is cooler than the operating temperature of the cells). In closed-loop systems this heat is either exhausted (to cool it) or transferred at a heat exchanger, where it flows to its application. In open-loop systems, this heat is used or exhausted before the fluid returns to the PV cells.

PV/T concentrator (CPVT)

A concentrating system has the advantage to reduce the amount of solar cells needed. It also can get very good solar thermal performance compared to flat PV/T collectors. The main obstacles are to provide good cooling of the solar cells and a durable tracking system.

4.13 Solar furnace

A solar furnace is a structure that captures sunlight to produce high temperatures, usually for industry. This is done with a curved mirror (or an array of mirrors) that acts as a parabolic reflector, concentrating light (Insolation) onto a focal point. The temperature at the focal point may reach 3500°C (6330°F), and this heat can be used to generate electricity, melt steel, make hydrogen fuel or nanomaterials.

The term 'solar furnace' has also evolved to refer to solar concentrator heating systems using parabolic mirrors or heliostats where 538°C is now commonly achieved. The largest solar furnace is at Odeillo in the Pyrénées-Orientales in France, opened in 1970. It employs an array of plane mirrors to gather sunlight, reflecting it onto a larger curved mirror. The rays are then focused onto an area the size of a cooking pot and can reach 3500°C, depending on the process installed, for example:

1. About 1000°C for metallic receivers producing hot air for the next generation solar towers as it will be tested at the Themis plant with the Pegase project.
2. About 1400°C to produce hydrogen by cracking methane molecules.
3. Up to 2500°C to test materials for extreme environment such as nuclear reactors or space vehicle atmospheric re-entry.
4. Up to 3500°C to produce nanomaterials by solar induced sublimation and controlled cooling, such as carbon nanotubes or zinc nanoparticles.

4.13.1 Modern uses

The solar furnace principle is being used to make inexpensive solar cookers and solar-powered barbecues, and for solar water pasteurisation. A prototype

Scheffler reflector is being constructed in India for use in a solar crematorium. This 50 m^2 reflector will generate temperatures of 700°C (1292°F) and displace 200–300 kg of firewood used per cremation. It has been suggested that solar furnaces could be used in space to provide energy for manufacturing purposes. Their reliance on sunny weather is a limiting factor as a source of renewable energy on earth but could be tied to thermal energy storage systems for energy production through these periods and into the night.

5
Solar photovoltaic systems

5.1 Introduction

Photovoltaic devices use semiconducting materials to convert sunlight directly into electricity. It was first observed in 1839 by the French scientist Becquerel who detected that when light was directed onto one side of a simple battery cell, the current generated could be increased. In the late 1950s, the space program provided the impetus for the development of crystalline silicon solar cells. The first commercial production of photovoltaic modules for terrestrial applications began in 1953 with the introduction of automated photovoltaic production plants.

Conventional photovoltaic cells are made of crystalline silicon that has atoms arranged in a three- dimensional array, making it an efficient semiconductor. While this material is most commonly used in converting light energy into electricity, it has associated drawbacks, like high material costs for silicon, costly processes for purifying silicon and manufacturing wafer, additional processes for assembly of modules, and bulky and rigid nature of the photovoltaic panels.

Photovoltaic cells convert sunlight directly into electricity without creating any air or water pollution. Photovoltaic cells are made of at least two layers of semiconductor material. One layer has a positive charge, the other negative. When light enters the cell, some of the photons from the light are absorbed by the semiconductor atoms, freeing electrons from the cell's negative layer to flow through an external circuit and back into the positive layer. This flow of electrons produces electric current. To increase their utility, a number of individual photovoltaic cells are interconnected together in a sealed, weatherproof package called a module (Fig. 5.1). When two modules are wired together in series, their voltage is doubled while the current stays constant. When two modules are wired in parallel, their current is doubled while the voltage stays constant. To achieve the desired voltage and current, modules are wired in series and parallel into what is called a PV array. The flexibility of the modular PV system allows designers to create solar power systems that can meet a wide variety of electrical needs, no matter how large or small.

Photovoltaic modules are usually installed on special ground or pole mounting structures. Modules may be mounted on rooftops provided that proper building and safety precautions are observed. For more output, modules are sometimes installed on a tracker—a mounting structure that moves to continually face the sun throughout the day.

The performance of photovoltaic modules and arrays are generally rated according to their maximum DC power output under standard test conditions (STC). Standard test conditions are defined by a module operating temperature of 250°C, and incident solar irradiance level of 1000 W/m^2 and under air mass 1.5 spectral distribution. Since these conditions are not always typical of how PV modules and arrays operate in the field, actual performance is usually 85–90 per cent of the STC rating.

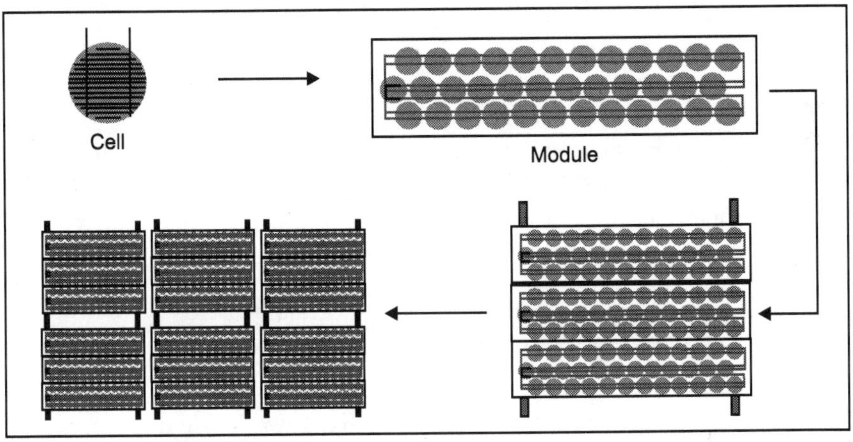

Figure 5.1 Photovoltaic cells, modules, panels and arrays

5.2 Fabrication of photovoltaic cells

5.2.1 Silicon-based photovoltaic cells

The process of fabricating conventional single and polycrystalline silicon photovoltaic cells begins with very pure semiconductor-grade polysilicon—a material processed from quartz and used extensively throughout the electronics industry. The polysilicon is then heated to melting temperature, and trace amounts of boron are added to the melt to create a p-type semiconductor material. Next, an ingot or block of silicon is formed, commonly using one of two methods: (i) by growing a pure crystalline silicon ingot from a seed crystal drawn from the molten polysilicon, and (ii) by casting the molten polysilicon in a block, creating a polycrystalline silicon material. Individual wafers are then sliced from the ingots using wire saws and then subjected to a surface etching process. After the wafers are cleaned, they are placed in a phosphorus diffusion furnace, creating a thin n-type semiconductor layer around the entire outer surface of the cell. Next, an anti-reflective coating is applied to the top surface of the cell, and electrical contacts are imprinted on the top (negative) surface of the cell. An aluminised conductive material is deposited on the back (positive) surface of each cell, restoring the p-type properties of the back surface by displacing the diffused phosphorus layer. Each cell is then electrically tested, sorted based on current output, and electrically connected to other cells to form cell circuits for assembly in PV modules.

5.2.2 Band gap energies of semiconductors

When light shines on crystalline silicon, electrons within the crystal lattice may be freed. But not all photons, only photons with a certain level of energy can free electrons in the semiconductor material from their atomic bonds to produce an electric current. This level of energy, known as the 'band gap energy', is the amount of energy required to dislodge an electron from its covalent bond and allow it to become part of an electrical circuit. To free an electron, the energy of a photon must be at least as great as the band gap energy. However, photons with more energy than the band gap energy will expend that extra amount as heat when freeing electrons. So, it is important for a photovoltaic cell to be 'tuned' through slight modifications to the silicon's molecular structure to optimise the photon energy. A key to obtaining an efficient PV cell is to convert as much sunlight as possible into electricity. Crystalline silicon has band gap energy of 1.1 eV. The band gap energies of other effective photovoltaic semiconductors range from 1.0 to 1.6 eV. In this range, electrons can be freed without creating extra heat. The photon energy of light varies according to the different wavelengths of the light. The entire spectrum of sunlight, from infrared to ultraviolet, covers a range of about 0.5 eV to about 2.9 eV. For example, red light has energy of about 1.7 eV, and blue light has energy of about 2.7 eV. Most PV cells cannot use about 55 per cent of the energy of sunlight, because this energy is either below the band gap of the material or carries excess energy.

5.2.3 Doping silicon to create n-type and p-type silicon

In a crystalline silicon cell, we need to contact p-type silicon with n-type silicon to create the built-in electrical field. The process of doping, which creates these materials, introduces an atom of another element into the silicon crystal to alter its electrical properties. The dopant, which is the introduced element, has either three or five valence electrons, which is one less or one more that silicon's four.

Phosphorus atoms, which have five valence electrons, are used in doping n-type silicon, because phosphorus provides its fifth free electron. A phosphorus atom occupies the same place in the crystal lattice formerly occupied by the silicon atom it replaces (Fig. 5.2). Four of its valence electrons take over the bonding responsibilities of the four silicon valence electrons that they replaced. But the fifth valence electron remains free, having no bonding responsibilities. When phosphorus atoms are substituted for silicon in a crystal, many free electrons become available.

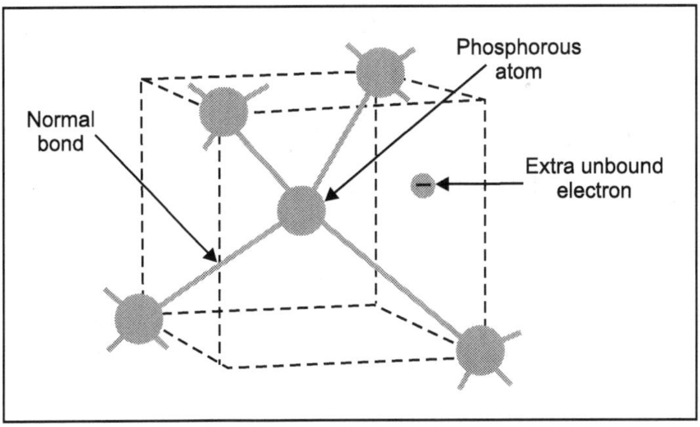

Figure 5.2 Phosphorus substituted n-type silicon

The most common method of doping is to coat a layer of silicon material with phosphorus and then heat the surface. This allows the phosphorus atoms to diffuse into the silicon. The temperature is then reduced so the rate of diffusion drops to zero. Other methods of introducing phosphorus into silicon include gaseous diffusion, a liquid dopant spray-on process, and a technique where phosphorus ions are precisely driven into the surface of the silicon.

The n-type silicon doped with phosphorus cannot form an electric field by itself. One also needs p-type silicon. Boron, which has only three valence electrons, is used for doping p-type silicon (Fig. 5.3). Boron is introduced during silicon processing when the silicon is purified for use in photovoltaic devices. When a boron atom takes a position in the crystal lattice formerly occupied by a silicon atom, a bond will be missing an electron. In other words, there is an extra positively charged hole.

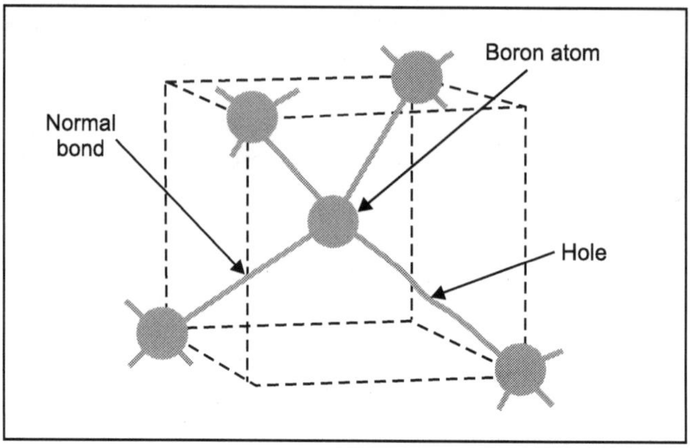

Figure 5.3 Boron substituted p-type silicon

5.2.4 Absorption and conduction

In a photovoltaic cell, photons are absorbed in the p-layer. And it is very important to 'tune' this layer to the properties of incoming photons to absorb as many as possible, and thus, to free up as many electrons as possible. Another challenge is to keep the electrons from meeting up with holes and recombining with them before they can escape from the photovoltaic cell. To do all this, the material is design to free the electrons as close to the junction as possible, so that the electric field can help send the free electrons through the conduction layer (the n-layer) and out into the electrical circuit. By optimising all these characteristics, one improves the photovoltaic cell's conversion efficiency, which is how much of the light energy is converted into electrical energy by the cell.

5.2.5 Electrical contacts

Electrical contacts are essential to a photovoltaic cell because they bridge the connection between the semiconductor material and the external electrical load, such as a light bulb. The back contact of a cell, i.e. on the side away from the incoming sunlight, i.e. is relatively simple. It usually consists of a layer of aluminium or molybdenum metal. But the front contact, on the side facing the sun, i.e. is more complicated. When sun rays fall on the photovoltaic cell, electron current flows all over its surface. If we attach contacts only at the edges of the cell, it will not work well because of the great electrical resistance of the top semiconductor layer. Only a small number of electrons would make it to the contact.

To collect the maximum current, one must place contacts across the entire surface of a photovoltaic cell. This is normally done with a 'grid' of metal strips or 'fingers'. However, placing a large grid, which is opaque, on the top of the cell shades active parts of the cell from the sun. The cell's conversion efficiency is thus significantly reduced. To improve the conversion efficiency, we must minimise these shading effects. Another challenge in cell design is to minimise the electrical resistance losses when applying grid contacts to the solar cell material. These losses are related to the solar cell material's property of opposing the flow of an electric current, which results in heating the material. Therefore, in designing grid contacts, we must balance shading effects against electrical resistance losses. The usual approach is to design grids with many thin, conductive fingers spreading to every part of the cell's surface. The fingers of the grid must be thick enough to conduct well (with low resistance), but thin enough not to block much of the incoming light. This kind of grid keeps resistance losses low while shading only about 3 per cent to 5 per cent of the cell's surface.

Grids can be expensive to make and can affect the cell's reliability. To make top-surface grids, we can either deposit metallic vapours on a cell through a mask or paint them on via a screen-printing method. Photolithography is the preferred method for the highest quality, but has the greatest cost. This process involves transferring an image via photography, as in modern printing. An alternative to metallic grid contacts is a transparent conducting oxide (TCO) layer such as tin oxide (SnO_2). The advantage of TCOs is that they are nearly invisible to incoming light, and they form a good bridge from the semiconductor material to the external electrical circuit. TCOs are very useful in manufacturing processes involving a glass superstrate, which is the covering on the sun-facing side of a PV module. Some thin-film PV cells, such as amorphous silicon and cadmium telluride, use superstrates. In this process, the TCO is generally deposited as a thin film on the glass superstrate before the semiconducting layers are deposited. The semiconducting layers are then followed by a metallic contact that will actually be the bottom of the cell. The cell is actually constructed 'upside down', from the top to the bottom. But the construction technique is not the only thing that determines whether a metallic grid or TCO is best for a certain cell design. The sheet resistance of the semiconductor is also an important consideration. In crystalline silicon, for example, the semiconductor carries electrons well enough to reach a finger of the metallic grid. Because the metal conducts electricity better than a TCO, shading losses are less than losses associated with using a TCO. Amorphous silicon, on the other hand, conducts very poorly in the horizontal direction. Therefore, it benefits from having a TCO over its entire surface.

5.2.6 Antireflective coating

Since, silicon is a shiny gray material and can act as a mirror, reflecting more than 30 per cent of the light that shines on it. To improve the conversion efficiency of a solar cell, to minimise the amount of light reflected so that the semiconductor material can capture as much light as possible to use in freeing electrons. Two techniques are commonly used to reduce reflection. The first technique is to coat the top surface with a thin layer of silicon monoxide (SiO). A single layer reduces surface reflection to about 10 per cent, and a second layer can lower the reflection to less than 4 per cent. A second technique is to texture the top surface. Chemical etching creates a pattern of cones and pyramids, which capture light rays that might otherwise to deflect away from the cell. Reflected light is redirected down into the cell, where it has another chance to be absorbed.

5.3 Photovoltaic module performance ratings

Generally, the performances rating of photovoltaic are expressed in terms of peak watt. The peak watt (Wp) rating is determined by measuring the maximum power of a PV module under laboratory conditions of relatively high light level, favourable air mass, and low cell temperature. But these conditions are not typical in the real world. Therefore, one may uses a different procedure, known as the normal operating cell temperature (NOCT) rating. In this procedure, the module first equilibrates with a specified ambient temperature so that maximum power is measured at a nominal operating cell temperature. This NOCT rating results in a lower watt value than the peak-watt rating, but it is probably more realistic. Neither of these methods is designed to indicate the performance of a solar module under realistic operating conditions. Another technique, the AMPM Standard, involves considering the whole day rather than 'peak' sunshine hours. This standard, which seeks to address the practical user's needs, is based on the description of a standard solar global-average day (or a practical global average) in terms of light levels, ambient temperature, and air mass. Solar arrays are designed to provide specified amounts of electricity under certain conditions. The following factors are usually considered when determining array performance: characterisation of solar cell electrical performance, determination of degradation factors related to array design and assembly, conversion of environmental considerations into solar cell operating temperatures, and calculation of array power output capability.

5.3.1 Power output

Power available at the power regulator, specified either as peak power or average power produced during one day.

5.3.2 Energy output

The energy is expressed as watt-hour or Wh. This indicates the amount of energy produced during a certain period of time. The parameters are output per unit of array area (Wh/m^2), output per unit of array mass (Wh/kg), and output per unit of array cost (Wh/\$).

5.3.3 Conversion efficiency

This parameter is defined as:

$$\frac{\text{Energy output from array}}{\text{Energy input from sun}} \times 100$$

This last parameter is often given as a power efficiency, equal to 'power output from array'/'power input from sun' × 100 per cent. Power is typically given in units of watts (W), and energy is typical in units of watt-hours (Wh). To ensure the consistency and quality of photovoltaic systems and increase consumer confidence in system performance, various groups such as the Institute of Electrical and Electronics Engineers (IEEE) and the American Society for Testing and Materials (ASTM) are working on standards and performance criteria for photovoltaic systems.

5.4 Reliability of photovoltaic systems

Reliability of photovoltaic arrays is an important factor in the cost of systems and in consumers accepting this technology. The photovoltaic cell itself is considered a 'solid-state' device with no moving parts, and therefore, it is highly reliable and long-lived. Therefore, reliability of photovoltaic usually focuses not on cells, but on modules and systems. One way to measure reliability is the rate of failure of particular parts. The failure of solar cells mostly involves cell cracking, interconnect failures (resulting in open circuits or short circuits), and increased contact resistance. Module-level failures include glass breakage, electrical insulation breakdown, and various types of encapsulate failures.

Fault-tolerant circuit design involves using various redundant features in the circuit to control the effect of partial failure on overall module yield and array power degradation. Degradation can be controlled by dividing the modules into a number of parallel solar cell networks called branch circuits. This type of design can also improve module losses due to broken cells and other circuit failures. Bypass diodes or other corrective measures can mitigate the effects of local cell hot-spots. Replacement of the entire module is a final option in dealing with photovoltaic array failures. However, today's component failure rates are low enough that, with multiple-cell interconnects, series/paralleling, and bypass diodes; it is possible to achieve high levels of reliability.

5.5 Classification of photovoltaic systems

Photovoltaic power systems are generally classified according to their functional and operational requirements, their component configurations, and how the equipment is connected to other power sources and electrical loads. The two principle classifications are grid-connected or utility-interactive systems and stand-alone systems. Photovoltaic systems can be designed to provide DC and/or AC power service, can operate interconnected with or independent of the utility grid, and can be connected with other energy sources and energy storage systems.

5.5.1 Grid-connected (utility-interactive) PV systems

Grid-connected or utility-interactive photovoltaic systems are designed to operate in parallel with and interconnected with the electric utility grid. The primary component in grid-connected photovoltaic systems is the inverter or power conditioning unit (PCU). The PCU converts the DC power produced by the photovoltaic array into AC power consistent with the voltage and power quality requirements of the utility grid, and automatically stops supplying power to the grid when the utility grid is not energised (Fig. 5.4). A bi-directional interface is made between the photovoltaic system AC output circuits and the electric utility network, typically at an on-site distribution panel or service entrance. This allows the AC power produced by the photovoltaic system to either supply on-site electrical loads or to back feed the grid when the photovoltaic system output is greater than the on-site load demand. At night and during other periods when the electrical loads are greater than the photovoltaic system output, the balance of power required by the loads is received from the electric utility This safety feature is required in all grid-connected photovoltaic systems, and ensures that the photovoltaic system will not continue to operate and feed back onto the utility grid when the grid is down for service or repair.

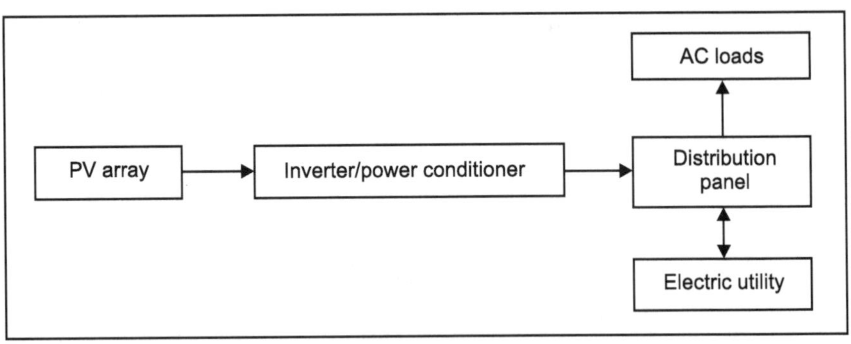

Figure 5.4 Diagram of grid-connected photovoltaic system

5.5.2 Stand-alone photovoltaic systems

Stand-alone photovoltaic systems are designed to operate independent of the electric utility grid, and are generally designed and sized to supply certain DC and/or AC electrical loads. These types of systems may be powered by a photovoltaic array only or may use wind, an engine-generator or utility power as an auxiliary power source in what is called a photovoltaic-hybrid system. The simplest type of stand-alone photovoltaic system is a direct-coupled system, where the DC output of a photovoltaic module or array is directly connected to a DC load (Fig. 5.5). Since there is no electrical energy storage (batteries) in direct-coupled systems, the load only operates during sunlight hours, making these designs suitable for common applications such as ventilation fans, water pumps, and small circulation pumps for solar thermal water heating systems. Matching the impedance of the electrical load to the maximum power output of the photovoltaic array is a critical part of designing well-performing direct-coupled system. For certain loads such as positive-displacement water pumps; a type of electronic DC–DC converter, called a maximum power point tracker (MPPT) is used between the array and load to help better utilise the available array maximum power output.

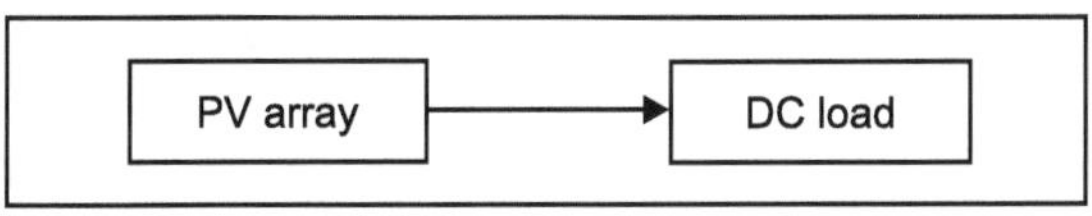

Figure 5.5 Simplest type of stand-alone PV system

In many stand-alone photovoltaic systems, batteries are used for energy storage. Figure 5.6 shows a diagram of a typical stand-alone PV system powering DC and AC loads. Figure 5.7 shows how a typical photovoltaic hybrid system might be configured.

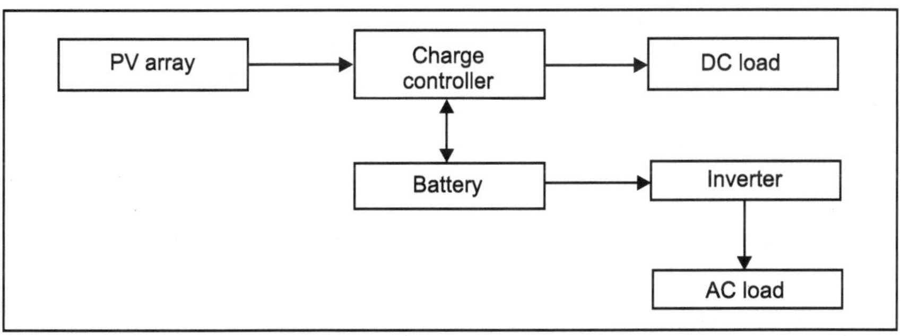

Figure 5.6 Diagram of stand-alone PV system with battery storage powering DC and AC loads

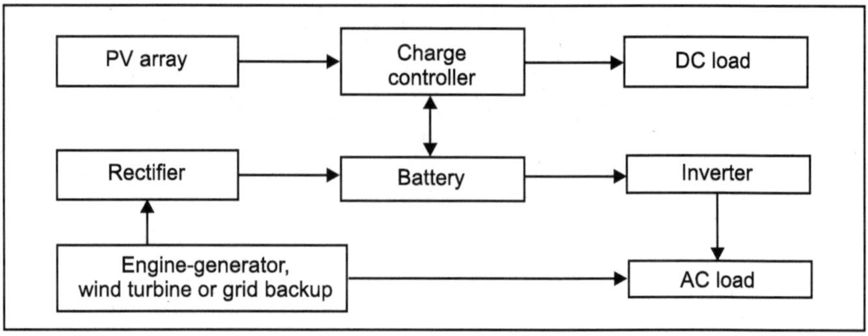

Figure 5.7 Diagram of photovoltaic hybrid system

5.6 Non-silicon-based photovoltaic systems

The alternative material and technology used in manufacturing photovoltaic components, termed as second and third generation photovoltaic technologies include less-costly raw material and manufacturing techniques. Second generation photovoltaic imply thin-film solar cells, that use amorphous silicon or other compounds with semi-conducting properties, which are deposited on flexible substrates ranging from glass to plastics and other polymers. Third generation technologies include organic, nano and spheral technologies. Most of these are presently in the process of development and are soon expected to be commercially produced.

5.6.1 Thin-film technology

Thin-film silicon solar cells offset many of the disadvantages of the conventional silicon cells by using a fraction of the pure silicon required in manufacturing solar cells. They are also easier to manufacture and easy to use in a variety of applications.

Thin-film solar cells are made by depositing a thin layer of semiconductor on a supporting material (substrates) such as glass, stainless steel or polyimide through a process called chemical vapour deposition. The materials selected for deposition are strong light absorbers, most commonly amorphous silicon (a-Si), cadmium telluride (CdTe) and copper indium (gallium) diselenide (CIS or CIGS). These materials are suitable for deposition over large substrate areas (up to 1 metre) and hence allow high volume manufacturing. In terms of costs, amorphous silicon thin film solar cells use less than 1 per cent of the silicon used in conventional cells, and the material costs are also lower for cells using CdTe or CIS technologies. These cells also do not require assembling and are flexible, hence having versatile applications. The efficiency levels of these cells range between 6 to 8 per cent.

5.6.2 Amorphous silicon (a-Si)

Used mostly in consumer electronic products, which require lower power output and cost of production, amorphous silicon has been the dominant thin-film PV material since it was first discovered in 1974. Amorphous silicon is a non-crystalline form of silicon, i.e. its silicon atoms are disordered in structure. A significant advantage of a-Si is its high light absorptivity, about 40 times higher than that of single-crystal silicon. Therefore only a thin layer of a-Si is sufficient for making PV cells (about 1 micrometre thick as compared to 200 or more micrometres thick for crystalline silicon cells). Also, a-Si can be deposited on various low-cost substrates, including steel, glass and plastic, and the manufacturing process requires lower temperatures and thus less energy. So the total material costs and manufacturing costs are lower per unit area as compared to those of crystalline silicon cells.

Despite the promising economic advantages, a-Si still has two major roadblocks to overcome. One is the low cell energy conversion efficiency, ranging between 59 per cent, and the other is the outdoor reliability problem in which the efficiency degrades within a few months of exposure to sunlight, losing about 10–15 per cent.

5.6.3 Cadmium telluride (CdTe)

As a polycrystalline semiconductor compound made of cadmium and tellurium, CdTe has a high light absorptivity level; only about a micrometre thick can absorb 90 per cent of the solar spectrum. Another advantage is that it is relatively easy and cheap to manufacture by processes such as high-rate evaporation, spraying or screen printing. The conversion efficiency for a CdTe commercial module is about 7 per cent, similar to that of a-Si. The instability of cell and module performance is one of the major drawbacks of using CdTe for PV cells. Another disadvantage is that cadmium is a toxic substance. Although very little cadmium is used in CdTe modules, extra precautions have to be taken in manufacturing process.

5.6.4 Copper indium diselenide ($CuInSe_2$ or CIS)

A polycrystalline semiconductor compound of copper, indium and selenium, CIS has been one of the major research areas in the thin-film industry. The reason for it to receive so much attention is that CIS has the highest 'research' energy conversion efficiency of 19.7 per cent in 2006 is not only the best among all the existing thin-film materials, but also came close to the 18 per cent research efficiency of the polycrystalline silicon PV cells. (A prototype CIS power module has a conversion efficiency of 10 per cent.) Being able to deliver such high energy conversion efficiency without suffering from the outdoor degradation problem, CIS has demonstrated that thin-film PV cells are a viable and competitive choice for the solar industry in the future.

CIS is also one of the most light-absorbent semiconductors; 0.5 micrometres can absorb 90 per cent of the solar spectrum. CIS is an efficient but complex material. Its complexity makes it difficult to manufacture. Also, safety issues might be another concern in the manufacturing process as it involves hydrogen selenide, an extremely toxic gas. So far, CIS is not commercially available yet although Siemens solar has plans to commercialise CIS thin-film PV modules.

5.6.5 Nanotechnology in photovoltaic

Various nanosize materials are under investigation; the major advantage of nanoparticle in the field of photovoltaics is increases in the charge transfer rate and tunability, which can be achieved by reducing the particle size. By controlling the particle size one can tune the band gap of the material so that it matches well with the solar spectrum and render the nanoparticles ideal for photovoltaic applications. Various attempts have been under investigation and it is believed that the appropriate photovoltaic system with maximum efficiency will be achieved in sooner.

5.6.6 Organic technology

Organic solar cells are based on the photosynthesis process in plants. The absorption of light in organic cells is done by the 'dye' which substitutes for the silicon in conventional cells. This light causes the dye molecules to excite and release electrons that are converted to electrical energy. The use of chemicals called dyes for the conversion process has led to organic cells also being known as 'dye-sensitised solar cells'. The absorption of light occurs in dye molecules that are in a highly porous film of titanium dioxide (TiO_2). This causes the electron to be injected into TiO_2 and is conducted to the transparent conductive oxide layer. The material and manufacturing costs of these cells are relatively much lower than conventional silicon photovoltaic cells. However, the low efficiency rates (3–5 per cent) result in an overall increase in the costs. This technology is presently being developed and expected to be produced commercially.

5.7 Asphalt roads as solar power producers

Researchers at the University of Rhode Island are starting research into using the miles and miles of roads that stretch across our country as solar power producers. It is a well-known phenomenon that roads collect a lot of heat from sunlight during the day. This heat collection is so extreme that it is blamed for the increased heat that is associated with urban centres as compared to areas with fewer roads. The researchers are exploring four ways to collect and harness that energy. The most basic approach is to attach photovoltaic cells to all the exposed areas of the roads. The flexible cells could be attached to the barriers on the roads and embedded in the roadway in areas like the shoulders where vehicles typically do not travel. The electricity produced could be collected and used locally to power street lights and signs. The great thing about current flexible PV cells is that they do not have to be pointed directly at the sun, so they could be placed anywhere where there is sunlight and still collect power during the day.

A second intriguing idea is embedding water filled piping into the road bed that would collect heat from the sun. As the water warms it would circulate into areas where ice and snow accumulate thus melting the snow and ice and reducing the need for salt to be used on the roads. The warm water could also be circulated to nearby buildings to assist in the heating needs of the structures as well as for hot water. This system would be very similar to the geothermal heat pumps which are already in use in many homes and buildings around the country, producing heating and hot water at minimal cost. And, of course, the heat could be used as steam to turn a steam turbine and produce electricity as well.

The third idea works on the thermal-electric principal that a small electric current is produced when a hot and cold location are connected using two types of semi-conductors. By putting semi-conductors in different places in the roadway, like in the sunny and the shady areas, a current could be produced. Although there are no practical applications of this idea yet, research suggests it could produce enough electricity to de-ice roads.

6

Energy consuming and converting equipments

6.1 Introduction

Energy is available in various forms in nature. Normally energy is utilised in a form different from the one in which it is available. We therefore require equipments which convert energy from one form to a more conveniently usable form. Broadly we can categorise the equipment as: (i) energy-converting equipment, and (ii) energy-using or -consuming equipment.

There is however a thin demarcation line between the two. In fact the energy-consuming equipments are also converting the energy into some from of useful activity to carry out a particular assignment. Under energy-converting equipment, we can identify the following major equipments and systems: (i) boilers, (ii) turbines (hydro steam, gas and nuclear), (iii) generators, (iv) batteries, (v) solar cells/fuel cells, and (vi) nuclear reactors.

Under energy-consuming equipments, we can identify the following equipments: (i) motor, (ii) fans, (iii) pumps, (iv) crushers/grinders, (v) heaters, and (vi) misc. appliances.

6.2 Boilers

The steam generator or boiler is a combination of systems and equipments for the purpose of converting chemical energy from fossil fuels into thermal energy and transferring the resulting thermal energy to a working fluid, usually water, for use in high-temperature processes or for partial conversion to mechanical energy in a turbine. In most modern large power plants, one boiler is used to supply steam to one steam-turbine generator unit. The boiler complex includes the air-handling equipment and ductwork, the fuel-handling system, the water-supply system, the steam drums and piping, the exhaust-gas system, and the pollution-control system.

Steam generators are the most common vaporisers. They fall into two general categories, namely, (i) industrial; and (ii) power-utility steam boilers.

Industrial steam boiler types include fire-tube boilers and waste-tube boilers. Fire-tube boilers (or steam generators) are characterised by the containment of the products of combustion within the tubes of the boiler. The water being vaporised surrounds the tubes and is contained by a larger shell with front and rear tube sheets. Almost all fire-tube boilers built today are of the modified Scotch marine type. This design utilises a central cylindrical combustion chamber with return of the gases through smaller tubes. Baffles are frequently provided to return the gases through more tubes for a third or fourth pass. These units are usually completed as a shop-erected package boiler with all combustion equipment and controls installed and ready to operate after electrical and piping connections are made at the place of utilisation. Fire-tube boilers are rated for capacity in terms of boiler horsepower. One boiler horsepower is defined as the ability to evaporate 34.5 lb water from and at 212°F. Such units are available in size from 5 to 600 hp (about 20,000 lb/hr steam) and in pressures to 250 lb/sq. inch. For the size and pressure range they are very competitive in price and installation costs. They do not lend themselves readily to the installation of superheaters and are thus limited to the generation of saturated steam only. They are applicable to oil and gas-fuel firing only.

Water-tube boilers are characterised by the containment of the vaporising water within the tubes while the products of combustion are on the outer surfaces. The furnace chamber may consist of containment by either refractory walls or special configurations of the water tubes. For economic reasons, size of 10,000 to 1,20,000 lb/hr are generally of 'package' shop-erected construction. These are currently available only for oil and gas-fuel firing, with active work underway to develop similar coal-fired units. A few coal-fired units have been built but full acceptance and utilisation have yet to develop. Larger-capacity oil and gas-fired units and almost all coal-fired units are generally 'field-erected' or constructed in place. Since the water tubes are of small diameter compared with the shells of fire-tube boilers, there is no similar limitation on their operating pressures.

Many older water-tube boilers are of the straight-water-tube construction, but modern units are almost universally of bent water-tube design. The use of bent tubes allows extreme flexibility in design. The shaping of the tubes to enclose the combustion space has sharply reduced the amount of refractory used and has permitted higher rates of burning within the combustion space. Superheaters may be readily fitted to such units. Industrial units in excess of 2,50,000 lb/hr capacity are rare and operating pressures in excess of 600 lb/sq. inch are also seldom encountered. Higher-pressure units are sometimes installed where electric power is generated within the plant, but the economic trends had been away from power generation with a few notable exceptions.

Utility boilers include conventional boilers and unconventional boilers. Conventional boilers used in power utility plants are those units operating below the critical pressure of the steam with widely accepted modes of operation. They may include features that yesterday were unconventional such as forced circulation, reheat, divided furnaces, once-through flow, and pressurised furnaces. They are usually custom-designed and built, with the trend to larger units in the multi-million pounds per hour capacity. Their superheated output is taken directly to a single turbine generator in recent designs for the production of electricity. They are universally of water-tube design, and all fuels are fired, with coal being the most common.

Unconventional boilers are those which are experimental or extremely rare in their application. Currently, mercury binary, combination gas turbine, supercritical, and nuclear boilers are in this category. Supercritical units are conventionally fired once-through forced-fluid-flow designs where the entire system is pressurised over the critical vapour pressure. Units of over 5000 lb/sq. inch operating pressure have been constructed and are in operation. Mercury-binary-system boilers impose a mercury-vapour topping system on a conventional steam cycle. The mercury is vaporised at a high saturated vapour temperature by conventional fuel firing. After electricity is produced in a mercury-vapour turbine generator the latent heat of the mercury is transferred to steam in a condenser-boiler. A number of such plants have been built and are in operation. Combination gas-turbine and steam-turbine plants have been built for many years (Velox boilers). New designs are currently being built which incorporate a more active part by the gas turbine. These designs include those where the exhaust gases of large gas turbines are re-heated by supplemental fuel firing to serve as the combustion gases of conventional steam boilers. Nuclear units have been built for power utility steam boilers in the United States and other parts of the world. It does not appear that they will compete economically with fossil-fuel plants until fundamental problems in fuel preparation and waste disposal are solved.

6.3 Turbines

A turbine is a device that converts the stored mechanical energy in a fluid into rotational mechanical energy. There are several different types of turbines, including steam turbines, gas turbines, water turbines and wind turbines or wind mills. The material presented in this section is applicable to only steam and gas turbines.

6.3.1 Steam turbines

'A steam turbine may be defined as a form of heat engine in which the energy of the steam is transformed into kinetic energy by means of expansion through nozzles, and the kinetic energy of the resulting jet is in turn converted into force doing work on rings of blading mounted on a rotating part.'

This definition may be restated as 'A steam turbine is a prime mover which converts the thermal energy of steam directly into mechanical energy of rotation.'

The gas turbine is an internal combustion engine differing in many respects from the standard reciprocating model. In the first place, the process by which a gas turbine operates involves steady flow; hence pistons and cylinders are eliminated. Secondly, each part of the thermodynamic cycle is carried out in a separate apparatus. The basic process involves compression of air in a compressor, introduction of the compressed air and fuel into the combustion chamber(s), and finally expansion of the gaseous combustion products in a power turbine.

A steam turbine is probably the most flexible driver available to industry. With the advent of modern precision gears, turbine speeds are seldom below 1200 rpm and may be as high as 25,000 rpm. Radial-flow units operating at still higher speeds (50,000 rpm) have been used in the refrigeration of process gases. Most designs are for a single speed either to drive electric generators or to operate over a limited speed range for industrial applications. Turbines developing full torque at reduced speeds may require special design considerations. Marine turbines are built with reversing stages to go from ahead to reverse rotation. Turbines can utilise steam at very low pressures or as high as 5000 lb/sq. inch. The exhaust can be any pressure lower than the inlet that will economically develop power, and in modern power plants this is frequently as low as 1 inch Hg absolute. Turbines can be used as pressure-reducing stations, taking steam at one level and exhausting it at a lower-pressure level, thus developing power and reducing steam temperature. In modernising old utility plants, a new high-pressure boiler and a 'topping turbine' can provide additional power and furnish steam to the original turbines with increased overall plant efficiency.

Design characteristics; sizes are based on shaft power, in terms of horsepower for mechanical-drive turbines, and in kilowatts for turboelectric generators. Frame size depends on the number and types of stages, blade lengths, and wheel diameters. Physical size alone does not indicate the power rating, since the power output of a frame is proportional to steam-pressure level.

Turbines are built in power outputs from fractional horsepower up to 5,00,000 kW. In the larger sizes, a turbine may consist of several different casings in series, i.e. a high-pressure turbine, an intermediate-pressure turbine, and a single or double-flow low-pressure turbine. Regardless of the number of casings, an installation is considered a single turbine as long as it is controlled by a single governor. Each unit of the turbine may have its own overspeed trips or special safety devices including those used for reheat installations.

Process plant drivers are seldom large enough or use a high enough steam pressure to justify more than one turbine casing. The following types of turbines are commonly used in process industries:

1. Straight-through turbines have only one inlet and one exhaust steam connection, either single or multi-stage.
2. Bleeder turbines are multi-stage turbines with outlets at one or more of the stages, so that steam may be bled and used for boiler feed water heating or process requirements.
3. Extraction turbines are bleeder turbines with the bleed pressure controlled and thus they constitute essentially two or more turbines in a single casing.
4. Mixed-pressure turbines have a governor admitting steam to the low-pressure end of a turbine and another governor admitting enough high-pressure steam to the high-pressure end to carry the load.
5. An extraction-induction turbine, is a combination mixed-pressure-extraction turbine, in that it can either supply or use low-pressure steam.

In all these casing arrangements the turbines may be condensing (exhausting to a condenser) or back-pressure exhausting to process or other uses.

Selection

Since all turbines have essentially the same operating characteristics, the user should be concerned primarily with guaranteed efficiencies, operating conditions and performance curves. Comparison of efficiencies and performance must, of course, be on an equivalent basis. Some manufacturers guarantee that turbine performance will be equal to or greater than the quoted value. Others allow a plus or minus variation. At part loads, reaction blades, because of their greater ability to utilise 'carry-over', may show better efficiencies.

Turbines for electric power generation are designed for operation at only one speed and under standard steam conditions. Process drives, usually known as mechanical-drive turbines, have to meet a wide variety of speed and steam conditions, and about the only standardisation that can be effected is in blading and casing patterns. The casings are usually selected from a series of high-pressure ends, intermediate spool pieces, and exhaust ends that can be bolted together in a variety of combinations.

The most frequently used steam turbine is the single-stage mechanical-drive turbine, which is low in cost and is made in several frame sizes, from 9- up to 24-inch wheels. Cast-iron casings are used for low steam temperatures and pressures and cast steel for the higher pressures and temperatures. A minimum of one nozzle can be used in each size, but the maximum nozzle area that can be installed increases rapidly with increased frame size.

The power that a turbine can develop is a function of size, steam conditions, and rpm. The lowest investment cost results from the use of the smallest possible turbine. For most applications, however, the additional investment in larger turbines will be paid for in a short time by reduced steam consumption.

The general-purpose turbines can usually operate from zero speed to full speed at any torque requirement within the temperature limits of the casing. The speed governors are of the simple direct-acting type, with at least a hand speed changer, although some models may include provisions for either pneumatic or electrically actuated speed reset.

Operation and safety consideration

By adjusting the throttling or governing valve the quantity of steam flow can be varied from zero to full flow. For each flow the turbine will have definite speed-torque characteristics, and a turbine will speed up until its torque is equal to that required by the driven equipment. If greater torque and speed are required the governor opening is increased to give more steam flow. Most turbines will burst at their 'runaway' speed, and as a result the user seldom changes the governing valve directly. The preferred procedure is to reset the speed setting of the governor and let it change the valve.

Turbines without governing valves are used under special situations. For example, where (i) the speed of a turbocharger turbine depends on the gas flow through an engine (i.e., engine rpm) and failure of the supercharging compressor will cause the engine to shut-down; (ii) turbines driving the lubrication-oil pumps of a larger turbine are frequently designed for 'runaway' speeds; and (iii) gas-turbine governors vary fuel rate and hence the speed.

Turbines can be installed with the necessary supervisory instrumentation to start up automatically and to shut-down in the event of malfunction of failures. Failure of lubricating oil automatically shuts the trip throttle valve. In addition, trips can be connected to vibration-detecting devices and temperature detectors in the inlet and exhaust piping.

A turbine has maximum available torque at zero speed and therefore the load does not have to be disconnected during start-up. The preferred starting procedure is to warm up the lubricating oil system by first placing it in operation. Then the turbine should be warmed up gradually by opening the throttle valve to bring the speed up slowly as the turbine warms up. This procedure minimises the possibility of turbine 'rubs'. Some of the smaller multi-stage turbines used as process drives can be started quickly when occasion demands it. The shut-down procedure of large turbines is critical and the rotors are turned slowly while the casing cools to prevent hot spots that might cause 'kinking' of the shaft.

A turbine can operate satisfactorily with the moisture that forms naturally in the stages, provided the casing is drained to prevent accumulation. The inlet-steam piping should be designed to provide dry steam to the turbine nozzle since a slug of moisture can knock out blades and/or thrust bearings.

Solids carried over from the boiler drum are deposited principally in turbine nozzles, resulting in lower turbine output. Their presence can be detected by higher than normal interstage pressures. These deposited principally in turbine nozzles, resulting in lower turbine output. These deposits can be washed out by injecting water into the turbine while it is running, but under conditions prescribed by the manufacturer to prevent turbine damage.

It is good practice to inspect turbines periodically, but the interval between shut-downs for thorough inspection is longer than those for other plant elements. Annual boiler inspections or the cleaning of heat exchangers in a chemical plant are often appropriate occasions. A steam turbine is generally considered to have 100 per cent availability, with longer than a year between shut-downs, provided its performance is periodically checked.

Most important is the checking of the safety features that have been incorporated to protect the turbine. For example, the operation of overspeed trips can be checked by using these to shut-down the turbine. Many continuous process drivers cannot be shut-down, and these safety features can only be checked during plant turnarounds. The only preventive maintenance required is to (i) keep the oil clean, at the correct temperature, and at the right level in the oil reservoir, (ii) inspect and check auxiliaries such as lubricating-oil pumps, coolers, and oil strainers, (iii) inspect turbine parts during the annual turnaround, and (iv) check safety devices.

Heat recovery

Since two-third of the heat entering the turbine leaves in the steam exhaust, with other losses minor, the major improvement in thermal efficiency results from the use of exhaust and/or extraction steam for heating, process or other purposes. The smallest amount of steam that can be used for heating purposes is that required for heating boiler feed water.

Turbine steam consumption can be decreased by superheating. The energy available (Btu/lb) for given inlet and exhaust pressure is proportional to the absolute inlet steam temperature. Frequently economics justify the addition of superheaters to waste-heat boiler installations.

Turbines that operate over a wide load change should have first-stage nozzle area control. Economics do not ordinarily justify controlling the nozzle areas of the other stages, except for extraction or mixed-pressure turbines where nozzle area control of the first stage in the low-pressure parts can be used. Nozzle area can be changed by hand valves or automatically by the governor.

Turbines that run at constant speed but at zero load for considerable periods will require less steam at this condition if the casing is under vacuum or if the speed is reduced. Under a vacuum and zero load, 5–10 per cent of full-load steam flow is required. Turbines that run at one load and at overloads for only short periods should be designed to carry overloads with an overload valve.

Combustion of gas turbines

Gas turbines are usually rated according to power output at sea level and 80°F ambient air (some European designs are rated at 60°F) and when burning a specified fuel. Power output and efficiency will be larger for those fuels which produce larger volumes of products of combustion since the compressor does not do any work on additional volume. Gas turbines are further classified by the physical arrangements of the component parts, such as (i) single shaft, (ii) two shaft, (iii) regenerative (a heat exchanger is used to recover exhaust losses and heat the air to the combustors), (iv) intercooled (heat removed between compressors), and (v) reheat (heat added between turbines).

The overall efficiency of a gas turbine is a function of compressor and turbine efficiencies, ambient air temperature, nozzle inlet temperature, and type of cycle used (i.e. reheat, etc.). The compressor and turbine are designed for high efficiency and the first-stage gas temperature establishes material and stress conditions for the first set of rotating blades. To the gas temperature at these blades is added the temperature drop across the first-stage nozzles to determine the inlet temperature of the turbine, which may vary from 1300° to 1500°F for industrial gas turbines and higher for aviation gas turbines. The higher values are usually used in impulse turbines.

In a simple cycle turbine, there is, for each turbine inlet temperature, an optimum pressure ratio producing the highest possible efficiency. The efficiency and optimum pressure ratio increases with increasing turbine inlet temperatures. These pressure ratios vary from 4 at 1300°F up to 6 for 1500°F turbine inlet temperature.

Regenerative cycles favour lower pressure ratios which results in low compressor discharge temperatures, thus allowing greater recovery, of heat from the turbine exhaust gases. High-ratio regenerative plants use intercoolers in the compressor circuit to lower the compressor discharge air temperature.

Turbine pumps

The term turbine pump is applied to units with mixed-flow (i.e. part axial and part centrifugal) impellers. Such units are available in capacities starting 100 gal/min upward for heads up to about 100 ft per stage. Turbine pumps are usually vertical.

A common form of turbine pump has the pump element mounted at the bottom of a column that serves as the discharge pipe. Such units are immersed in the liquid to be pumped and are commonly used for wells, condenser circulating water, large volume drainage, etc. Another form of the pump has a shell surrounding the pumping element which is connected to the intake pipe. In this form the pump is used on condensate service in power plants and for process work in oil refineries and elsewhere.

Regenerative pumps, also sometimes referred to as turbine pumps because of the shape of the impeller, employ a combination of mechanical impulse and centrifugal force to produce heads of several hundred feet at low volumes (usually less than 100 gal/min). The impeller, which rotates at high speed with small clearances, has many short radial passages milled on each side at the periphery. Similar channels are milled in the mating surfaces of the casing. Upon entering, the liquid is directed into the impeller passages and proceeds in a spiral pattern around the periphery, passing alternately from impeller to the casing, receiving successive impulses as it does so.

These pumps are particularly useful where it is required to handle low volumes of low-viscosity liquids at higher pressures than are normally available with centrifugal pumps. Close clearances limit their use to clean liquids. For very high heads multiple-stage units are available.

6.4 Diesel generating sets (DG sets)

Many industries operate DG sets on continuous basis for various reasons like economy, reliability of power, etc. Many industries have processes which emits lots of energy to atmosphere. One must be aware that huge amount of heat energy is rejected to the atmosphere through exhaust gases and jacket cooling system (the heat energy rejected to atmosphere is to the tune of 60 to 67 per cent of input energy from the fuel burnt in the engine generator set). Process like incineration, sulphonation, cement plant kilns, glass furnaces, steel plant furnaces emit huge amount of heat energy to atmosphere.

This waste heat can be converted into useful forms like:

1. Steam
2. Hot oil
3. Hot air

4. Hot water
5. Chilling
6. Refrigeration
7. Electricity
8. Waste-water recycling

The fuel cost of DG sets is a substantial part of running cost for all process industries. The prices of petroleum fuels in India have increased considerably. This has substantially increased the running cost of these DG sets.

Waste heat recovery system on existing resources in this era of 'self-reliance' is essential when our import bill of petroleum products is scaling to new heights. It is not only in individual company's interest, but also in national interest, that, by recovering the waste heat we reduce our fuel import bill.

This section highlights on waste heat recovery from furnace oil/gas fired DG sets, process and furnace with its heat recovery units.

Few salient features of exhaust heat recovery systems are listed below:

1. Payback period varies in the range of 6 months to 18 months.
2. Waste heat recovery boiler (WHRB) can be provided with economiser/ feed water preheated to improve the performance.
3. The equipment can be designed within the back pressure constraints of primary equipment and does not affect the performance of source equipment in any way if designed properly.
4. Variety of designs is possible to suit different types of exhaust gases.
   (a) Smoke tube type
   (b) Water tube type (bare tube/finned tube construction)
5. Suitable for wide range of temperature and pressures.
6. Waste heat recovery systems can be provided with isolation and bypass facility on gas side by fully automatic three-way pneumatically operated diverter valve.
7. Particulate deposited from dirty gases can be removed with on-line soot blowing valve.
8. Location of heat recovery boiler can be indoor as well as outdoor. Following safety instruments are provided for safe and reliable operation of WHRB as well as engine genset.
   (a) High steam pressure.
   (b) High exhaust back pressure.
   (c) Low water level in boiler.
9. Economiser/feed water pre-heater for WHRB can be provided with unique control circuit for operation against cold end corrosion due to sulphur in fuel (only applicable for heavy oil fired engine genset).

6.5 Motors

Electric motors are the prime movers of any industry. Electric motor efficiency is the measure of its ability to convert kilowatt of electric power supplied to the motor terminal and the horsepower of mechanical energy taken out of the motor at the rotating shaft.

Conservation of electrical energy is one of the prime requirements of today and future times. Due to drastic increase in electricity tariff, use of energy efficient motors for all the applications is inevitable. Indian industry/OEMs normally look at initial costs rather than running cost, leading to purchase of standard motors with low efficiency at full load/at partial loads. During life of motors, they are rewound ineffectively leading to further reduction of efficiency and leading to consumption of more power. The gap between power generated and demand is widening. Due to the above reasons now it is a 'National Priority', since 70 per cent of the energy consumed is used for driving electric motors. There exists a national BIS standard for three phase EE induction motors up to 37 kW. Energy efficient motors are about 15 per cent more expensive. With this increase in price the payback period is only about 1 year for 1500 working hours per annum.

Although induction motors require a minimum of attention in service, they should be inspected at regular intervals to check for excessive: (i) dirt, (ii) moisture, (iii) friction, and (iv) vibration, which account for 90 per cent of all motor failures.

There are two basic types of bearings used in integral horsepower induction motors, sleeve bearings and ball bearings. For modern motors, ball bearings are more popular. With proper design, application, and maintenance, each type will give good service.

Any of the following drive methods may be used, depending on the particular motors application: (i) V-belts drive, (ii) chain drive, (iii) gear drive, and (iv) direct drive.

Be sure the motor is connected as shown on the name-plate diagram, and that the power supply (voltage, frequency, and number of phases) corresponds with the nameplate data.

Run the motor without load to check the connections and direction of rotation. The motor will operate satisfactorily with a 10 per cent variation in voltage, a 5 per cent variation in frequency or a combined voltage and frequency variation of 10 per cent, but not necessarily in accordance with the standards of performance established for operation at normal rating.

Features of energy efficient motor

1. Highest efficiency
2. Lower operating cost

3. Lower demand charges
4. Has higher overload capacity
5. Suitable for operations at higher ambient temperature
6. Fewer power factor correction
7. Lower branch circuit losses
8. Reduced air-conditioned load
9. Saving increases with time
10. Confirmation with NEMA standards of protection and control
11. Cooler and quieter operation
12. Longer insulation life: EEM windings run about 20°C cooler which increases insulation life by four times
13. Improved bearing life: EEM bearings run about 10°C cooler than standard motor (STM) bearings, which doubles the life
14. Less starting thermal stress
15. Higher service factors
16. Better suited for energy management systems
17. EEM performs better under adverse conditions of abnormal voltage conditions like unbalanced voltages.
18. Efficiency of EEM remains almost constant from 50 to 100 per cent of load.

6.6 Pumps

Necessity for the movement of liquids by pumping occurs often throughout a manufacturing plant. Raw materials may be delivered to the plant as liquids. Many materials are liquids during all or part of the manufacturing processes. Water and sanitary services may require pumps as do steam-generating equipments, cooling and washing equipments, etc.

Pumps may be classified on the basis of the applications they serve, the materials from which they are constructed, the liquids they handle, and even their orientation in space. All such classifications, however, are limited in scope and tend to substantially overlap each other. A more basic system of classification, the one used, first defines the principle by which energy is added to the fluid, goes on to identify the means by which this principle is implemented, and finally delineates specific geometries commonly employed. This system is therefore related to the pump itself and is unrelated to any consideration external to the pump or even to the materials from which it may be constructed.

Under this system, all pumps may be divided into two major categories: (i) dynamic, in which energy is continuously added to increase the fluid velocities within the machine to values in excess of those occurring at the discharge such that subsequent velocity reduction within or beyond the pump produces a pressure increase; and (ii) displacement, in which energy is periodically added by application of force to one or more movable boundaries of any desired number of enclosed, fluid-containing volumes, resulting in a direct increase in pressure up to the value required to move the fluid through valves or ports into the discharge line. Dynamic pumps may be further subdivided into several varieties of centrifugal and other special-effect pumps. Displacement pumps are essentially divided into reciprocating and rotary types, depending on the nature of movement of the pressure-producing members.

Energy savings in pumping systems account for 40 per cent power in the industrial sector and 15 per cent of the national consumption of India.
Centrifugal pump consists of a set of rotating vanes, enclosed within a housing or casing and used to impart energy to a fluid through centrifugal force. Thus, stripped of all refinements, a centrifugal pump has two main parts: (i) a rotating element, including an impeller and a shaft; and (ii) a stationary element made up of a casing, stuffing box, and bearings.

In a centrifugal pump the liquid is forced, by atmospheric or other pressure, into a set of rotating vanes. These vanes constitute an impeller which discharges the liquid at its periphery at a higher velocity. This velocity is converted to pressure energy by means of a volute or by a set of stationary diffusion vanes surrounding the impeller periphery. Pumps with volute casings are generally called volute pumps, while those with diffusion vanes are called diffuser pumps. Diffuser pumps were once quite commonly called turbine pumps, but this term has recently been more selectively applied to the vertical deep-well centrifugal diffuser pumps usually referred to as vertical turbine pumps.

Centrifugal pump is selected on the basis of discharge requirements and total head of the system. While selecting operating speed, there are certain limitations from site conditions like available NPSH or available submergence, etc. It is recommended that the speed should be so selected that pump has maximum achievable efficiency. As per hydraulic institute standard, maximum efficiency can be achieved when the specific speed is around 2500 NsUS.

Reciprocating displacement pump is one in which a plunger or piston displaces a given volume of fluid for each stroke. The basic principle of a reciprocating pump is that a solid will displace an equal volume of liquid. For example, when an ice cube is dropped into a glass of water, the volume of water that spills out of the glass is equal to the submerged volume of the ice cube. All reciprocating pumps have a fluid-handling portion, commonly called the liquid end.

Power pump is a constant-speed, constant-torque, and nearly constant-capacity reciprocating machine whose plungers or pistons are driven through a crankshaft from an external source. The pump's capacity fluctuates with the number of plungers or pistons. In general, the higher the number, less is the capacity variation at a given rpm. The pump is designed for a specific speed, pressure, capacity, and horsepower. The pump can be applied to horsepower conditions less than the specific design point, but at a sacrifice of the most economical operating condition.

Diaphragm pumps are displacement pumps with flexible membranes clamped at their peripheries in sealing engagement with a stationary housing. The central portion moves in a reciprocating manner through mechanical means, such as a crank or an eccentric cam or by fluid means, such as compressed air or liquid under alternating pressure. An inlet check valve and an outlet check valve control the flow of pumped liquid into and out of the pumping chamber. A distinguishing feature of all diaphragm pumps is that they have no seals or packing and can be used in applications requiring zero leakage. They are also self-priming and can run dry without damage. In single-diaphragm pumps, the pumped liquid can have a lot of inertia if the suction and discharge lines are relatively long. A pumping cycle consists of an inlet stoke and a discharge stroke. A simple accumulator on the suction (inlet) side of the pump will allow the pump to draw liquid from the accumulator while it simultaneously draws liquid through the suction line. If the discharge line from the pump is relatively long, the inertia of the liquid can be great, as mentioned above, and can impose severe loads on the diaphragm and cranking means as the diaphragm enters the discharge stroke of a complete cycle.

Screw pumps are a special types of rotary positive displacement pumps in which the flow through the pumping elements is truly axial. The liquid is carried between screw threads on one or more rotors and is displaced axially as the screws rotate and mesh. In all other rotary pumps the liquid is forced to travel circumferentially, thus giving the screw pump with its unique axial flow pattern and low internal velocities a number of advantages in many applications where liquid agitation or churning is objectionable.

The applications of screw pumps cover a diversified range of markets: navy, marine, and utilities fuel oil service; marine cargo; industrial oil burners; lubricating oil service; chemical processes; petroleum and crude oil industries; power hydraulics of navy and machine tools; and many others. The screw pump can handle liquids in a range of viscosities, from molasses to gasoline, as well as synthetic liquids in a pressure range from 50 to 5000 lb/in^2 (3.5 to 350 bar) and flows up to 5000 gal/min. (1135 m^3/hr). Because of the relatively low inertia of their rotating parts, screw pumps are capable of operating at higher speeds than other rotary or reciprocating pumps of comparable displacement. Some turbine-attached lubricating oil pumps operate at 10,000 rpm and even higher. Screw pumps, like other rotary positive displacement pumps, are self-priming and have a delivery flow characteristics which is essentially independent of pressure.

Rotary pumps are rotary positive displacement pumps in which the main pumping action is caused by relative movement between the rotating and stationary elements of the pump. The rotary motion of these pumps distinguishes them from reciprocating positive displacement pumps, in which the main motion of moving elements is reciprocating. The positive displacement nature of the pumping action of rotary pumps distinguishes them from the general class of centrifugal pumps, in which liquid displacement and pumping action depend in large part on developed liquid velocity. It is characteristic of rotary pumps, as positive displacement pumps, that the amount of liquid displaced by each revolution is independent of speed. Also, it is characteristic of rotary pumps that a time-continuous liquid seal of sorts is maintained between the inlet and outlet ports by the action and position of the pumping elements and the close running clearances of the pump. Hence rotary pumps generally do not require inlet and outlet valve arrangements, as reciprocating pumps do.

Rotary pumps are useful in handling both fluids and liquids, where fluid is a general term that includes liquids, gases, vapours, and mixtures thereof, and sometimes solids in suspension, and where liquid is a more specific term that is limited to true liquids which are relatively incompressible and relatively free of gases, vapours, and solids.

6.6.1 Agricultural pumps

With enhanced irrigation and mechanisation in agricultural sector, the share of energy consumption— especially electricity has been rapidly increasing during the past decade. A major share of this energy is consumed by irrigation pumping systems, which operate at low energy efficiency.

The consumption of direct commercial energy in the agricultural sector is largely for two activities, viz. (i) mechanised land preparation, and (ii) mechanised life irrigation. The mechanised life irrigation comprises operation of electric and diesel powered pumps for extracting groundwater for irrigation purposes. These pumping systems include besides the pump, other components like motor, suction and delivery pipes, valves, pipe fittings, etc. The pumps and their appurtenances are manufactured by different agencies and hence have varying operational characteristics. In view of the fact that the energy consumption by these pumps is directly related to their operational efficiency, it is necessary to improve their efficiency through suitable rectification measures. As a first step towards achieving this objective, it is necessary to assess the current operational features of the agricultural pumping systems, the geological and agricultural factors affecting their performance, and the institutional arrangements for rectification of these pumping systems.

Factors influencing energy consumption include: (i) groundwater status, (ii) agricultural situation, (iii) performance status of pumping system, (iv) level of electricity supply, and (v) awareness among farmers.

6.7 Belt drives

Belts can be used to transmit power from one shaft to another, and at the same time obtain a reduction or increase in rpm. As a rule the shafts are parallel; however, by proper design and at lower loads they can be placed at an angle. Most flat belts used in industry are made of either duck or leather. Canvas or duck belts are protected from erosion by a rubber covering. For special conditions, flat belts can be made from almost any material that can be bent around pulleys.

Under normal conditions, flat belts operate at speeds below 3000 ft/min. However, carefully designed installations have been operated successfully at belt speeds as high 6000 ft/min. Belts experience a critical speed phenomenon which is evidenced by the pulleys 'riding' from side to side or by violet 'flapping' of the slack side of the belt. The critical speed of an installation is a function of the geometry of the installation, belt characteristics, and tension. Thus, when critical conditions occur they are corrected by changing some of these parameters.

7

Nanotechnology and solar power

7.1 Introduction

Nanotechnology provides essential improvement potentials for the development of both conventional energy sources (fossil and unclear fuels) and renewable energy sources like geothermal energy, sun, wind, water, tides or biomass. Breakthroughs in nanotechnology open up the possibility of moving beyond our current alternatives for energy supply by introducing technologies that are more efficient, inexpensive, and environmentally sound. According to various researchers and authors, nanotechnologies provide the potential to enhance energy efficiency across all branches of industry and to economically leverage renewable energy production through new technological solutions and optimised production technologies. Nanotechnology innovations could impact each part of the value-added chain in the energy sector.

Nanotechnology has the potential for significant impact at all stages of the energy value chain. A number of oil companies are investigating ways in which nanotechnology can improve the effectiveness of oil drilling. One of the main applications for nanotechnology in energy generation is solar photovoltaics, with thin film solar technologies, using CIGS and CdTe, and dye-sensitised solar cells. Energy storage is also affected by nanotechnology. Novel electrode materials are already commercially available and are increasing the power density and stability of batteries.

The primary driver—and it is of immense importance—is the increasing awareness of the threat of climate change, and a policy climate which reflects that. Companies which are developing energy technologies that do not produce significant quantities of CO_2; generating energy from the sun, the wind, water and vegetable matter; know that they will have access to a supportive policy environment and an investor community which is very much focused on developing these technologies. Currently two applications of nanotechnology for energy are considered; solar photovoltaics, and batteries.

Solar photovoltaic (PV) cells convert sunlight to electricity. As an indication of the market potential, some projections call for 12 per cent of Europe's electricity needs to be satisfied by solar generation by 2020, a total of 420 terawatt hours (TWh). Solar PV applications can be divided into grid-connected, off-grid systems, and portable applications. A number of companies have commercial products and development activities focused on nanotechnology-based solar technologies. The battery market is expected to grow substantially in the coming years. This is partially driven by growth in existing battery applications—particularly laptops, 'netbooks' and mobile devices. However, the primary driver will be new applications, particularly in the automotive industry. Nanotechnology for batteries is a very active field of research and development. Nanotechnology can provide an improved solution to important functional requirements such as power density, charge/discharge efficiency, and self-discharge rate, as well as making the batteries themselves less volatile. The nanorods behave as wires because when they absorb light of a specific wavelength they generate electrons. These electrons flow through the nanorods until they reach the aluminium electrode where they are combined to form a current and are used as electricity. Another potential feature of these solar cells is that the nanorods could be 'tuned' to absorb various wavelengths of light. This could significantly increase the efficiency of the solar cell because more of the incident light could be utilised (Fig. 7.1).

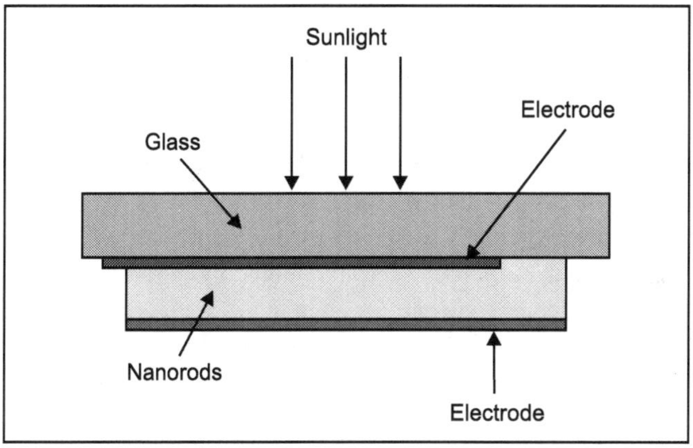

Figure 7.1 Diagram of a nano solar cell

7.2 Conserve energy with nanotechnology solar panel

With the evident effects of environmental neglect, energy all over the world is becoming more and more depleted. With energy sources becoming scarce, it generally results to a stiff competition to obtain the limited sources for energy resulting to high power charges. There are several sources of energy, and some of them are even natural resources such as wind, solar, and hydro. But even if the sources of energy are natural resources and should be delivered to the people at a lower cost, still, power costs are high due to expenses incurred while generating electricity.

For several years now, scientists and researches has been looking at other means to obtain the following:

1. Minimise generation costs of energy
2. Provide energy to the people without causing any damages to the environment
3. Generate renewable and recyclable energy resources
4. Provide energy and electricity with the least possible charges
5. Deliver electricity and energy to the public at the lowest possible cost

7.2.1 Achieving these goals using nanotechnology solar panel

Nanotechnology aims to build important things with the use of tiny machines or by producing highly advanced, essential, and quality products using tools and techniques provided by nanotechnology.

Part of the researches being conducted is to come up with a solar panel that would accurately capture solar energy that can be utilised at home, in the office, and even in large-scale industrial areas. Although there are already solar panels available in the market today, these solar panels that are usually made of silicon panels only capture 67.4 per cent light incident or solar energy emitted by the sun. Whereas, with nanotechnology, industrial and electrical manufacturers can now produce a coating for solar panels using nanorods that can capture 96.7 per cent light incident. Nanorods used in nanotechnology solar panel looks like multi-layer funnels designed to capture light incident. The first layer absorbs the light that hits the panel at wide outer angles and turns it to a slightly narrower angle. The succeeding layers become narrower, like they are passing the light to the next layer even further as it funnels the light till it reaches the active region of the panel at a 90° angle. This process makes the solar panel securely capture the sun's energy without having to rotate with the sun. The energy market encompasses a vast range of economic activity, from the extraction of fossil fuels, generation of energy by burning coal to generate electricity, for example; distribution, and consumption. A simple value chain is shown in Fig. 7.2.

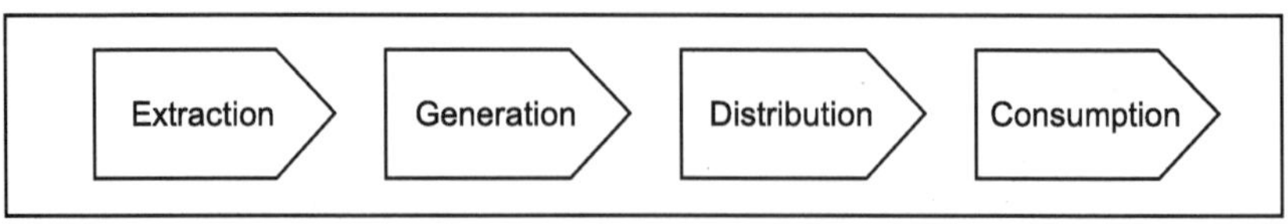

Figure 7.2 A simple value chain

The threat of climate change has highlighted the environmental damage caused by a build up of carbon dioxide in the atmosphere. A substantial proportion of CO_2 is generated by activities in this energy value chain; burning fossil fuels in a power plant, a factory or a vehicle. Therefore, and with varying degrees of success, renewable energy sources are being introduced.

The macro-level consumption figures obscure growth in renewable energies, albeit from a fairly low base. For example, 2008 saw wind energy capacity grow by 28.8 per cent to 120.8 GW, with 27 GW coming on stream during the year. The market value of the wind industry is €36.5 billion, and it employs 0.4 Millions people. A third of new capacity was installed in Asia, doubling China's capacity to 12.2 GW. The European Photovoltaics Industry Association estimated that installed solar PV capacity in 2007 was 2.25 GW, and that the amount installed per year would increase to 7.3 GW by 2012. Hydroelectric power generation accounts for energy consumption of 29,728 quadrillion BTUs. This is fractionally more than the amount of energy produced by nuclear generation.

Efforts to price in the effect of CO_2 emissions, including the EU Emission Trading Scheme, increase the costs of fossil fuel usage, further driving adoption of alternative technologies. There are also other pressures beyond climate change on the current energy economy. The world's oil resources are finite; as oil becomes more challenging to extract, prices will rise. Post-extraction, there is a global shortage in refining capacity to turn crude oil into usable petroleum products.

7.3 Nanotechnology for energy extraction

Technologies include sensors that can be injected into oil reservoirs in order to provide information about the location, consistency and presence of oil deposits. Nanotechnology could also be used in the process of converting crude oil to petroleum; a filter which is composed of carbon nanotubes was found to filter large hydrocarbons from oil. One of the main applications for nanotechnology in energy generation is solar photovoltaics. Thin film solar technologies, using CIGS and CdTe, now account for 10 per cent new solar installations, and are a very active area of development; especially in production optimisation. Dye-sensitised solar cells are another promising technology area, and dye-solar has already given rise to commercial products, albeit in far smaller quantities than thin film approaches. Nanotechnology has the potential to generate step changes in the efficiency and production cost of solar. Nanotechnology can also assist in other areas of energy generation. Mechanical methods; wind turbines, and gas turbines, can benefit from improved efficiency due to material innovation. CNT composite material for wind turbine blades can prevent shearing and increase the lifetime of the turbine. Sensors on a turbine blade could detect build up of fouling and increase preventative maintenance.

7.3.1 Supply

Energy storage is also affected by nanotechnology. Novel electrode materials are already commercially available and are increasing the power density and stability of batteries. Retrofitting nano-enabled batteries to hybrid 'Toyota Prius' is increasing their range (and therefore their attractiveness). More generally, improvements in battery technology will increase the speed with which electric vehicles and hybrid-electric vehicles can take over from conventional combustion engines. Fuel cells are another area that will benefit from nanotechnology innovations. The use of carbon nanotubes in the catalyst support structure as a replacement for metal membranes allows for higher temperature operation.

7.3.2 Consumption

Finally and more indirectly, nanotechnology can also reduce energy consumption. Vehicles whose chassis is made from more lightweight materials will consume less energy; machinery which employs nanotechnology in low friction bearings or lubricants will be more energy efficient to operate. Solid state lighting as a replacement for incandescent lighting reduces energy consumption, and also provides a positive feedback loop in which lower power requirements can be satisfied with local renewable energy sources—such as a roof mounted solar cell.

7.4 Drivers and barriers to innovation

7.4.1 Drivers for innovation

The primary driver—and it is of immense importance—is the increasing awareness of the threat of climate change, and a policy climate which reflects that. Companies which are developing energy technologies that do not produce significant quantities of CO_2; generating energy from the sun, the wind, water and vegetable matter; know that they will have access to a supportive policy environment and an investor community which is very focused on developing these technologies. Subsidies decrease the cost of renewable energies; cap and trade schemes increase the price of energy produced with fossil fuels. However, there remains a price gap in which fossil fuels are substantially lower cost for important uses such as electricity generation. An important driver for the development of new technologies is therefore improvements in their cost-effectiveness; in the case of photovoltaics, this comes from reducing material costs, increasing production efficiency, and increasing the efficiency of PV cells in use.

7.4.2 Barriers to innovation

The largest energy companies are still those which extract and process fossil fuels. Whilst they may still invest in new technologies—and some are quite supportive of renewable energy—there is not as much substantial industrial investment in energy technology as there is in ICT, for example. However, this is offset by the large amounts of public funding that are being directed towards energy research.

7.5 Solar photovoltaics

The dominant current technology for the production of solar PV is amorphous or single crystal silicon. This accounts for over 90 per cent of current solar PV production. The technology sector identifies several nanotechnology-based approached to solar PV, which are differing stages of market readiness:

Thin-film: This technology area is often described as 'second generation' solar PV. Materials used include amorphous silicon (aSi), copper indium gallium sellenide (CIGS), and cadmium telluride (CdTe). Thin-film solar PV is currently in production.

Emerging approaches: The third generation of solar PV includes emerging approaches such as dye-sensitive solar cells (DSSC) and quantum dots. DSSC and conducting polymer PV cells are currently in production, though seemingly in much lower quantities that crystalline silicon and thin film.

The impact of nanotechnology may be to lower the cost of solar PV by reducing material requirements or by introducing more efficient manufacturing methods. It may also enable novel applications of solar PV technology, such as integration with building materials or even clothing.

7.5.1 Short application description

Solar photovoltaic (PV) cells convert sunlight to electricity. Applications range from very large grid-connected solar power plants to small solar cells for portable devices. A combination of factors, including government subsidies, cost reductions, and the increasing cost of alternatives contributes to substantial growth in the solar PV market. As an indication of the market potential, some projections call for 12 per cent of Europe's electricity needs to be satisfied by solar generation by 2020, a total of 420 terawatt hours (TWh). Solar PV applications can be divided into grid-connected, off-grid systems, and portable applications.

Grid-connected solar PV

The application category includes solar PV installations that are connected to the electricity grid. These range from solar power plants which are capable of generating sufficient electricity to power thousands of homes, to installations in industrial facilities, retail spaces and private homes. A number of companies have invested in the production of solar power plants. Grid-connected solar replaces other sources of electricity generation; typically coal or gas-fired power stations. Industrial and home installations may be used to satisfy local electricity demand, and to supply additional power at times of peak demand; the classic example being their use to power air conditioning units on hot, sunny summer days. Grid connectivity also enables excess local electricity production to be supplied to the power grid, although this also requires billing and metering mechanisms and a device to convert the current. Grid-connected solar PV has been the engine of solar growth over the last decade, and accounts for a very high proportion of all solar cells sold. The quantities involved render this a large-scale, manufacturing intensive business. Many of the larger companies in this space, such as Sharp or Kyocera, have experience in mass production of glass or displays, and have applied this experience to the production of crystalline silicon PV.

Off-grid systems

Although this category includes a range of applications, off-grid systems are generally smaller scale than grid-connected PV. Off-grid solar PV may be used to provide power in places which are not connected to the electricity grid, in developing countries or in rural areas of developed countries. Electrifying rural areas is not only an economic opportunity, and also has wider societal benefits.

A significant application is the use of off-grid solar to power utilities, such as cell towers for mobile telephone networks. A combination of the power requirement of telecom base stations, and their location on high ground, make off-grid solar PV an attractive power source. Similarly, solar PV cells may be used to power street lighting, traffic lights, and signage. This application area reduces the cost and complexity of connecting these utilities to the electricity grid.

Portable devices

The most ubiquitous application of solar PV is their use to replace batteries as a power source for portable devices such as calculators and watches. This application area currently accounts for a very small proportion of the total global PV market. Despite their ubiquity, these are a small, commoditised component of a low cost device, and therefore the value is low:

Category	Existing application examples
Grid-connected	Solar power plants: industrial, retail and domestic applications
Off-grid	Rural electrification: utility in lighting, signage and telecoms
Portable	Devices: watches, calculators

Nanotechnology-enabled novel applications

The current dominant technology, silicon, limits the form factor of solar PV to inflexible, fragile flat panels. Whilst this is adequate for most current applications, especially large-scale grid connected solar; there are novel applications that would be enabled by new form factors.

Reduced material costs

Until recently, limited supply and commensurate high cost of silicon has been a major factor in increasing the cost of crystalline silicon PV. There is some evidence that this pressure will ease in coming years, though it may also be the case that the bottleneck simply shifts down the value chain to wafer production. Whilst this factor will not alter if one costly element (silicon) is replaced with another (indium), several new approaches either employ less rare materials or reduce the quantities needed.

Low-cost manufacturing

There are two elements at work here. Production of crystalline silicon is being scaled up, with major manufacturers opening larger and larger facilities; Solar World has expanded its main German facility to production capacity of 1GW. This should result in manufacturing economies of scale.

Meanwhile, production processes for thin-film solar are being improved. Nanosolar, which builds thin-film CIGS cells, claims that it has optimised several different elements of the production process including applying CIGS as a printable ink using a roll to roll process. The key question is to what extent rapid, incremental reductions in the cost of manufacturing crystalline silicon cells are outweighed by the inherent efficiencies in print-like production.

Improved efficiency

The efficiency of a PV cells is a measure of how well it converts absorbed light into electrical energy, and is typically expressed as a percentage. Whilst this is an important measure amongst the research community, it is just one element of the key metric, cost per kilowatt hour. Currently the highest efficiencies are obtained by concentrator PV (CPV) cells, which are capable of generating conversion efficiencies of around 40 per cent (a European record of 39.7 per cent was achieved by Fraunhofer ISE in 2008). Efficiencies achieved by other technology approaches are shown in Table 7.1.

The conversion efficiency of the specific technology used is just one element of the overall system efficiency, it being further affected by the packaging and other elements of the photovoltaic system.

20 year + installed lifetime: Clearly one of the most important factors in calculating the life-cycle cost of PV cells is how its efficiency will change over time (and how long it takes until it simply fails to operate). The target lifetime for a PV cell is at least 20 years, in a temperature range of –40° to 85°C, and many manufactures issue warrantees to this effect. Durability is a particular issue with organic approaches; degradation of conducting polymers is a well-studied phenomenon.

Table 7.1 Comparison of efficiency levels achieved by solar PV technologies.

Technology	Highest recorded efficiency (%)
Concentrator PV	40
Crystalline silicon	24
Thin-film–CIGS	19
Thin-film–CdTe	16
Thin-film–aSi	12
Organic PV	5

Drivers for market size increase of solar PV

The growth of solar PV has been driven by the factors; reduced costs (via subsidy), and an increase in the cost of alternatives due to rising fuel costs and emission pricing.

Drivers affecting market share of nanotechnology-enabled solar PV

Two factors will impact the extent to which nanotechnology-based approaches become price competitive with crystalline silicon; cost of materials, and cost of production.

Silicon shortage in 2008 may give way to oversupply by 2012. Renewable Energy Corporation, an integrated producer of silicon, wafers, cells and modules, projects that polysilicon production capacity will increase from under 50,000 MT in 2008 to 125–1,60,000 MT in 2012. This would exceed their projection of demand for polysilicon in 2012, potentially leading to a price drop. This is in part because the solar industry currently competes with the semiconductor industry for high grade silicon, but in future 'solar grade' silicon will emerge with a distinct supply chain.

Restrictions on capital expenditure will drive low cost manufacturing. Manufacturing expenses account for a small proportion of total solar PV cost on an ongoing basis (materials are the largest cost item). However, the costs of building a manufacturing plant are substantial, and during a period where capital will be difficult to access, lower cost manufacturing methods could gain market share.

Effect of nanotechnology-enabled products on the solar PV value chain

The current value chain for crystalline silicon solar PV is shown in Fig. 7.3.

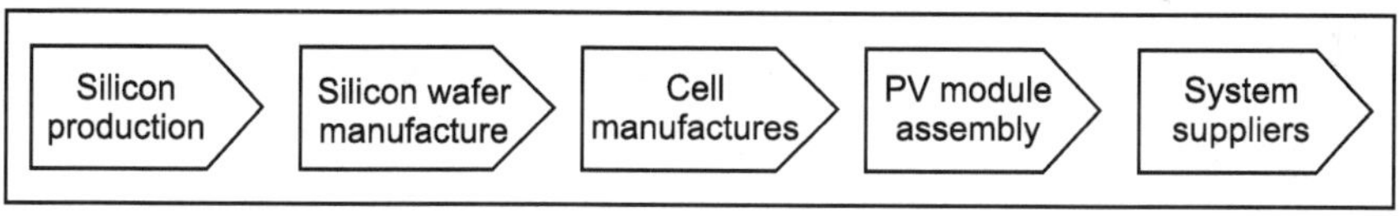

Figure 7.3 Value chain for crystalline silicon solar PV

Thin-film solar PV will reduce the importance of silicon production and wafer manufacture, given the flexibility to apply onto a number of substrates. However approaches which continue to use rare metals (such as CIGS) could simply replace the silicon supply bottleneck with an Indium shortage.

8
Wind energy

8.1 Introduction

Wind is simply air in motion. It is caused by the uneven heating of the earth's surface by the sun. Because the earth's surface is made of very different types of land and water, it absorbs the sun's heat at different rates. One example of this uneven heating can be found in the daily wind cycle.

During the day, the air above the land heats up more quickly than the air over water. The warm air over the land expands and rises, and the heavier, cooler air rushes in to take its place, creating wind. At night, the winds are reversed because the air cools more rapidly over land than over water.

In the same way, the atmospheric winds that circle the earth are created because the land near the earth's equator is heated more by the sun than the land near the North and South Poles. Today, wind energy is mainly used to generate electricity. Wind is a renewable energy source because the wind will blow as long as the sun shines.

Wind is a form of solar energy. Winds are caused by the uneven heating of the atmosphere by the sun, the irregularities of the earth's surface, and rotation of the earth. Wind flow patterns are modified by the earth's terrain, bodies of water, and vegetative cover. This wind flow or motion energy, when 'harvested' by modern wind turbines, can be used to generate electricity.

The terms 'wind energy' or 'wind power' describe the process by which the wind is used to generate mechanical power or electricity. Wind turbines convert the kinetic energy in the wind into mechanical power. This mechanical power can be used for specific tasks (such as grinding grain or pumping water) or a generator can convert this mechanical power into electricity to power homes, businesses, schools, and the like.

Wind turbines: Wind turbines, like aircraft propeller blades, turn in the moving air and power an electric generator that supplies an electric current. Simply stated, a wind turbine is the opposite of a fan. Instead of using electricity to make wind, like a fan, wind turbines use wind to make electricity. The wind turns the blades, which spin a shaft, which connects to a generator and makes electricity.

Modern wind turbines fall into two basic groups; the horizontal-axis variety, like the traditional farm windmills used for pumping water, and the vertical-axis design, like the eggbeater-style Darrieus model, named after its French inventor. Most large modern wind turbines are horizontal-axis turbines.

Turbine components: Horizontal turbine components include:

1. Blade or rotor, which converts the energy in the wind to rotational shaft energy.
2. A drive train, usually including a gearbox and a generator.
3. A tower that supports the rotor and drive train.
4. Other equipment, including controls, electrical cables, ground support equipment, and interconnection equipment.

Wind turbines are often grouped together into a single wind power plant, also known as a wind farm, and generate bulk electrical power. Electricity from these turbines is fed into a utility grid and distributed to customers, just as with conventional power plants.

Wind turbines are available in a variety of sizes, and therefore power ratings. The largest machine has blades that span more than the length of a football field, stands 20 building stories high, and produces enough electricity to power 1400 homes. A small home-sized wind machine has rotors between 8 and 25 ft in diameter and stands upwards of 30 ft and can supply the power needs of an all-electric home or small business. Utility-scale turbines range in size from 50 to 750 kW. Single small turbines, below 50 kW, are used for homes, telecommunications dishes or water pumping.

A renewable non-polluting resource: Wind energy is a free, renewable resource, so no matter how much is used today, there will still be the same supply in the future. Wind energy is also a source of clean, non-polluting, electricity.

Cost issues: The technology requires a higher initial investment than fossil-fuelled generators. Roughly 80 per cent of the cost is the machinery, with the balance being site preparation and installation. If wind generating systems are compared with fossil-fuelled systems on a 'life-cycle' cost basis (counting fuel and operating expenses for the life of the generator), however, wind costs are much more competitive with other generating technologies because there is no fuel to purchase and minimal operating expenses.

Environmental concerns: Although wind power plants have relatively little impact on the environment compared to fossil fuel power plants, there is some concern over the noise produced by the rotor blades, aesthetic (visual) impacts, and birds and bats having been killed (avian/bat mortality) by flying into the rotors. Most of these problems have been resolved or greatly reduced through technological development or by properly citing wind plants.

Supply and transport issues: The major challenge to using wind as a source of power is that it is intermittent and does not always blow when electricity is needed. Wind cannot be stored (although wind-generated electricity can be stored, if batteries are used), and not all winds can be harnessed to meet the timing of electricity demands. Further, good wind sites are often located in remote locations far from areas of electric power demand (such as cities). Finally, wind resource development may compete with other uses for the land, and those alternative uses may be more highly valued than electricity generation. However, wind turbines can be located on land that is also used for grazing or even farming.

8.2 Wind power

Wind power is the conversion of wind energy into a useful form of energy, such as using wind turbines to make electricity, windmills for mechanical power, wind pumps for water pumping or drainage, or sails to propel ships. The total amount of economically extractable power available from the wind is considerably more than present human power use from all sources. Wind power, as an alternative to fossil fuels, is plentiful, renewable, widely distributed, clean, and produces no greenhouse gas emissions during operation, and the cost per unit of energy produced is similar to the cost for new coal and natural gas installations.

At the end of 2010, worldwide nameplate capacity of wind-powered generators was 197 gigawatts (GW). Wind power now has the capacity to generate 430 TWh annually, which is about 2.5 per cent of worldwide electricity usage. Over the past five years the average annual growth in new installations has been 27.6 per cent. Wind power market penetration is expected to reach 3.35 per cent by 2013 and 8 per cent by 2018. Several countries have already achieved relatively high levels of wind power penetration, such as 21 per cent of stationary electricity production in Denmark, 18 per cent in Portugal, 16 per cent in Spain, 14 per cent in Ireland and 9 per cent in Germany in 2010. As of 2011, 83 countries around the world are using wind power on a commercial basis.

Although a variable source of power, the intermittency of wind seldom creates problems when using wind power to supply up to 20 per cent of total electricity demand, but as the proportion rises, increased costs, a need to use storage such as pumped-storage hydroelectricity, upgrade the grid or a lowered ability to supplant conventional production may occur. Power management techniques such as excess capacity, storage, dispatchable backing supply (usually natural gas), exporting and importing power to neighbouring areas or reducing demand when wind production is low, can mitigate these problems.

A large wind farm may consist of several hundred individual wind turbines which are connected to the electric power transmission network. Offshore wind power can harness the better wind speeds that are available offshore compared to on land, so offshore wind power's contribution in terms of electricity supplied is higher. Small onshore wind facilities are used to provide electricity to isolated locations and utility companies increasingly buy back surplus electricity produced by small domestic wind turbines. The construction of wind farms is not universally welcomed because of their visual impact, but any effects on the environment from wind power are generally less problematic than those of any other power source.

8.3 Wind energy

The earth is unevenly heated by the sun, such that the poles receive less energy from the sun than the equator; along with this, dry land heats up (and cools down) more quickly than the seas do. The differential heating drives a global atmospheric convection system reaching from the earth's surface to the stratosphere which acts as a virtual ceiling. Most of the energy stored in these wind movements can be found at high altitudes where continuous wind speeds of over 160 km/hr (99 mph) occur. Eventually, the wind energy is converted through friction into diffuse heat throughout the earth's surface and the atmosphere.

The total amount of economically extractable power available from the wind is considerably more than present human power use from all sources. The most comprehensive study as of 2008 found the potential of wind power on land and near-shore to be 72 TW, equivalent to 54,000 MToE (million tons of oil equivalent) per year or over five times the world's current energy use in all forms. The potential takes into account only locations with mean annual wind speeds ≥6.9 m/s at 80 m. The study assumes six 1.5 MW, 77 m diameter turbines per square kilometre on roughly 13 per cent of the total global land area (though that land would also be available for other compatible uses such as farming). The practical limit to exploitation of wind power will be set by economic and environmental factors, since the resource available is far larger than any practical means to develop it.

8.3.1 Distribution of wind speed

Distribution of wind speed (red) and energy (blue) for all of 2007 at the Lee Ranch facility in Colorado. The histogram shows measured data, while the curve is the Rayleigh model distribution for the same average wind speed.

The strength of wind varies, and an average value for a given location does not alone indicate the amount of energy a wind turbine could produce there.

To assess the frequency of wind speeds at a particular location, a probability distribution function is often fit to the observed data. Different locations will have different wind speed distributions. The Weibull model closely mirrors the actual distribution of hourly wind speeds at many locations. The Weibull factor is often close to 2 and therefore a Rayleigh distribution can be used as a less accurate, but simpler model.

8.3.2 Electricity generation

In a wind farm, individual turbines are interconnected with a medium voltage (often 34.5 kV), power collection system and communications network. At a substation, this medium-voltage electric current is increased in voltage with a transformer for connection to the high voltage electric power transmission system. The surplus power produced by domestic microgenerators can, in some jurisdictions, be fed into the network and sold to the utility company, producing a retail credit for the microgenerators' owners to offset their energy costs.

Grid management

Induction generators, often used for wind power, require reactive power for excitation so substations used in wind-power collection systems include substantial capacitor banks for power factor correction. Different types of wind turbine generators behave differently during transmission grid disturbances, so extensive modelling of the dynamic electromechanical characteristics of a new wind farm is required by transmission system operators to ensure predictable stable behaviour during system faults. In particular, induction generators cannot support the system voltage during faults, unlike steam or hydro turbine-driven synchronous generators.

Doubly fed machines generally have more desirable properties for grid interconnection. Transmission systems operators will supply a wind farm developer with a grid code to specify the requirements for interconnection to the transmission grid. This will include power factor, constancy of frequency and dynamic behaviour of the wind farm turbines during a system fault.

Capacity factor

Since wind speed is not constant, a wind farm's annual energy production is never as much as the sum of the generator nameplate ratings multiplied by the total hours in a year. The ratio of actual productivity in a year to this theoretical maximum is called the capacity factor. Typical capacity factors are 20–40 per cent, with values at the upper end of the range in particularly favourable sites. For example, a 1 MW turbine with a capacity factor of 35 per cent will not produce 8760 MW·h in a year (1 × 24 × 365), but only 1 × 0.35 × 24 × 365 = 3066 MW·h, averaging to 0.35 MW. Online data is available for some locations and the capacity factor can be calculated from the yearly output.

Unlike fuelled generating plants, the capacity factor is affected by several parameters, including the variability of the wind at the site, but also the generator size—having a smaller generator would be cheaper and achieve higher capacity factor, but would make less electricity (and money) in high winds. Conversely a bigger generator would cost more and generate little extra power and, depending on the type, may stall out at low wind speed. Thus an optimum capacity factor can be used, which is usually around 20–35 per cent.

Penetration

Wind energy 'penetration' refers to the fraction of energy produced by wind compared with the total available generation capacity. There is no generally accepted 'maximum' level of wind penetration. The limit for a particular grid will depend on the existing generating plants, pricing mechanisms, capacity for storage or demand management, and other factors. An interconnected electricity grid will already include reserve generating and transmission capacity to allow for equipment failures; this reserve capacity can also serve to regulate for the varying power generation by wind plants. Studies have indicated that 20 per cent of the total electrical energy consumption may be incorporated with minimal difficulty. These studies have been for locations with geographically dispersed wind farms, some degree of dispatchable energy or hydropower with storage capacity, demand management, and interconnection to a large grid area export of electricity when needed. Beyond this level, there are few technical limits, but the economic implications become more significant. Electrical utilities continue to study the effects of large (20 per cent or more) scale penetration of wind generation on system stability and economics.

Intermittency and penetration limits

Electricity generated from wind power can be highly variable at several different time-scales: from hour to hour, daily, and seasonally. Annual variation also exists, but is not as significant. Related to variability is the short-term (hourly or daily) predictability of wind plant output. Like other electricity sources, wind energy must be 'scheduled'. Wind power forecasting methods are used, but predictability of wind plant output remains low for short-term operation. Because instantaneous electrical generation and consumption must remain in balance to maintain grid stability, this variability can present substantial challenges to incorporating large amounts of wind power into a grid system. Intermittency and the non-dispatchable nature of wind energy production can raise costs for regulation, incremental operating reserve, and (at high penetration levels) could require an increase in the already existing energy demand management, load shedding or storage solutions or system interconnection with HVDC cables. At low levels of wind penetration, fluctuations in load and allowance for failure of large generating units requires reserve capacity that can also regulate for variability of wind generation. Wind power can be replaced by other power stations during low wind periods. Transmission networks must already cope with outages of generation plant and daily changes in electrical demand. Systems with large wind capacity components may need more spinning reserve (plants operating at less than full load).

Pumped-storage hydroelectricity or other forms of grid energy storage can store energy developed by high-wind periods and release it when needed. Stored energy increases the economic value of wind energy since it can be shifted to displace higher cost generation during peak demand periods. The potential revenue from this arbitrage can offset the cost and losses of storage; the cost of storage may add 25 per cent to the cost of any wind energy stored, but it is not envisaged that this would apply to a large proportion of wind energy generated.

Capacity credit and fuel saving

Many commentators concentrate on whether or not wind has any 'capacity credit' without defining what they mean by this and its relevance. Wind does have a capacity credit, using a widely accepted and meaningful definition, equal to about 20 per cent of its rated output (but this figure varies depending on actual circumstances). This means that reserve capacity on a system equal in MW to 20 per cent of added wind could be retired when such wind is added without affecting system security or robustness. But the precise value is irrelevant since the main value of wind is its fuel and CO_2 savings.

Installation placement

Good selection of a wind turbine site is critical to economic development of wind power. Aside from the availability of wind itself, other factors include the availability of transmission lines, value of energy to be produced, cost of land acquisition, land use considerations, and environmental impact of construction and operations. Offshore locations may offset their higher construction cost with higher annual load factors, thereby reducing cost of energy produced. Wind farm designers use specialised wind energy software applications to evaluate the impact of these issues on a given wind farm design.

8.4 Energy from the wind

Windmills have been used for many centuries for pumping water and milling grain. The discovery of the internal combustion engine and the development of electrical grids caused many windmills to disappear in the early part of this century.

However, in recent years there has been a revival of interest in wind energy and attempts are underway all over the world to introduce cost-effective wind energy conversion systems for this renewable and environmentally benign energy source.

In developing countries, wind power can play a useful role for water supply and irrigation (wind pumps) and electrical generation (wind generators). These two variants of windmill technology are discussed in separate technical briefs. This brief gives a general overview of the resource and of the technology of extracting energy from the wind.

8.4.1 Energy availability in the wind

The power in the wind is proportional to the cube of wind velocity. The general formula for wind
power is:

$$\text{Power} = \frac{\text{Energy output from array}}{\text{Energy input from sun}} \times 100$$

$$P = \tfrac{1}{2}\rho.A.v^3$$

If the velocity (v) is in m/s, then at sea level (where the density of air is 1.2 kg/m^3) the power in the wind is:

Power = 0.6 × v^3 watts per m^2 of rotor swept area.

This means that the power density in the wind will range from 10 W/m^2 at 2.5 m/s (a light breeze) to 41,000 W/m^2 at 40 m/s (a hurricane). This variability of the wind power resource strongly influences virtually all aspects of wind energy conversion systems design, construction, citing, use and economy.

8.5 Wind resource

Unfortunately, the general availability and reliability of wind speed data is extremely poor in many regions of the world. Large areas of the world appear to have mean annual wind speeds below 3 m/s, and are unsuitable for wind power systems, and almost equally large areas have wind speeds in the intermediate range (3–4.5 m/s) where wind power may or may not be an attractive option. In addition, significant land areas have mean annual wind speeds exceeding 4.5 m/s where wind power would most certainly be economically competitive.

8.6 Principles of wind energy conversion

There are two primary physical principles by which energy can be extracted from the wind; these are through the creation of either drag or lift force (or through a combination of the two). The difference between drag and lift is illustrated (Fig. 8.1) by the difference between using a spinaker sail, which fills like a parachute and pulls a sailing boat with the wind, and a bermuda rig, the familiar triangular sail which deflects with wind and allows a sailing boat to travel across the wind or slightly into the wind. Drag forces provide the most obvious means of propulsion, these being the forces felt by a person (or object) exposed to the wind. Lift forces are the most efficient means of propulsion but being more subtle than drag forces are not so well understood.

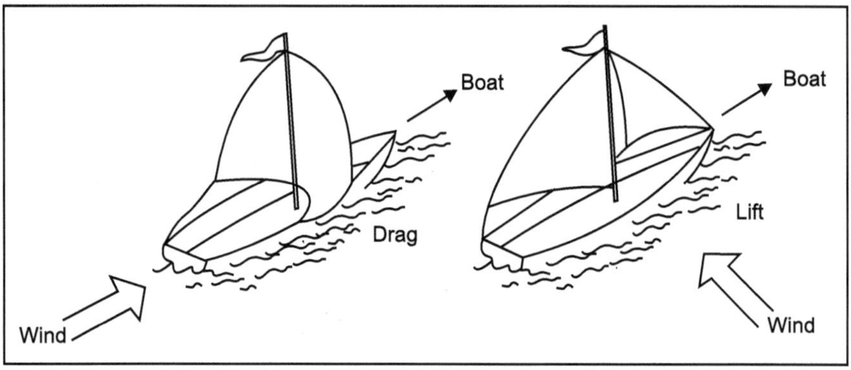

Figure 8.1 Drag and lift forces

The basic features that characterise lift and drag are:

1. Drag is in the direction of airflow.
2. Lift is perpendicular to the direction of airflow.
3. Generation of lift always causes a certain amount of drag to be developed.
4. With a good aerofoil (Fig. 8.2), the lift produced can be more than thirty times greater than the drag.
5. Lift devices are generally more efficient than drag devices.

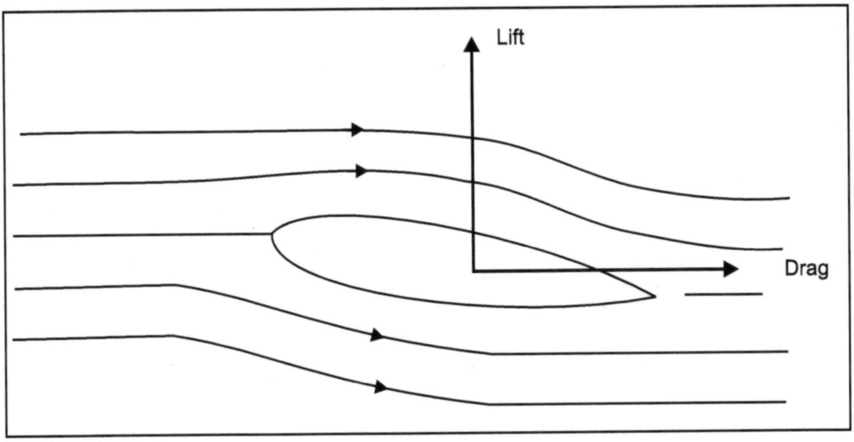

Figure 8.2 Aerofoil

8.7 Types and characteristics of windmill rotors

There are two main families of windmills: vertical axis machines and horizontal axis machines. These can in turn use either lift or drag forces to harness the wind. Of these types the horizontal axis lift device represents the vast majority of successful wind machines, either ancient or modern. In fact other than a few experimental machines virtually all windmills come under this category.

There are several technical parameters that are used to characterise windmill rotors. The tip-speed ratio is defined as the ratio of the speed of the extremities of a windmill rotor to the speed of the free wind. It is a measure of the 'gearing ratio' of the rotor (Fig. 8.3). Drag devices always have tip-speed ratios less than one and hence turn slowly, whereas lift devices can have high tip-speed ratios and hence turn quickly relative to the wind.

$$\text{Tip speed ratio} = \frac{\text{Bade tip speed}}{\text{Wind speed}}$$

The proportion of the power in the wind that the rotor can extract is termed the coefficient of performance (or power coefficient or efficiency; symbol C_p) and its variation as a function of tip-speed ratio is commonly used to characterise different types of rotor. It is physically impossible to extract all the energy from the wind, without bringing the air behind the rotor to a standstill. Consequently, there is a maximum value of C_p of 59.3 per cent (known as the Betz limit); although in practice, real wind rotors have maximum C_p values in the range of 25–45 per cent.

Solidity is usually defined as the percentage of the circumference of the rotor which contains material rather than air.

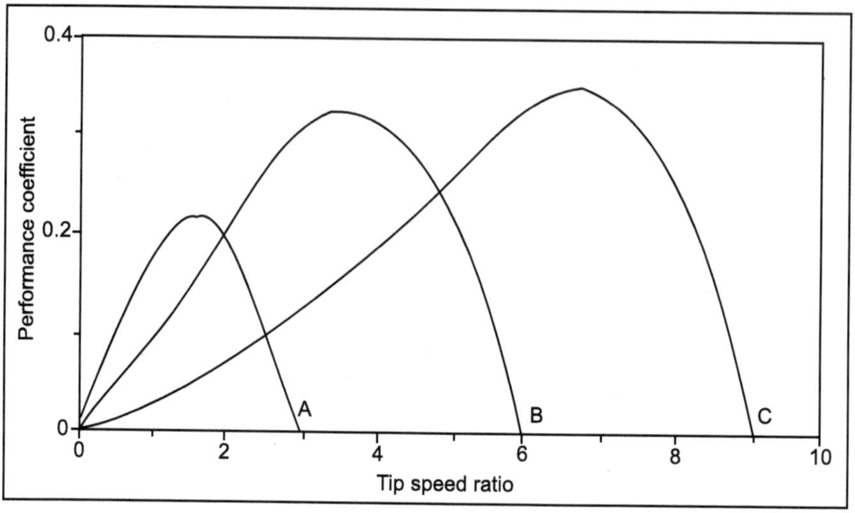

Figure 8.3 Tip speed ratio and the performance coefficient

High-solidity machines carry a lot of material and have coarse blade angles. They generate much higher starting torque than low-solidity machines but are inherently less efficient than low-solidity machines as shown in Fig. 8.4. The extra materials also cost more money. However, low-solidity machines need to be made with more precision which leads to little difference in costs. The choice of rotor is dictated largely by the characteristic of the load and hence of the end use. These aspects are discussed separately in the technical briefs on wind pumps and wind generators. Table 8.1 compares different rotor types.

Table 8.1 Comparison of rotor types.

Type	Speed	Torque	Manufacture	C_p	Solidity %
Horizontal axis					
Cretan sail	Low	Medium	Simple	0.05–0.15	50
Cambered plate fan	Low	High	Moderate	0.15–0.30	50–80
Moderate speed aero-generator	Moderate	Low	Moderate	0.20–0.35	5–10
High speed aero-generator	High	Very low	Precise	0.30–0.45	<5
Vertical axis					
Panemone	Low	Medium	Crude	>0.10	50
Savonius	Moderate	Medium	Moderate	0.15	100
Darrieus	Moderate	Very low	Precise	0.25–0.35	10–20
Variable geometry	Moderate	Very low	Precise	0.20–0.35	15–40

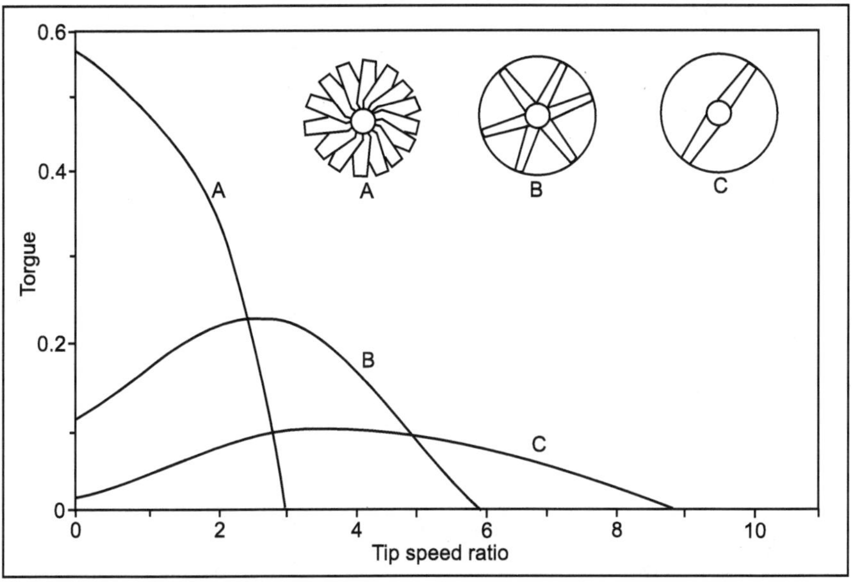

Figure 8.4 Solidity and torque

8.8 Windmill performance

Although the power available is proportional to the cube of wind speed, the power output has a lower order dependence on wind speed. This is because the overall efficiency of the windmill (the product of rotor Cp, transmission efficiency and pump or generator efficiency) changes with wind speed. There are four important characteristic wind speeds:

1. The cut-in wind speed: when the machine begins to produce power.
2. The design wind speed: when the windmill reaches its maximum efficiency.
3. The rated wind speed: when the machine reaches its maximum output power.
4. The furling wind speed: when the machine furls to prevent damage at high wind speeds.

Performance data for windmills can be misleading because they may refer to the peak efficiency (at design wind speed) or the peak power output (at the rated wind speed). The data could also refer to the average output over a time period (e.g. a day or a month).

Because the power output varies with wind speed, the average output over a time period is dependent in the local variation in wind speed from hour to hour. Hence to predict the output for a given windmill one needs to have output characteristics of the windmill and the wind speed distribution curve of the site (duration at various wind speeds). Multiplying the values of both graphs for each wind speed interval and adding all the products gives the total energy output of that windmill at that site.

8.9 Wind speed

Wind speed or wind velocity is a fundamental atmospheric rate. Wind speed affects weather forecasting, aircraft and maritime operations, construction projects, growth and metabolism rate of many plant species, and countless other implications. Wind speed is now commonly measured with an anemometer but can also be classified using the older Beaufort scale which is based on people's observation of specifically defined wind effects.

8.9.1 Factors affecting wind speed

Wind speed is affected by a number of factors and situations, operating on varying scales (from micro to macro scales). These include the pressure gradient, Rossby waves and jet streams, and local weather conditions. There are also links to be found between wind speed and wind direction, notably with the pressure gradient and surfaces over which the air is found.

Pressure gradient is a term to describe the difference in air pressure between two points in the atmosphere or on the surface of the earth. It is vital to wind speed, because the greater the difference in pressure, the faster the wind flows (from the high to low pressure) to balance out the variation. The pressure gradient, when combined with the Coriolis effect and friction, also influences wind direction.

Rossby waves are strong winds in the upper troposphere. These operate on a global scale and move from West to East (hence being known as Westerlies). The Rossby waves are themselves a different wind speed from what we experience in the lower troposphere.

Local weather conditions play a key role in influencing wind speed, as the formation of hurricanes, monsoons and cyclones as freak weather conditions can drastically affect the velocity of the wind.

8.10 Wind turbine

A wind turbine is a device that converts kinetic energy from the wind into mechanical energy. If the mechanical energy is used to produce electricity, the device may be called a wind generator or wind charger. If the mechanical energy is used to drive machinery, such as for grinding grain or pumping water, the device is called a windmill or wind pump. Developed for over a millennium, today's wind turbines are manufactured in a range of vertical and horizontal axis types. The smallest turbines are used for applications such as battery charging or auxiliary power on sailing boats; while large grid-connected arrays of turbines are becoming an increasingly large source of commercial electric power (Fig. 8.5).

Figure 8.5 Wind farm

Wind turbines in locations with constantly high wind speeds bring best return on investment. With a wind resource assessment it is possible to estimate the amount of energy the wind turbine will produce.

A quantitative measure of the wind energy available at any location is called the wind power density (WPD). It is a calculation of the mean annual power available per square metre of swept area of a turbine, and is tabulated for different heights above ground. Calculation of wind power density includes the effect of wind velocity and air density. Colour-coded maps are prepared for a particular area described, for example, as 'mean annual power density at 50 m'.

8.10.1 Types of wind turbines

Wind turbines can rotate about either a horizontal or a vertical axis, the former being both older and more common.

Horizontal axis

Horizontal-axis wind turbines (HAWT) have the main rotor shaft and electrical generator at the top of a tower, and must be pointed into the wind. Small turbines are pointed by a simple wind vane, while large turbines generally use a wind sensor coupled with a servo motor. Most have a gearbox, which turns the slow rotation of the blades into a quicker rotation that is more suitable to drive an electrical generator.

Since a tower produces turbulence behind it, the turbine is usually positioned upwind of its supporting tower. Turbine blades are made stiff to prevent the blades from being pushed into the tower by high winds. Additionally, the blades are placed a considerable distance in front of the tower and are sometimes tilted forward into the wind a small amount.

Downwind machines have been built, despite the problem of turbulence (mast wake), because they don't need an additional mechanism for keeping them in line with the wind, and because in high winds the blades can be allowed to bend which reduces their swept area and thus their wind resistance. Since cyclical (that is repetitive) turbulence may lead to fatigue failures, most HAWTs are of upwind design.

Subtypes

(a) *12th-century windmills:* These squat structures, typically (at least) four bladed, usually with wooden shutters or fabric sails, were developed in Europe. These windmills were pointed into the wind manually or via a tail-fan and were typically used to grind grain. In the Netherlands they were also used to pump water from low-lying land, and were instrumental in keeping its polders dry.

(b) *19th-century windmills:* The Eclipse windmill factory was set up around 1866 in Beloit, Wisconsin and soon became successful building mills for pumping water on farms and for filling railroad tanks. Other firms like Star, Dempster, and Aeromotor also entered the market. Hundreds of thousands of these mills were produced before rural electrification and small numbers continue to be made. They typically had many blades, operated at tip speed ratios not better than one, and had good starting torque. Some had small direct-current generators used to charge storage batteries, to provide power to lights or to operate a radio receiver.

(c) *Modern wind turbines:* Turbines used in wind farms for commercial production of electric power are usually three-bladed and pointed into the wind by computer-controlled motors. These have high tip speeds of up to six times the wind speed, high efficiency, and low torque ripple, which contribute to good reliability. The blades are usually coloured light gray to blend in with the clouds and range in length from 20 to 40 m (65 to 130 ft) or more. The tubular steel towers range from 200 to 300 ft (60 to 90 m) tall. The blades rotate at 10–22 revolutions per minute. A gear box is commonly used to step up the speed of the generator, although designs may also use direct drive of an annular generator. Some models operate at constant speed, but more energy can be collected by variable-speed turbines which use a solid-state power converter to interface to the transmission system. All turbines are equipped with shutdown features to avoid damage at high wind speeds.

Advantages of HAWT

1. Variable blade pitch, which gives the turbine blades the optimum angle of attack. Allowing the angle of attack to be remotely adjusted gives greater control, so the turbine collects the maximum amount of wind energy for the time of day and season.
2. The tall tower base allows access to stronger wind in sites with wind shear. In some wind shear sites, every 10 m up, the wind speed can increase by 20 per cent and the power output by 34 per cent for every 10 m in elevation.
3. High efficiency, since the blades always move perpendicularly to the wind, receiving power through the whole rotation. In contrast, all vertical axis wind turbines, and most proposed airborne wind turbine designs, involve various types of reciprocating actions, requiring airfoil surfaces to backtrack against the wind for part of the cycle. Backtracking against the wind leads to inherently lower efficiency.
4. The face of a horizontal axis blade is struck by the wind at a consistent angle regardless of the position in its rotation. This results in a consistent lateral wind loading over the course of a rotation, reducing vibration and audible noise coupled to the tower or mount.

Disadvantages of HAWT

1. HAWTs have difficulty operating in near ground, turbulent winds.
2. The tall towers and blades up to 90 m long are difficult to transport. Transportation can now cost 20 per cent of equipment costs.
3. Tall HAWTs are difficult to install, needing very tall and expensive cranes and skilled operators.

4. Massive tower construction is required to support the heavy blades, gearbox, and generator.
5. Reflection on tall HAWTs may affect side lobs of radar installations creating signal clutter, although filtering can suppress it.
6. Their height makes them obtrusively visible across large areas, disrupting the appearance of the landscape and sometimes creating local opposition.
7. Downwind variants suffer from fatigue and structural failure caused by turbulence when a blade passes through the tower's wind shadow (for this reason, the majority of HAWTs use an upwind design, with the rotor facing the wind in front of the tower).
8. HAWTs require an additional yaw control mechanism to turn the blades toward the wind.
9. In order to minimise fatigue loads due to wake turbulence, wind turbines are usually sited a distance of 5 rotor diameters away from each other, but the spacing depends on the manufacturer and the turbine model.

Vertical axis wind turbines

Vertical axis wind turbines (or VAWTs) have the main rotor shaft arranged vertically. Key advantages of this arrangement are that the turbine does not need to be pointed into the wind to be effective. This is an advantage on sites where the wind direction is highly variable. VAWTs can utilise winds from varying directions. With a vertical axis, the generator and gearbox can be placed near the ground, so the tower doesn't need to support it, and it is more accessible for maintenance. Drawbacks are that some designs produce pulsating torque. Drag may be created when the blade rotates into the wind.

It is difficult to mount vertical-axis turbines on towers, meaning they are often installed nearer to the base on which they rest, such as the ground or a building rooftop. The wind speed is slower at a lower altitude, so less wind energy is available for a given size turbine. Air flow near the ground and other objects can create turbulent flow, which can introduce issues of vibration, including noise and bearing wear which may increase the maintenance or shorten the service life.

However, when a turbine is mounted on a rooftop, the building generally redirects wind over the roof and this can double the wind speed at the turbine. If the height of the rooftop mounted turbine tower is approximately 50 per cent of the building height, this is near the optimum for maximum wind energy and minimum wind turbulence.

Subtypes

Darrieus wind turbine: 'Eggbeater' turbines or Darrieus turbines, were named after the French inventor, Georges Darrieus. They have good efficiency, but produce large torque ripple and cyclical stress on the tower, which contributes to poor reliability. They also generally require some external power source or an additional Savonius rotor to start turning, because the starting torque is very low. The torque ripple is reduced by using three or more blades which results in greater solidity of the rotor. Solidity is measured by blade area divided by the rotor area. Newer Darrieus type turbines are not held up by guy-wires but have an external superstructure connected to the top bearing.

Giromill: A subtype of Darrieus turbine with straight, as opposed to curved, blades. The cycloturbine variety has variable pitch to reduce the torque pulsation and is self-starting. The advantages of variable pitch are: high starting torque; a wide, relatively flat torque curve; a lower blade speed ratio; a higher coefficient of performance; more efficient operation in turbulent winds; and a lower blade speed ratio which lowers blade bending stresses. Straight, V or curved blades may be used.

Savonius wind turbine: These are drag-type devices with two (or more) scoops that are used in anemometers, Flettner vents (commonly seen on bus and van roofs), and in some high-reliability low-efficiency power turbines. They are always self-starting if there are at least three scoops. They sometimes have long helical scoops to give a smooth torque.

Advantages of vertical axis wind turbines

1. A massive tower structure is less frequently used, as VAWTs are more frequently mounted with the lower bearing mounted near the ground.
2. Designs without yaw mechanisms are possible with fixed pitch rotor designs.
3. The generator of a VAWT can be located nearer the ground, making it easier to maintain the moving parts.
4. VAWTs have lower wind start-up speeds than HAWTs. Typically, they start creating electricity at 6 mph (10 km/hr).
5. VAWTs may be built at locations where taller structures are prohibited.
6. VAWTs situated close to the ground can take advantage of locations where mesas, hilltops, ridgelines, and passes funnel the wind and increase wind velocity.
7. VAWTs may have a lower noise signature.

Disadvantages of vertical axis wind turbines

1. A VAWT that uses guy-wires to hold it in place puts stress on the bottom bearing as all the weight of the rotor is on the bearing. Guy wires attached to the top bearing increase downward thrust in wind gusts. Solving this problem requires a superstructure to hold a top bearing in place to eliminate the downward thrusts of gust events in guy wired models.
2. The stress in each blade due to wind loading changes sign twice during each revolution as the apparent wind direction moves through 360 degrees. This reversal of the stress increases the likelihood of blade failure by fatigue.
3. While VAWTs' components are located on the ground, they are also located under the weight of the structure above it, which can make changing out parts very difficult without dismantling the structure, if not designed properly.
4. Having rotors located close to the ground where wind speeds are lower due to the ground's surface drag, VAWTs may not produce as much energy at a given site as a HAWT with the same footprint or height.

8.10.2 Turbine design and construction

Wind turbines are designed to exploit the wind energy that exists at a location. Aerodynamic modelling is used to determine the optimum tower height, control systems, number of blades and blade shape. Wind turbines convert wind energy to electricity for distribution. Conventional horizontal axis turbines can be divided into three components.

1. The rotor component, which is approximately 20 per cent of the wind turbine cost, includes the blades for converting wind energy to low speed rotational energy.
2. The generator component, which is approximately 34 per cent of the wind turbine cost, includes the electrical generator, the control electronics, and most likely a gearbox (e.g. planetary gearbox, adjustable-speed drive or continuously variable transmission) component for converting the low speed incoming rotation to high speed rotation suitable for generating electricity.
3. The structural support component, which is approximately 15 per cent of the wind turbine cost, includes the tower and rotor yaw mechanism.

8.10.3 Small wind turbines

Small wind turbines may be as small as a fifty-watt generator for boat or caravan use. The US Department of Energy's National Renewable Energy Laboratory (NREL) defines small wind turbines as those smaller than or equal to 100 kW. Small units often have direct drive generators, direct current output, aeroelastic blades, lifetime bearings, and use a vane to point into the wind.

Larger, more costly turbines generally have geared power trains, alternating current output, flaps and are actively pointed into the wind. Direct drive generators and aeroelastic blades for large wind turbines are being researched.

8.10.4 Rotor

To increase the efficiency of wind turbines, rotor planning principles are very important for wind turbines with horizontal axis. It is not possible to expect the maximum efficiency from a wind turbine that is installed without doing the optimisation processes. At this point, rotor blades aerodynamic features are very important. It is necessary to put forward the power value that can be obtained from rotor blades. It is possible to conduct power tests indoors or under natural conditions. But in both methods, real rotor blades have to be produced. And this means cost. For this reason, first miniature blades can be produced and the tests can be conducted on these miniature rotor blades.

The main forces in atmosphere that produce wind and effect its speed is: pressure grading force, diverting force, centrifugal force and friction force. Pressure gradient force acts to move the air from high pressure to low pressure. Diverting force affects the air from two ways: one is as diverting force of earth's rotation for movements from the equator to poles or in the opposite direction. Winds, in general, are under the effect of a force that wants to divert them from their centre because they curl around a centre. This force is called centrifugal force. Friction force tries to decrease the speed of wind. The effect of this forces the greatest when near to earth's surface.

Importance of rotor

Rotor is the organ that transforms the kinetic energy of wind to mechanic energy. For this reason it is very important for wind turbines. It is very important for rotor and rotor blades to have optimum features, because these have a direct effect on the efficiency of wind turbines.

A flow mass has a kinetic energy because of its speed.

$$E = \frac{M}{2} V^2 \qquad \text{... (8.1)}$$

The achievement derives from the energy per time. Calculation of the mass flow:

$$m = \rho \cdot V \cdot A \qquad \text{... (8.2)}$$

Into these conditions we can compute the power with the following equation:

$$P = \frac{\rho}{2} . A . V^3 \qquad \text{... (8.3)}$$

Power equity gives the theoretic power that can be obtained from the kinetic energy that is stored in wind as Watt. This theoretic power has to be transformed into useful power by the help of turbine rotor. At this point a constant, which is effected by wind speed, turbine shaft speed and blade choice has to be taken into consideration. This constant is called ideal power constant (Cp). Blade types, blade form, inclination angle and speed of blade's tip are effective factors in here. In Fig. 8.6 the diagram of the ideals power coefficient. Theoretically ideal power constant cannot exceed 0.59. This constant is called Betz constant. In practice, this value is even smaller, because mechanical losses (η) come into effect in practice. But mechanic efficiency value can be neglected in calculations, because it is close to 1. Thus the equation can be written as below:

$$P = \frac{\rho}{2} . A . V^3 . C_p . \eta \qquad \text{... (8.4)}$$

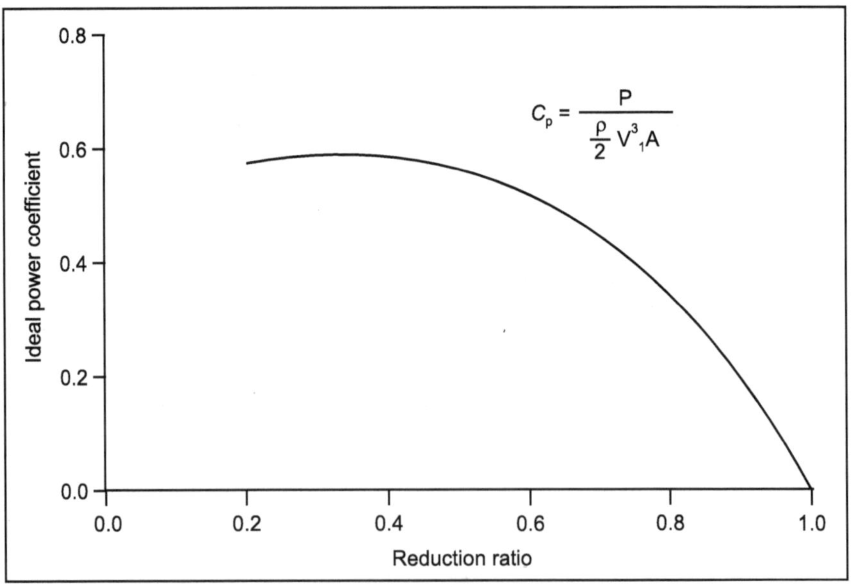

Figure 8.6 Power coefficient (Cp) according to Betz

One thing that must not be forgotten in here is that air density is 1.225 kg m^{-3} under standard meteorological conditions (temperature: 15°C and air pressure: 1013.3 hPa). The changes in air temperature and air pressure will change air density.

Obtaining maximum energy production from a wind turbine depends on various factors. These are factor like the height of wind turbine; wind turbine blade's sweep area and aerodynamic structure, air density and wind speed. The most important ones of these factors are the height of wind turbine and aerodynamic structure of wind turbine. The height of wind turbine is important because wind speed increases as we go away from earth's surface. Aerodynamic structure of wind turbine blade is important, because it can transform maximum 59 per cent of the kinetic energy that wind has to useful energy.

Rotor aerodynamic of wind turbine: In transformation of wind energy, which is formed by heating of different points of the atmosphere by the main energy source sun, to electric energy; wind rotor, which is the first ring in transformation chain, can be designed according to Betz or Glaubert–Schmitz for the purpose of transferring the existing wind power with minimum loss.

Wind turbine rotor height: As wind speed gets away from earth's surface, it frees itself from the friction effect cause by the roughness of earth's surface. Thus it moves more freely. As it gets away from obstacles that decelerate its speed, its speed increases. It is assumed that winds that are 1000 m above the earth's surface, namely geostrophic winds are not affected by the roughness of earth's surface and friction losses. At the light of these thoughts, we can say that there is a relation between wind speed and wind height. This is the reason why wind turbines are built as high as possible.

Wind turbine rotor diameter: Besides determining turbine height, the diameter of the wind that rotor blade sweeps has to be determined. The diameter of the wind that rotor blade sweeps has a direct effect on the power that will obtained from turbine.

The wind potential in where the wind turbine rotor will operate: Wind potential in where the wind turbine will installed is very important. For this reason it is one of the parameters that have to be considered in rotor design. Wind speed potential in where the wind turbine will installed has to be observed at least for 6 months. Wind potentials is directly effective on the efficiency of the wind turbine rotor. Wind speed is the most important factor about the energy of the wind. The power that will be obtained from wind is directly proportional to wind speed's third power.

When wind speed has this much effect on the power that will be obtained from the wind turbine, the wind potential in where the turbine will be built is very important. The main objective is to find the point where the wind speed is maximum and install the turbine there.

Land structure of the place where the wind turbine rotor will operate: There is a relation between wind speed and wind height. This relation is dependent on some conditions. These conditions originate from land shapes. Same conditions do not apply in a flat land surface and surface with obstacles. In a flat land surface as height increases, speed also increase in a direct ratio, but his will not be true in a land surface with obstacles. In a surface with obstacles, wind will have to climb over obstacles to resume its course and this will cause a pause in wind's speed.

If the most important criteria for obtaining energy from wind are wind speed, then it is very important to find the areas where wind speed is high and install wind turbine rotor there. But we don't have to make a false assumption that when we increase the height we will always catch winds that contain more energy. The factors that wind speed is affect depending on height are: Von Karman constant, surface friction speed and roughness length.

The performance demanded from wind turbine rotor: Before choosing a wind turbine for an enterprise, first we have to determine how much electric power our enterprise needs. This way, power of the wind turbine that will be installed can be determined. The performance demanded from the rotor has to be determined according to the enterprise's installed power.

To increase the efficiency of wind turbines, rotor planning principles are very important for wind turbines with horizontal axis. It is not possible to expect the maximum efficiency from a wind turbine that is installed without doing the optimisation processes.

Like in every topic, to be successful when making use of wind energy, basic values must be based on scientific data. When the topic is considered from this point of view, rotor design parameters have to be put forward in wind turbines. Starting from here, the most suitable systems can be developed by taking source data as a basis. The basic strategy is to put into practice high performance systems that can compete.

8.11 Wind electricity basics

Small wind-electric systems can provide electricity on remote, off-grid sites or right in town connected to the utility grid. Although wind systems require more maintenance and need more attention than solar-electric or microhydro-electric systems, if you invest up front in good equipment, design, and installation, wind-electric systems can make economic and environmental sense. They also bring a great deal of satisfaction—there's nothing quite like watching your wind generator convert a summer breeze or a winter storm into electrical energy.

Boiled down to its simplest principles, a wind generator's rotating blades convert the wind's kinetic energy into rotational momentum in a shaft. The rotating shaft turns an alternator, which makes electricity. This electricity is transmitted through wiring down the tower to its end use.

The blades use engineered airfoils, matched to the alternator, that capture the wind's energy. Most modern wind generators use three blades, the best compromise between the highest efficiency possible (one blade) and the balance that comes with multiple blades. Together, the blades and the hub they are attached to are termed the rotor, which is the collector of the system, intercepting winds that pass by. Most turbines on the market today are upwind machines—their blades are on the windward side of the tower. A few downwind machines are available, but neither configuration has a clear performance advantage over the other.

In most small-scale designs, the rotor is connected directly to the shaft of a permanent magnet alternator, which creates wild, three-phase AC. Wild, three-phase electricity means that the voltage and frequency vary continuously with the wind speed. They are not fixed like the 60 Hz and 120 VAC electricity coming out of common household outlets. The wild output is rectified to DC to either charge batteries or feed a grid-synchronous inverter. In most designs (up to 15 kW in peak capacity), the rotor is usually connected directly to the alternator, which eliminates the additional maintenance of gears. In systems 20 kW and larger, as well as some smaller wind systems (like the Endurance, Tulipo or Aircon), a gearbox is used to increase alternator speed from a slower turning rotor.

The blades must turn to face the wind, so a yaw bearing is needed, allowing the wind turbine to track the winds as they shift direction. The tail directs the rotor into the wind. Some sort of governing system limits the rotor rpm as well as generator output to protect the turbine from high winds. A shutdown mechanism is also useful to stop the machine when necessary, such as during an extreme storm, when you do not need the energy or when you want to service the system.

How wind turbines are rated: Wind turbine rating is a tricky affair. While solar-electric module or microhydro-electric turbine production can be predicted fairly realistically based on rated output, this number is very misleading with wind turbines. Why? Because rated output is pegged to a particular wind speed, and different manufacturers use different wind speeds to determine rated output. Also, the power available in the wind varies with the cube of its speed, so small increases in wind speed result in large increases in power available to the rotor. A 10 per cent increase in wind speed yields a 33 per cent increase in power available in the wind. Conversely, this means that a turbine rated at 1000 watts at 28 mph might produce only 125 watts or less at half that wind speed, 14 mph.

So what's a wind turbine buyer to do? Ignore the peak output and the power curve. Look for the monthly or annual energy numbers for the turbine, estimated for the average wind speed you expect or measure at your site. These will be given in kWh per month (or year) in the manufacturer's specifications for each turbine. Energy is what you are after, not peak power. If, for example, you are looking for a turbine that can produce 300 kWh per month, and you know that you have a 10 mph average wind speed at the proposed turbine height, you can shop for a turbine that is predicted to generate that much energy in that average wind speed.

If you can not get energy production estimates from the manufacturer or a turbine owner, look for a different manufacturer. This is basic information that any manufacturer should supply. However, knowing a turbine's swept area may also help you calculate the annual energy output for the wind turbine. All other things being equal, 'there is no replacement for displacement'.

A turbine's revolutions per minute (rpm) at its rated wind speed can give you some idea of the relative aerodynamic sound of the machine, and also speaks to longevity. Slower-turning wind turbines tend to be quieter and last longer. High rpm machines wear out components, such as bearings, much faster. In addition, the faster blades move through the air, the greater the possibility that they will waste some of that energy as sound from the blades.

How to choose a wind turbine: Trying to keep an inexpensive wind generator running can be an uphill battle that you'll soon tire of. But expect to pay more for a better machine—it is a tough job to design and manufacture a long-lasting, small-scale wind generator.

The bottom line: Buy a turbine that has a very good track record and a good warranty—five years is preferable but not always available in the small wind industry. A warranty is one indication of the manufacturer's confidence in their product, and their intention to stand behind it.

Real-world reports from users carry even more weight than a warranty, so search for people who own the model of turbine you are considering buying, and get the straight scoop from them about performance, durability, reliability, and maintenance issues.

Some manufacturers make only battery-charging machines, and may offer a variety of turbine voltages. Others produce machines intended to connect to grid-synchronous inverters without batteries. One machine even includes an inverter integrated with the turbine itself. Make sure you are buying a machine that is appropriate for your intended use.

When you look at prices, keep in mind that just buying a wind turbine will not get you any wind-generated electricity. You will also need most or all of the components mentioned elsewhere; also budget for equipment rental, like a backhoe and crane, concrete and rebar, electrical components, shipping, and sales tax. Unless you do all of the work yourself, also factor in installation labour expenses. These costs can add up significantly, so make sure that you research and understand all of the associated expenses before committing to a purchase. Many people are quite surprised to learn that the wind turbine cost can range from only 10 per cent to as much as 40 per cent of the entire wind system's expenses. Small-scale wind energy is not for the half-hearted, uninvolved or uncommitted, and probably not for folks who never change the oil in their vehicles (or are willing to spend the bucks to hire someone to do the tower work).

8.12 Wind-electric system types

8.12.1 Off-grid wind-electric systems

Off-grid wind-electric systems are battery based. People generally choose these systems because their home or other energy use is not connected to the grid, and connection would be expensive. Others prefer the independence of off-grid systems or live where utilities and governments make it difficult to tie a renewable energy system to the grid.

Off-grid systems are limited in capacity by the size of the generating sources (wind turbine, solar-electric array, fuel-fired generator, etc.), the resources available, and the battery bank size. Off-grid homeowners have to learn to live within the limitations of their system capacity (Fig. 8.7).

8.12.2 Grid-tied wind-electric system with battery backup

Connecting a wind-electric system to the utility grid with battery backup gives you the best of both worlds. You have the unlimited capacity of the grid at your disposal, and you can send your surplus wind energy to the grid. When the grid is down, you can still use your system, within the limitations of the battery bank and turbine. Wind-electric systems can be a much better match for utility backup than solar-electric systems, since many grid outages are caused by high winds. The drawback is that this is the most expensive type of wind-electric system you can install (Fig. 8.8).

8.12.3 Battery-less grid-tied wind-electric system

Connecting to the grid without batteries is the most cost-effective and environmentally friendly way to go. You eliminate batteries, which are costly, require maintenance, and carry a significant efficiency penalty. The only drawback of battery-less systems is that when the grid is down, your system shuts down. But in most grid-serviced areas, utility outages are only a few hours a year—a small inconvenience to endure for the efficiency, environmental friendliness, and thriftiness of these systems. Battery-less grid-tie systems may see increased performance (sometimes dramatically) from the wind turbine compared to battery-based systems.

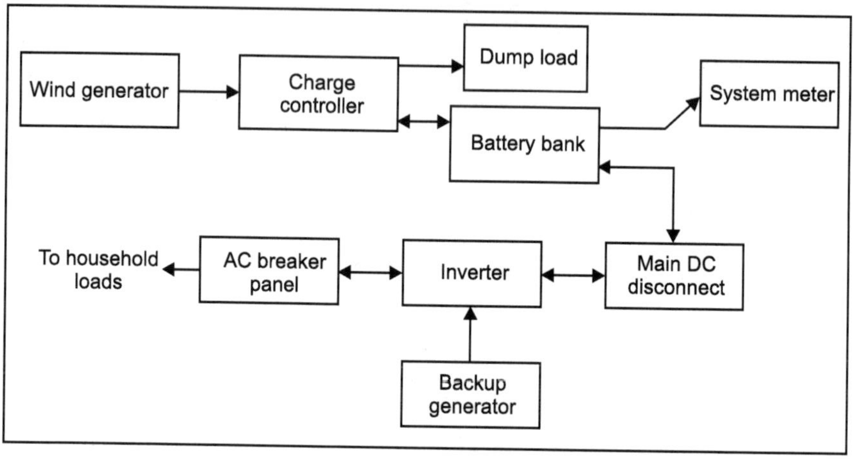

Figure 8.7 Off-grid wind-electric system

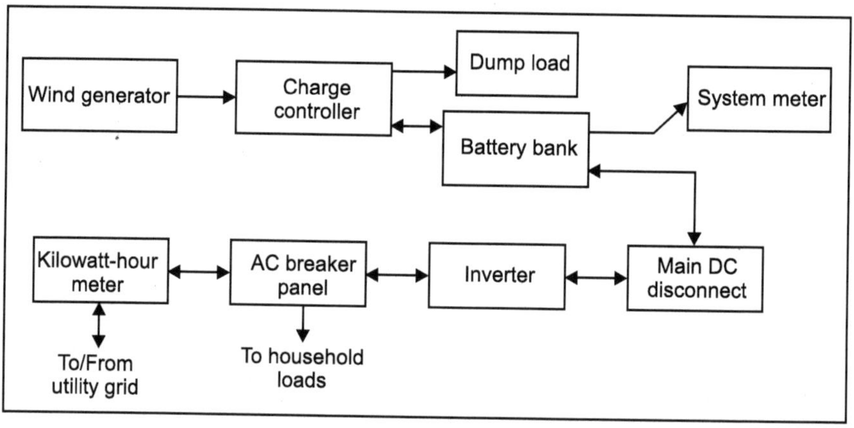

Figure 8.8 Grid-tied wind-electric system with battery backup

This is because the inverter's electronics can match the wind's load more exactly, running the turbine at optimum speed, and extracting the maximum energy (Fig. 8.9).

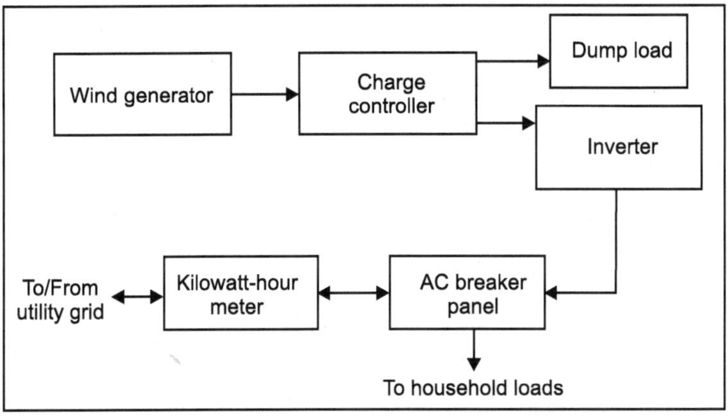

Figure 8.9 Battery-less grid-tied wind-electric system

8.12.4 Direct-drive battery-less wind-electric system

These are the least common wind-electric systems, typically used for water pumping. A turbine is matched to a pump, often through an electronic controller. When the wind blows, water is pumped to an elevated tank, a stock-watering tank or directly to the land to irrigate. These systems can be simple and cost effective in the right situation. Direct-drive systems are also used for heating, which can be a good match, since it's normally colder when it's windy. But heating is a big load, so large turbines are needed (Fig. 8.10).

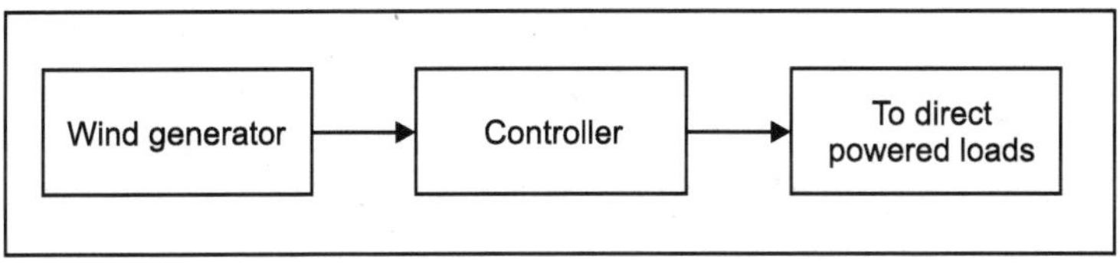

Figure 8.10 Direct-drive battery-less wind-electric system

8.12.5 Wind-electric system components

Here are some brief descriptions of the common equipment used in grid-intertied and off-grid wind-electric systems. Systems vary—not all equipments are necessary for every system type.

1. Wind generator
2. Tower
3. Brake
4. Charge controller

5. Dump load
6. Battery bank
7. System meter
8. Main DC disconnect
9. Inverter
10. AC breaker panel
11. Kilowatt-hour meter
12. Backup generator

Wind generator

The wind generator is what actually generates electricity in the system. Most modern wind generators are upwind designs (blades are on the side of the tower that faces into the wind), and couple permanent magnet alternators directly to the rotor (blades). Three-bladed wind generators are most common, providing a good compromise between efficiency and rotor balance.

Small wind turbines protect themselves from high winds (governing) by tilting the rotor up or to the side or by changing the pitch of the blades. Electricity is transmitted down the tower on wires, most often as three-phase wild alternating current (AC). It is called 'wild' because the voltage and frequency vary with the rotational speed of the wind turbine. The output is then rectified to direct current (DC) to charge batteries or to be inverted for grid connection.

Tower

A wind generator tower is very often more expensive than the turbine. The tower puts the turbine up in the 'fuel'—the smooth strong winds that give the most energy. Wind turbines should be sited at least 30 ft (9 m) higher than anything within 500 ft (152 m).

Three common types of towers are tilt-up, fixed-guyed, and freestanding. Towers must be specifically engineered for the lateral thrust and weight of the turbine, and should be adequately grounded to protect your equipment against lightning damage.

Brake

Most wind turbines have some means of stopping the turbine for repairs, in an emergency, for routine maintenance, or when the energy is not needed. Many turbines have 'dynamic braking', which simply shorts out the three electrical phases and acts as a disconnect. Others have mechanical braking, either via a disc or drum brake, activated by a small winch at the base of the tower. Still others have mechanical furling, which swings the rotor out of the wind. Mechanical braking is usually more effective and reliable than dynamic braking.

Charge controller

A wind-electric charge controller's primary function is to protect your battery bank from overcharging. It does this by monitoring the battery bank—when the bank is fully charged, the controller sends energy from the battery bank to a dump (diversion) load.

Many wind-electric charge controllers are built into the same box as the rectifiers (AC-to-DC converters). Overcurrent protection is needed between the battery and controller/dump load.

In battery-less grid-tie systems, there is no controller in normal operation, since the inverter is selling whatever energy the turbine is generating. But there will be some control function in the case of grid failure, and there may be electronics before the inverter to regulate the input voltage.

Dump load

Solar-electric modules can be turned off—open circuited—with no damage. Most wind generators should not run unloaded. They will run too fast and too loud, and may self-destruct. They must be connected to a battery bank or load. So normally, a charge controller that has the capability of being a diversion controller is used. A diversion controller takes surplus energy from the battery bank and sends it to a dump load. In contrast, a series controller (commonly used in PV systems), actually opens the circuit.

A dump load is an electrical resistance heater, and it must be sized to handle the full generating capacity of the wind generator used. These dump loads can be air or water heaters, and are activated by the charge controller whenever the batteries or the grid cannot accept the energy being produced.

Battery bank

Your wind generator will produce electricity whenever the wind blows above the cut-in speed. If your system is off-grid, you'll need a battery bank—a group of batteries wired together—to store energy so you can have electricity when it is not windy. For off-grid systems, battery banks are typically sized to keep household electricity running for one to three calm days. Grid-intertied systems also can include battery banks to provide emergency backup during blackouts—perfect for keeping critical electric loads operating until the grid is up again.

Use only deep-cycle batteries in wind-electric systems. Lead-acid batteries are the most common battery type. Flooded lead-acid batteries are usually the least expensive, but require adding distilled water occasionally to replenish water lost during the normal charging process. Sealed absorbent glass mat (AGM) batteries are maintenance free and designed for grid-tied systems where the batteries are typically kept at a full state of charge. Sealed gel-cell batteries can be a good choice to use in unheated spaces due to their freeze-resistant qualities.

System meter

System meters can measure and display several different aspects of your wind-electric systems performance and status—tracking how full your battery bank is, how much electricity your wind generator is producing or has produced, and how much electricity is in use. Operating your system without metering is like running your car without any gauges—although possible to do, it is always better to know how much fuel is in the tank.

Main DC disconnect

In battery-based systems, a disconnect between the batteries and inverter is required. This disconnect is typically a large, DC-rated breaker mounted in a metal enclosure. This breaker allows the inverter to be quickly disconnected from the batteries for service, and protects the inverter-to-battery wiring against electrical fires.

Inverter

Inverters transform the electricity produced by your wind generator into the AC electricity commonly used in most homes for powering lights and appliances. Grid-tied inverters synchronise the electricity they produce with the grid's 'utility grade' AC electricity, allowing the system to feed wind electricity to the utility grid.

Grid-tie inverters are either designed to operate with or without batteries. Battery-based inverters for off-grid or grid-tie systems often include a battery charger, which is capable of charging a battery bank from either the grid or a backup generator during cloudy weather.

AC breaker panel

The AC breaker panel is the point at which all of a home's electrical wiring meets with the provider of the electricity, whether that is the grid or a solar-electric system. This wall-mounted panel or box is usually installed in a utility room, basement, garage or on the exterior of the building. It contains a number of labelled circuit breakers that route electricity to the various rooms throughout a house. These breakers allow electricity to be disconnected for servicing, and also protect the building's wiring against electrical fires.

Just like the electrical circuits in your home or office, an inverter's electrical output needs to be routed through an AC circuit breaker. This breaker is usually mounted inside the building's mains panel, which enables the inverter to be disconnected from either the grid or from electrical loads if servicing is necessary, and also safeguards the circuit's electrical wiring.

Kilowatt-hour meter

Most homes with a grid-tied wind-electric system will have AC electricity both coming from and going to the electric utility grid. A bidirectional kWh meter can simultaneously keep track of how much electricity you are using and how much your system is producing. The utility company often provides intertie-capable meters at no cost.

Backup generator

Off-grid wind-electric systems can be sized to provide electricity during calm periods when the wind doesn't blow. But sizing a system to cover a worst-case scenario, like several calm weeks during the summer, can result in a very large, expensive system that will rarely get used to its capacity and will run a huge surplus in windy times. To spare your pocketbook, go with at least two sources of energy. Wind-PV hybrid systems are often an excellent fit with local renewable resources. But a backup, fuel-powered generator still may be necessary.

Engine-generators can be fuelled with biodiesel, petroleum diesel, gasoline or propane, depending on the design. Most generators produce AC electricity that a battery charger (either stand-alone or incorporated into an inverter) converts to DC energy, which is stored in batteries. Like most internal combustion engines, generators tend to be loud and stinky, but a well-designed renewable energy system will require running them only 50 to 200 hours a year or less.

8.13 Wind farm

A wind farm is a group of wind turbines in the same location used for production of electric power. Individual turbines are interconnected with a

medium voltage power collection system and communications network. At a substation, this medium voltage electrical current is increased in voltage with a transformer for connection to the high voltage transmission system. Near shore turbine installations are on land within 5 miles of a shoreline or on water within ten miles. These areas are good sites for turbine installation, because of wind produced by convection due to differential heating of land and sea each day. Wind speeds in these zones share the characteristics of both onshore and offshore wind, depending on the prevailing wind direction (Fig. 8.11).
Offshore wind turbines are less obtrusive than turbines on land, as their apparent size and noise is mitigated by distance. Because water has less surface roughness than land (especially deeper water), the average wind speed is usually considerably higher over open water.

Spain, Denmark, and Germany are Europe's main wind energy producers. A large wind farm may consist of a few dozen to several hundred individual wind turbines, and cover an extended area of hundreds of square miles, but the land between the turbines may be used for agricultural or other purposes. A wind farm may be located offshore to take advantage of strong winds blowing over the surface of an ocean or lake. The United States is behind in developing sea-based wind farms for many reasons: economic and regulatory uncertainties, local opposition (not in my backyard), and even the relative bounty of cheaper land-based wind power resources have all conspired to slow any drive to develop wind power resources on the sea.

Figure 8.11 Wind farm in England

8.13.1 Location planning

A quantity called the wind power density (WPD) is used to select locations for wind energy development. The WPD is a calculation relating to the effective force of the wind at a particular location, frequently expressed in term of the elevation above ground level over a period of time. It takes into account velocity and mass. Colour coded maps are prepared for a particular area describing, for example, 'Mean Annual Power Density, at 50 m'. The results of the above calculation are used in an index developed by the National Renewable Energy Lab and referred to as 'NREL CLASS'. The larger the WPD calculation the higher it is rated by class.

Wind farm citing can be highly controversial, particularly when sites are picturesque or environmentally sensitive, such as having substantial bird life or requiring roads to be built through pristine areas. These areas are generally non-residential due to the noise concerns and setback requirements. Access to the power grid must be taken into mind. The further from the power grid, there will be need for more transmission lines to span from the farm directly to the power grid or transformers will have to be built on the premises depending upon the types of turbines being used.

Wind speed

As a general rule, wind generators are practical if wind speed is 10 mph (16 km/hr or 4.5 m/s) or greater. An ideal location would have a near constant flow of non-turbulent wind throughout the year with a minimum likelihood of sudden powerful bursts of wind. An important factor of turbine citing is also access to local demand or transmission capacity. Usually sites are preselected on basis of a wind atlas, and validated with wind measurements. Meteorological wind data alone is usually not sufficient for accurate citing of a large wind power project. Collection of site specific data for wind speed and direction is crucial to determining site potential. Local winds are often monitored for a year or more, and detailed wind maps constructed before wind generators are installed. To collect wind data a meteorological tower is installed with instruments at various heights along the tower. All towers include anemometers to determine the wind speed and wind vanes to determine the direction. The towers generally vary in height from 30 to 60 m. The towers primarily are guyed steel-pipe structures which are left to collect data for one to two years and then disassembled. Data is collected by a data logging device which stores and transmits data for analysis. Great attention must be paid to the exact positions of the turbines (a process known as micro-citing) because a difference of 30 m can nearly double energy production. For smaller installations where such data collection is too expensive or time consuming, the normal way of prospecting for wind-power sites is to directly look for trees or vegetation that are permanently 'cast' or deformed by the prevailing winds. Another way is to use a wind-speed survey map or historical data from a nearby meteorological station, although these methods are less reliable.

Altitude

The wind blows faster at higher altitudes because of the reduced influence of drag. The increase in velocity with altitude is most dramatic near the surface and is affected by topography, surface roughness, and upwind obstacles such as trees or buildings. Typically, the increase of wind speeds with increasing height follows a wind profile power law, which predicts that wind speed rises proportionally to the seventh root of altitude. Doubling the altitude of a turbine, then, increases the expected wind speeds by 10 per cent and the expected power by 34 per cent.

Wind park effect

The 'wind park effect' refers to the loss of output due to mutual interference between turbines. Wind farms have many turbines and each extracts some of the energy of the wind. Where land area is sufficient, turbines are spaced three to five rotor diameters apart perpendicular to the prevailing wind, and five to ten rotor diameters apart in the direction of the prevailing wind, to minimise efficiency loss. The loss can be as low as 2 per cent of the combined nameplate rating of the turbines.

In a large wind park, due to 'multifractal' effects between individual rotors, the behaviour deviates significantly from Kolmogorov's turbulence scaling for individual turbines.

Environmental and aesthetic impacts

Compared to the environmental effects of traditional energy sources, the environmental effects of wind power are relatively minor. Wind power consumes no fuel, and emits no air pollution, unlike fossil fuel power sources. The energy consumed to manufacture and transport the materials used to build a wind power plant is equal to the new energy produced by the plant within a few months of operation.

Danger to birds and bats has been a concern in many locations. Some dismiss the number of birds killed by wind turbines as negligible when compared to the number that die as a result of other human activities, and

especially the environmental impacts of using non-clean power sources.
Effect on power grid

Utility-scale wind farms must have access to transmission lines to transport energy. The wind farm developer may be obligated to install extra equipment or control systems in the wind farm to meet the technical standards set by the operator of a transmission line. The company or person that develops the wind farm can then sell the power on the grid through the transmission lines and ultimately chooses whether to hold on to the rights or sell the farm or parts of it to big business like GE, for example.

8.13.2 Types

Onshore

Onshore turbine installations in hilly or mountainous regions tend to be on ridgelines generally 3 km or more inland from the nearest shoreline. This is done to exploit the so-called topographic acceleration as the wind accelerates over a ridge. The additional wind speeds gained in this way make a significant difference to the amount of energy that is produced. Great attention must be paid to the exact positions of the turbines (a process known as micro-citing) because a difference of 30 m can sometimes mean a doubling in output.

Nearshore

Nearshore turbine installations are on land within 3 km of a shoreline or on water within 10 km of land. These areas are good sites for turbine installation, because of wind produced by convection due to differential heating of land and sea each day. Wind speeds in these zones share the characteristics of both onshore and offshore wind, depending on the prevailing wind direction.

Offshore

Offshore wind development zones are generally considered to be 10 km or more from land. Offshore wind turbines are less obtrusive than turbines on land, as their apparent size and noise is mitigated by distance. Because water has less surface roughness than land (especially deeper water), the average wind speed is usually considerably higher over open water. Capacity factors (utilisation rates) are considerably higher than for onshore and nearshore locations.

Transporting large wind turbine components (tower sections, nacelles, and blades) is much easier over water than on land, because ships and barges can handle large loads more easily than trucks/lorries or trains. On land, large goods vehicles must negotiate bends on roadways, which fixes the maximum length of a wind turbine blade that can move from point to point on the road network; no such limitation exists for transport on open water. Offshore wind turbines will probably continue to be the largest turbines in operation, since the high fixed costs of the installation are spread over more energy production, reducing the average cost.

Turbine components (rotor blades, tower sections) can be transported by barge, making large parts easier to transport offshore than on land, where turn clearances and underpass clearances of available roads limit the size of turbine components that can be moved by truck. Similarly, large construction cranes are difficult to move to remote wind farms on land, but crane vessels easily move over water. Offshore wind farms tend to be quite large, often involving over 100 turbines.

8.14 Environmental impact of wind power

Compared to the environmental impact of traditional energy sources, the environmental impact of wind power is relatively minor. Wind power consumes no fuel, and emits no air pollution, unlike fossil fuel power sources. The energy consumed to manufacture and transport the materials used to build a wind power plant is equal to the new energy produced by the plant within a few months. While a wind farm may cover a large area of land, many land uses such as agriculture are compatible, with only small areas of turbine foundations and infrastructure made unavailable for use.

There are reports of bird and bat mortality at wind turbines as there are around other artificial structures. The scale of the ecological impact may or may not be significant, depending on specific circumstances. Prevention and mitigation of wildlife fatalities, and protection of peat bogs, affect the citing and operation of wind turbines. There are anecdotal reports of negative effects from noise on people who live very close to wind turbines.

8.14.1 Carbon dioxide emissions and pollution

Wind power consumes no fuel and no water for continuing operation, and has no emissions directly related to electricity production. Wind turbines produce no carbon dioxide, sulphur dioxide, mercury, particulates or any other type of air pollution, unlike fossil fuel power sources. Wind power plants consume resources in manufacturing and construction. During manufacture of the wind turbine, steel, concrete, aluminium and other materials will have to be made and transported using energy-intensive processes, generally using fossil energy sources.

The wind turbine manufacturer Vestas states that initial carbon dioxide emissions 'pay back' is within about 9 months of operation for offshore turbines.

8.14.2 Ecology

Land use

Wind farms are often built on land that has already been impacted by land clearing. The vegetation clearing and ground disturbance required for wind farms is minimal compared with coal mines and coal-fired power stations. If wind farms are decommissioned, the landscape can be returned to its previous condition, although the concrete foundations will be left in place and access tracks may also be retained.

Impact on wildlife

Environmental assessments are routinely carried out for wind farm proposals, and potential impacts on the local environment (e.g. plants, animals, soils) are evaluated. Turbine locations and operations are often modified as part of the approval process to avoid or minimise impacts on threatened species and their habitats. Any unavoidable impacts can be offset with conservation improvements of similar ecosystems which are unaffected by the proposal.

Birds

A study estimates that wind farms are responsible for 0.3 to 0.4 fatalities per gigawatt-hour (GWh) of electricity while fossil-fuelled power stations are responsible for about 5.2 fatalities per GWh. The study therefore states that fossil fuel based electricity causes about 10 times more fatalities than wind farm based electricity, primarily due to habitat alteration from pollution and mountain-top removal for coal mining.

Bats

Bats may be injured by direct impact with turbine blades, towers or transmission lines. Recent research shows that bats may also be killed when suddenly passing through a low air pressure region surrounding the turbine blade tips. The numbers of bats killed by existing onshore and near-shore facilities has troubled bat enthusiasts. Migratory bat species appear to be particularly at risk, especially during key movement periods (spring and more importantly in fall).

Climate change

Wind farms may affect weather in their immediate vicinity. Spinning wind turbine rotors generate a lot of turbulence in their wakes like the wake of a boat. This turbulence increases vertical mixing of heat and water vapour that affects the meteorological conditions downwind.

8.14.3 Impact on people

Safety

Operation of any utility-scale energy conversion system presents safety hazards. Wind turbines do not consume fuel or produce pollution during normal operation, but still have hazards associated with their construction, operation and maintenance.

With the installation of industrial sized wind turbines numbering in the thousands, there have been at least 40 fatalities of workers due to the construction, operation, and maintenance of wind turbines, and other injuries and deaths attributed to the wind power life-cycle. Most worker deaths involve falls or becoming caught in machinery while performing maintenance inside turbine housings.

Aesthetics

Newer wind farms have larger, more widely spaced turbines, and have a less cluttered appearance than older installations. Wind farms are often built on land that has already been impacted by land clearing and they coexist easily with other land uses (e.g. grazing, crops). They have a smaller footprint than other forms of energy generation such as coal and gas plants. Wind farms may be close to scenic or otherwise undeveloped areas, and aesthetic issues are important for onshore and near-shore locations.

Aesthetic issues are subjective and some people find wind farms pleasant and optimistic or symbols of energy independence and local prosperity. While some tourism officials predict wind farms will damage tourism, some wind farms have themselves become tourist attractions, with several having visitor centres at ground level or even observation decks atop turbine towers.

Noise

Modern wind turbines produce significantly less noise than older designs. Turbine designers work to minimise noise, as noise reflects lost energy and output. Noise levels at nearby residences may be managed through the citing of turbines, the approvals process for wind farms, and operational management of the wind farm. Renewable UK, a wind energy trade organisation, has said that the noise measured 350 m from a wind farm is less than that from normal road traffic or in an office; some physicians and acoustic engineers have reported problems from wind turbine noise, including sleep deprivation, headaches, dizziness, anxiety and vertigo.

9

Status of wind energy in India

9.1 Introduction

Nearly two decades ago the Indian economy was snatched back from the brink of a composite economic crisis. The Indian government undertook some hard-hitting liberalisation measures that would have been unthinkable in a business as usual political landscape. Largely as a result of those actions, today India is in a position to be counted as one of the 'emerging economies'.

Successive governments have looked towards locking in an average economic growth rate of at least 6–8 per cent, up from 3.5 per cent from the 1950s through the 1980s. The original objective of the 11th Five-year plan (2007–2012) was to achieve a GDP growth rate of 9 per cent over this period. This was revised to 8.1 per cent last year by the planning commission. Given the plans for rapid economic growth, the requirement for energy services and supporting infrastructure is simultaneously escalating.

Electricity demand has continuously outstripped production, and a peak energy shortage of around 12.7 per cent prevailed in 2009–10. To meet this shortfall as well as the National Electricity Policy target of 'Electricity for All by 2012', the cleanest options available to India are Renewable Energy Technologies (RETs). For the government to seriously consider meeting its promise of electricity for all by 2012, renewable energy options including wind power will have to play a crucial role in India's emerging energy mix. Not only are they environmentally sound but also their project gestation periods are significantly shorter than those for thermal or nuclear power plants.

According to the Ministry of New and Renewable Energy (MNRE), today the share of renewable based capacity is 10.9 per cent (excluding large hydro) of the total installed capacity of 170 GW in the country, up from 2 per cent at the start of the 10th plan period (2002–2007). This includes 13,065.78 MW of wind, 2939 MW of small hydro power, 1562 MW of (bagasse based) cogeneration, 997 MW of biomass, 73.46 MW of 'waste to power' and 17.80 MW of solar PV for grid connected renewables at the end of 2010.

The originally stated cumulative target for the current plan period was to add 92 GW of new capacity of which about 14 GW was to come from renewable sources. Given the right mix of regulatory and institutional support, renewable sources could meet the proposed capacity addition of 14 GW from renewable energy before the end of the 11th five-year plan period (2007–2012). This would bring the total share of renewable energy sources up to 15 per cent of the new installed capacity in the 11th plan period.

Over the next decade, India will have to invest in options that not only provide energy security but also provide cost-effective tools for eradicating energy poverty across the board. India is a signatory to the United Nations Framework Convention on Climate Change (UNFCCC) and has as part of its obligations released a National Action Plan on Climate Change (released in June 2008) by Prime Minister Manmohan Singh which has laid out his government's vision for a sustainable and green future for India's economy. India's developmental needs will be challenged by climate change impacts. This requires a timely pre-emptive shift towards achieving an energy efficient and green economy. Over the next couple of decades renewable energy will play a major role in delivering that shift.

9.2 Wind power scenarios

There are several published scenarios that examine the future role of wind power globally as a part of the necessary energy system overhaul towards a clean energy future. The Global Wind Energy Council (GWEC) developed its scenarios in collaboration with Greenpeace international and the German Aerospace Centre (DLR). These scenarios are updated biennially. The resultant publication—the Global Wind Energy Outlook (GWEO)—first looks toward 2020, and then onwards to 2030 and 2050. Some of the other prominent scenarios are the World Energy Outlook (2010) from the International Energy Agency (IEA) and the Energy [R]evolution: A Sustainable World Energy Outlook by Greenpeace (2010).

There are many variables that will determine the path of development and growth of wind energy. Table 9.1 lists the assumptions underlying the GWEO scenarios and associated assumptions for wind power development.

9.3 GWEO scenario results

The GWEO scenarios show that even with the continuation of current policy measures to encourage wind power development and serious government efforts to meet existing targets, the resulting 'Moderate' scenario growth will put the development of wind power on a dramatically different trajectory from the IEA-based 'Reference' scenario. Table 9.2 lists the global cumulative wind power capacity.

The global wind markets have grown by an average 28 per cent per year in terms of total installed capacity during the last decade. The IEA's Reference scenario suggests that growth rates for wind power would decrease substantially in the coming years, and that 2013 would see an addition of only 26.8 GW. However, in reality the global wind industry added 35.8 GW during the year. Table 9.3 lists the cumulative wind power capacity of India.

Table 9.1 Global wind energy outlook: scenario and assumptions.

Reference scenario (IEA based)	Moderate scenario	Advanced scenario
The most conservative of all, the'reference' scenario is based on the projections in the 2009 World Energy extent to which this industry could outlook from the IEA. This takes into account only existing policies and measures, but includes assumptions such as continuing electricity and gas market reform, the liberalisation of cross-border energy trade and recent policies aimed at combating pollution. The IEA's figures only go out to the year 2030, but based on these assumptions, DLR has extrapolated both the overall reference scenario and the growth of wind power up to 2050.	The 'Moderate' scenario takes into account all policy and measures to already enacted or in planning stages around the world. It also assumes that the targets set by many countries for either renewable, emission reductions and/or wind energy are successfully implemented, as well as the modest implementation of new policies aimed at reducing pollution and carbon emissions. It also takes into account environmental and energy policy measures that were part of many governments economic stimulus packages implemented since late 2008. Up to 2014 the figures for installed capacity are closer to being forecasts than scenarios.	The most ambitious scenario, the 'advanced' version examines the support renewable energy either grow in a best case 'wind energy vision'. The assumption here is a clear and unambiguous commitment to renewable energy as per the industry's recommendations along with the political will necessary to carry it forward.

Table 9.2 Global cumulative wind power capacity (MW).

Scenario	2007	2008	2009	2010	2015	2020	2030
Reference	93,864	1,20,297	1,58,505	1,85,258	2,95,783	4,15,433	5,72,733
Moderate	93,864	1,20,297	1,58,505	1,98,717	4,60,364	8,32,251	17,77,550
Advanced	93,864	1,20,297	1,58,505	2,01,657	5,33,233	10,71,415	23,41,984

Table 9.3 India: cumulative wind power capacity (MW).

Year	Reference	Moderate	Advanced
2009	10,926	10,926	10,926
2010	12,276	12,629	12,833
2015	19,026	24,747	29,151
2020	24,026	46,104	65,181
2030	30,526	1,08,079	1,60,741

The Indian market grew by almost 68 per cent on a year-on-year basis with 2139 MW of new capacity installed between January and December 2010. This made India the third largest annual market after China and the USA for 2010. With more than 13 GW of total installed capacity at the end of 2010, India ranks fifth in the world in terms of cumulative installed capacity. The IEA projects that 327 GW of power generation capacity will be needed in India by 2020, which would imply a yearly addition of about 16 GW. This is reflected in the stated target for new capacity addition by the Indian government under its 11th Five-Year Plan. The plan envisages an addition of 78.7 GW by 2012 from traditional sources (coal, nuclear and large hydro) and an additional 9 GW by 2013 (revised from 10.5 GW) from new wind generation capacity.

During the first three years of the 11th Plan period ending March 2010, India added 4.6 GW of wind power capacity. With over a year to go before the current plan period is over it is very likely that Indian wind power installations will meet and exceed the 11th plan-period target, which will be a record of sorts as historically the targets have never been met through conventional thermal and hydro projects within a plan period.

Under the IEA's Reference scenario, India's wind power market is shown to shrink considerably to only about 600 MW per year by 2030. This translates into a total installed capacity of merely 24 GW by 2020 and 30.5 GW by 2030. Wind power would then produce close to 60 TWh every year by 2020 and 75 TWh by 2030, and save 35 MT of CO_2 in 2020 and 45 MT in 2030. Investment in wind power in India would drop to about US$910 million by 2030 (at 2010 $ value).

However under the GWEO scenarios, it is expected that by the end of 2015, between 24.7 GW and 29 GW will be installed in India. Under the moderate scenario this would reach almost 46 GW by 2020 and 108 GW by 2030. In this scenario, about US$9 billion would be invested in Indian wind power development every year by 2020, representing a quadrupling of the 2009 investment figures. Employment in the sector would grow from the currently estimated 28,000 jobs to over 84,000 by 2020 and 1,13,000 by 2030. The GWEO advanced scenarios show that wind power development in India could go much further depending upon adequate regulatory support and political will. By 2020, India could have 65 GW of wind power in operation, employing 1,70,000 people and saving 173 MT of CO_2 emissions each year. Investment by then would be to the tune of US$10.4 billion per year. The World Institute for Sustainable Energy (WISE) estimates deploying just the current generation of wind turbines could yield a potential onshore wind power capacity of 65–100 GW.

The Ministry of New and Renewable Energy (MNRE) has so far underplayed the potential of renewable energy (RE) sources in India. WISE did a revised estimate of the true potential of grid-connected RE in India as given in Table 9.4.

WISE sees its own numbers as a conservative estimation. With the present level of momentum established in India's wind sector, the ten years between 2020 and 2030 could see spectacular growth if some of the systemic barriers are addressed in a timely manner. With the political will geared towards fully exploiting the country's wind resource and reaping the accompanying economic, environmental and energy security benefits, the 'Advanced scenario' could be reached, which would see substantial wind power growth in many regions of the country. Wind power would then be instrumental in achieving a genuine energy revolution, putting India on the path to a sustainable energy future. India is now at a crossroads for making these decisions, which will determine the future of her energy system. As well as, to a great extent, the future of the planet (Table 9.5).

Table 9.4 Grid connected renewable energy potential in India.

Energy source	Capacity (MW)	Assumed plant load factor	Annual energy generation [billion kWh]
Wind (onshore)	1,00,000	25	219
Small hydro	15,000	45	46
Bagasse	5000	60	26.3
Biomass	16,881	60	88.72
Large hydro (existing and future)	1,00,000	60	525.6
Large hydro in Bhutan	16,000	60	84.1
Waste to energy	5000	60	26.28
Concentrated solar based power generation	2,00,000	35	613.2
Solar photovoltaic based power generation	2,00,000	20	350.4
Geothermal	10,000	80	70.1
Total	6,62,881		2049.70

Table 9.5 Comparative wind power development across few Indian states.

Particulars	AP	Gujarat	Karnataka	Kerala	MP	Maharashtra	Rajasthan	TN
Identified sites	32	40	26	17	7	39	8	45
Potential districts	7	9	9	3	5	13	5	11
Annual mean wind speed (m/sec) at 50 m mast height	4.86–6.61	4.33–6.97	5.19–8.37	4.41–8.12	5.0–6.25	4.31–6.58	4.02–5.73	4.47–7.32
Monitoring stations (Oct. 2010)	63	69	49	27	37	112	36	68
Operating stations (Dec. 2010)	2	6	5	–	4	22	1	1
Wind potential (MW)	5394	10,609	8591	790	920	5439	5005	5374
Inst. capacity (MW) (Dec. 2010)	176.8	2005.30	1576.20	28	230.8	2201.60	1353.40	5502.90
Untapped potential (MW) (Dec. 2010)	5217.20	8603.80	7014.90	762	689.2	3237.40	3671.70	128.9*

*Based on government estimated potential for wind power. Currently, at least another 3000 MW of projects are in the pipeline in Tamil Nadu.

9.4 Estimated wind power resource

The Centre for Wind Energy Technology (C-WET) published the Indian Wind Atlas in 2010, showing large areas with annual average wind power densities of more than 200 Watts/m^2 at 50 metres above ground level (MAGL). This is considered to be a benchmark criterion for establishing wind farms in India as per CWET and the MNRE. The potential sites have been classified according to annual mean wind power density ranging from 200 W/m^2 to 500 W/m^2. Most of the potential assessed sites have an annual mean wind power density in the range of 200–250 W/m^2 at 50 MAGL. The Wind Atlas has projected Indian wind power installable potential (name plate rating) as 49,130 MW at 2 per cent land availability. This is seen as a conservative estimate of wind power potential in India (Table 9.6).

Table 9.6 State-wise generation and installed capacity (up to 31st March 2010).

State	Cumulative generation (MU)	Cumulative installed capacity (MW)
Andhra Pradesh	1451	138.4
Gujarat	8016	1934.6
Karnataka	9991	1517.2
Madhya Pradesh	554	230.8
Maharashtra	11,790	2108.1
Rajasthan	3938	1095.6
Tamil Nadu	41,100	5073.1
Kerala	110	28
Total	76,950	12,125.8

With the improvement in technology and increase in the hub height of the wind turbine it has become possible to generate more electricity than assumed in earlier estimates. Based on the resource assessment carried out by C-WET, wind speeds in India are in the low to moderate range except in few pockets like coastal southern Tamil Nadu and the Rann of Katch (Gujarat). Further India's as yet unassessed offshore wind potential was not included in the C-WET study.

9.5 Offshore wind power development

A long coastline and relatively low construction costs could make India a favoured destination for offshore wind power. Offshore wind development is a relatively new phenomenon, and Europe is the only sizeable market at present, with a total offshore capacity of 3 GW. The global offshore wind turbine segment has been dominated by two established players, Vestas and Siemens. However, there are other manufacturers active in the market such as REPower, Sinovel, Areva and Bard, with strong interest from GE Wind, Gamesa, XEMC and WinWinD (Table 9.7).

Special construction requirements make offshore wind power 1.5–2.5 times more expensive than onshore, making large-scale offshore deployment difficult in developing regions. The current average rated capacity of offshore wind turbines is 2.5 MW, as compared to average onshore wind turbine capacity of 1.06 MW (BTM ApS, 2010). It should be noted that most of the 4–6 MW turbines currently in the testing or early deployment stage are designed for offshore operation. If the government supported small capacity offshore demonstration projects, it could build confidence and bring in public and private investment in this sector in the years to come.

To examine the feasibility of offshore wind farms, C-WET conducted the first phase of its study at Dhanushkodi in Tamil Nadu. So far, the area around Dhanushkodi has shown good potential, where wind power density of 350–500 W/m^2 has been recorded. For the next stage, C-WET is currently awaiting approval from various government agencies.

Based on a study carried out by WISE on the clearances required for offshore projects, it is understood that more than 20 central and state ministries and departments would need to be involved in the process. As this technology is in its nascent stage in India, there is a need for specific policy framework for offshore wind power generation.

On the corporate side, there have been a few early moves on offshore wind in India. Oil and Natural Gas Corporation (ONGC) announced its plans to tap offshore wind power. Further, in June 2010, global majors like Areva, Siemens and GE announced their plans to explore offshore wind power opportunities in the country. Tata Power is the first private sector player to submit a formal request to the Government of Gujarat and Gujarat Maritime Board for approval of an offshore project in India.

Table 9.7 Challenges for repowering.

Challenge	Issues
Turbine ownership	Repowering will reduce the number of turbines, and there may not be one-to-one replacement. Thus, the issue of ownership needs to be handled carefully.
Land ownership	Multiple owners of wind farm land may create complications for repowering projects.
Power purchase agreement	PPAs were signed with the state utility for 10, 13 or 20 years and the respective electricity board may not be interested in discontinuing or revising the PPA before its stipulated time.
Electricity evacuation facilities	The current grid facilities are designed to support present generation capacities and may require augmentation and upgrading.
Additional costs	The additional decommissioning costs for old turbines (such as transport charges) need to be assessed.
Disposal of old turbines	There are various options such as scrapping, buy-back by the government or manufacturer or export. Local capacity may need to be developed.
Incentives	One of the primary barriers to repowering is the general lack of economic incentive to replace the older WTGs. In order to compensate for the additional cost of repowering, appropriate incentives are necessary.
Policy package	A new policy package should be developed which would cover additional project cost and add-on tariff by the State Electricity Regulatory Commissions (SERCs) and include a repowering incentive (on the lines of the recently introduced generation-based incentive scheme by MNRE).

9.6 Wind turbine installations

Wind turbine generator (WTG) capacity addition in India has taken place at a CAGR of 24.67 per cent for the period of 1992–2010. The installed capacity increased from a modest base of 41.3 MW in 1992 to reach 13,065.78 MW by December 2010. The official installation figures show that amongst the states, Tamil Nadu ranks the highest both in terms of installed capacity and in terms of energy generation from wind, with shares of 41.8 and 53.4 per cent respectively. Other states like Gujarat, Maharashtra and Rajasthan have seen significant growth in wind capacity over the last 4–5 years, also due to a stable policy and regulatory regime (Table 9.8).

Table 9.8 Average size of WTG (kW) installed each year.

Country	2004	2005	2006	2007	2008	2009
China	771	897	931	1079	1220	1360
Denmark	2225	1381	1875	850	2277	2368
Germany	1715	1634	1848	1879	1916	1977
India	767	780	926	986	999	1117
Spain	1123	1105	1469	1648	1837	1897
Sweden	1336	1126	1138	1670	1738	1974
UK	1695	2172	1953	2049	2256	2251
USA	1309	1466	1667	1669	1677	1731

9.7 Repowering potential

Repowering is the process of replacing older, smaller wind turbines with modern and more powerful machines, which would reap considerably more power from the same site. In India, about 46 per cent of the WTGs were rated below 500 kW in 2010, adding up to 2331.3 MW (about 18 per cent of cumulative installed capacity). Figure 9.1 shows state-wise repowering potential as of March 2009. Amongst the states with good wind potential Tamil Nadu leads with a repowering potential of more than 800 MW followed by Gujarat, Maharashtra, Andhra Pradesh and Karnataka. A special drive for repowering of old wind farms undertaken by the central government would encourage the industry to take this up on a larger scale. This could be done by way of creating suitable mechanisms and offering support along with financial incentives, to make new repowering projects viable. Currently, neither the states nor the central government provides dedicated policy support or incentives to encourage Indian wind power developers or investors to repower their old projects. However, there are some challenges to be addressed before a comprehensive repowering attempt in India. All these issues related to repowering can no doubt be resolved by learning from the experiences in other markets such as Denmark and Germany, although they are still at the early stages of their own repowering. These markets have introduced various incentive mechanisms and policies to encourage repowering, and done away with provisions that initially hampered repowering. If a sensible policy package is developed, many old sites can provide two to three times their current electricity generation after repowering.

9.8 Technology development trends

Modern wind power technology has come a long way in the last two decades, and both globally and in India, improved technology has slowly and steadily improved capacity utilisation. A key trend in the Indian industry is the development of multi-megawatt turbines installed at greater hub heights. Larger diameter rotors mean that a single wind power generator can capture more energy or more 'power per tower'. This allows WTGs to take advantage of higher altitudes with stronger winds and less turbulence (wind speed generally increases with height above the ground). Subsequently, larger machines have resulted in a steady increase in the capacity factor on average from 10–12 per cent in 1998 to 20–22 per cent in 2010.

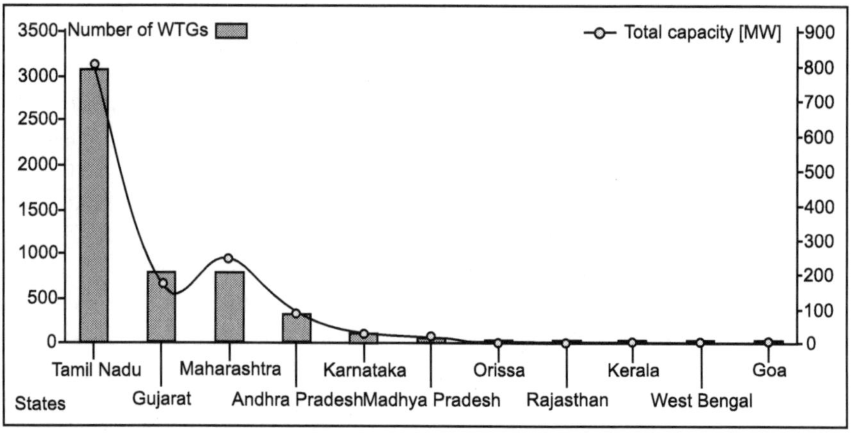

Figure 9.1 State-level repowering potential in India

For two decades now, global average WTG power ratings have grown almost linearly, with current commercial machines rated on average in the range of 1.5–2.1 MW. The average size of WTGs installed in India has gradually increased from 767 kW in 2004 to 1117 kW in 2009. The shift in India to larger WTGs is a result of improved infrastructure available to handle bigger turbines and improved economics of the sector. As generator size increases, fixed overall project costs fall on a 'per unit of output' basis. Given that finding sites and establishing transmission corridors is a significant investment, developers need to maximise the use of available sites for wind power generation. Installing fewer high capacity turbines, versus installing a greater number of smaller turbines, reduces overall capital investment by lowering installation, maintenance and potentially real estate costs. For example, instead of siting ten 600 kW turbines on acres of land, developers can instead site only three 2.0 MW WTGs.

9.9 Investment in wind power sector

The Government of India has outlined ambitious capacity expansion and investment plans for the current plan period (2007–2012) and wind power projects form the majority of the proposed capacity addition. The total investments on development of RE during the plan period is expected to be in excess of US$15 billion (~Rs. 60,000 crores). The majority of this investment is being raised through domestic private investors, concessional financing from specialised government agencies and multilateral financial institutions.

Due to growing awareness of the benefits of wind power and evolving government priorities more banks and lending institutions are showing interest in funding these projects. On top of the financing spectrum is IREDA, the Indian Renewable Energy Development Agency, the apex nodal agency for renewable energy development in India and a funding arm of MNRE. The other government agencies that actively fund renewable energy projects are the Power Finance Corporation (PFC) and Rural Electrification Corporation (REC). Figure 9.2 shows annual sanctioned and actual installed capacity of SWTs and wind-solar hybrid systems.

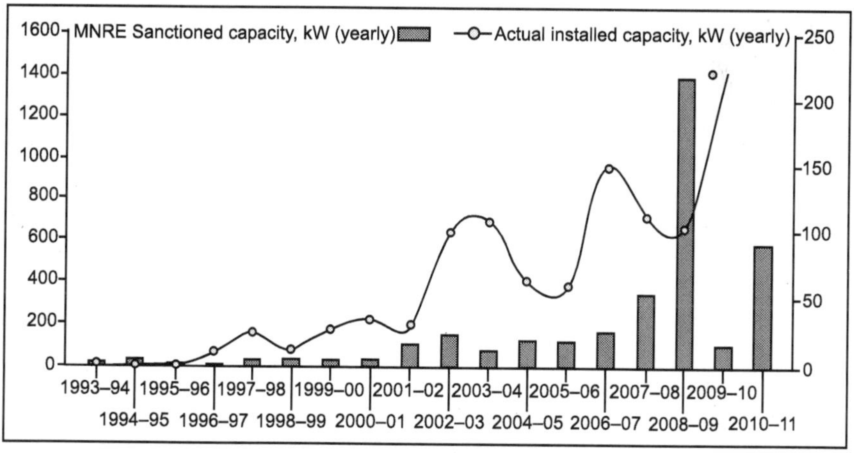

Figure 9.2 Annual sanctioned and actual installed capacity of SWTs and wind-solar hybrid systems

The multilateral agencies such as the World Bank, the International Finance Corporation (IFC), and the Asian Development Bank (ADB), as well as bilateral agencies such as KFW (German Development Bank) have also stepped up their assistance to the sector in the last few years. Prominent domestic banks that fund renewable projects are IDBI, ICICI, IFCI, SBI and PNB among others. Foreign banks such as Standard Chartered, RBS India (formerly known as ABN Amro) and Rabobank are also providing renewable energy project financing. Currently, the market in India for the RE business is growing at an annual rate of 15 per cent. The scope for private investment in RE is estimated at about US$3 billion per annum. Given the evolving regulatory and policy regime, the business outlook is generally positive at this time. Proposed policy guidance and regulations are also coming into place to further strengthen this rate of growth.

9.10 Small wind and hybrid systems

The global market for small wind turbines (SWTs) has been on the upswing over the last two to three years. This is driven by rapidly growing energy demand, higher fossil fuel prices and improved SWT technology, which can be deployed for a diverse pool of applications, both in 'grid-tied' and 'standalone' modes. With the increasing shortfall in power supply and energy across the country, India could benefit significantly from exploiting the potential of micro-generation technologies that can meet energy needs under the distributed generation mode, so as to provide long-term solutions. WISE estimates India's micro-generation potential at about 83 GW. However, costs are a major hurdle and policy support needs to be oriented towards promoting mass manufacturing and early adoption of these microgeneration options. Although a small annual market for such systems (~150–200 kW) currently exists in India, it is largely driven by the capital subsidy program of the MNRE. Most of the current installations are of the standalone type. Figure 9.3 shows state-wise distribution of wind-solar hybrid systems.

The 'small wind energy and hybrid systems' program initiated in 1994 by the MNRE focused solely on small wind energy and hybrid systems. The objective of the program is to develop technology and promote applications of water pumping windmills and aero-generators/wind-solar hybrid systems. Although the program helped to promote awareness of small wind systems in India, it created interest only among select users and has yet to make a real impact. The implementation of the program was extended in April 2010 to the fiscal year 2011–12. The physical annual target was set to installed 500 kW aero-generator/wind-solar hybrid systems and 25 water pumping windmills with estimated financial budget of Rs. 50 million over 2010–2012.

The program is implemented through State Nodal Agencies (SNA) mainly in Andhra Pradesh, Assam, Bihar, Gujarat, Karnataka, Kerala, Maharashtra, Rajasthan, Sikkim, Tamil Nadu, Goa, and West Bengal and the Andaman and Nicobar Islands. Manufacturers of water-pumping windmills, aerogenerators, and wind-solar hybrid systems are also eligible to market the systems directly to users. The program is being extended to other potential states.

An aggregate capacity of 1.07 MW of aerogenerators or hybrid systems was installed under the program up to December 2010. Interestingly, almost 57 per cent of the total cumulative installations in the country are in Maharashtra followed by Goa, Karnataka, West Bengal, Manipur and Tamil Nadu. Almost all the projects sanctioned by the MNRE and those actually commissioned availed themselves of capital subsidy benefits from the Ministry.

9.11 Barriers to higher growth

The low utilisation of the country's wind power potential so far is attributable to several factors, including lack of an appropriate regulatory framework to facilitate purchase of renewable energy from outside the host state, inadequate grid connectivity; high wheeling and open access charges in some states, delays in acquiring land and obtaining statutory clearances.

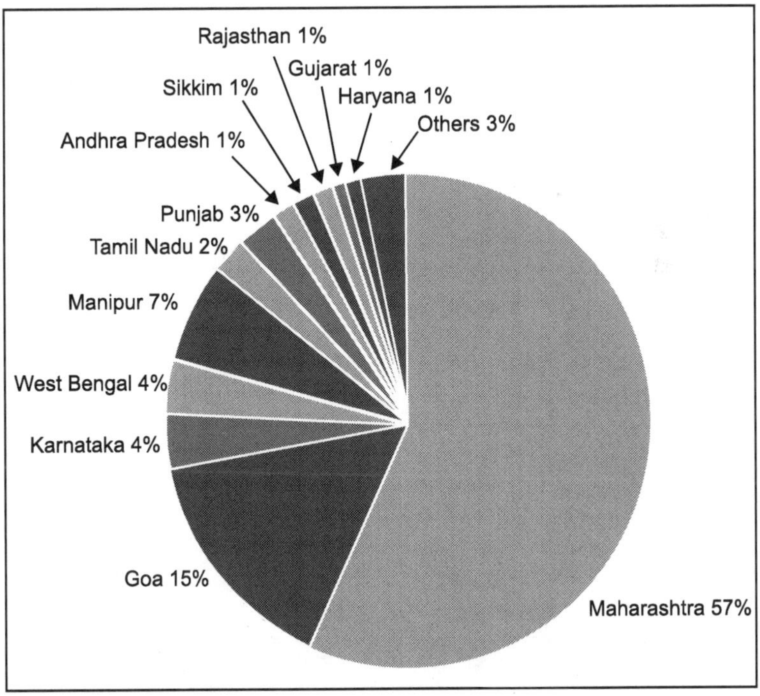

Figure 9.3 State-wise distribution of wind–solar hybrid systems

In 2010, India installed a record 2.1 GW of new wind power capacity. For this growth to be maintained it is essential that the industry is supported by a predictable policy and regulatory environment. Proposed amendments to India's tax laws (such as the Direct Tax Code; Goods and Services Tax) will have an impact on the investment portfolio of wind power.

Besides these there are other potential barriers to achieving higher growth rates in the short-to-medium term. The main reason for the growth of wind power has been the availability of accelerated depreciation (AD), providing the facility to offset taxes on income from other sources. With the possible introduction of the DTC from the next fiscal year (2012–13), the quantum of this benefit could be affected, which could have an impact on the investments in the Indian wind sector.

The Generation Based Incentive (GBI) scheme has not attracted as many Independent power producers as envisaged, since the investors are of the opinion that the current rate of Re. 0.5/kWh is not adequate or in line with the fiscal benefits offered under the AD scheme, and the two are mutually exclusive. Further, the multitude of regulatory agencies adds to the confusion—there is the Central Electricity Regulatory Commission (CERC) and each state also has a State Electricity Regulatory Commission (SERC). The CERC issues guidelines for determining the feed-in-tariff from RE sources and these are applicable to central government power generating stations and those who transmit power in the interstate corridor. However, this is applicable to a very small number of power producers and the vast majority is still covered by the tariff determined by the SERCs. This duality is not useful. For example, an SERC could determine the tariff which may or may not be equivalent to the tariff determined by the CERC. This has a major impact on the project developers.

Inadequate grid infrastructure is another key issue that needs to be addressed urgently. Across most of those states with significant wind potential, the grid does not have sufficient spare capacity to be able to evacuate ever increasing amount of wind power. As a result, the state distribution utilities are reluctant to accept more power and on a merit order basis prefer thermal power. Thus, there is an urgent need to augment the grid capacity and the regional Southern Grid needs to be connected with the rest of the country on a realtime basis. This requires better forecasting of power demand across the nation, and a modernisation of the grid.

In most of the states availability of land for wind farms is a contentious issue. Even if private lands are available, conversion of land use status from agricultural to non-agricultural is a time consuming process. Further, if the land is close to a protected area or forest lands then obtaining clearance from forest authorities for using the forest land for wind power generation is also time consuming.

Current and projected growth rates for wind power development in India are putting increasing strain on the WTG manufacturing sector, and the component supply chain needs to be improved. It would be beneficial for the small and medium enterprises (SMEs) to have access to concessional financing to bear the risks related to production capacity augmentation. As the industry grows, there will be demand for trained manpower and accordingly, the academic curriculum may need to be modified.

10

Hydroelectric power

10.1 Introduction

It is a form of energy—a renewable resource. Hydropower provides about 96 per cent of the renewable energy in the United States. Other renewable resources include geothermal, wave power, tidal power, wind power, and solar power. Hydroelectric power plants do not use up resources to create electricity nor do they pollute the air, land or water, as other power plants may. Hydroelectric power has played an important part in the development of world's electric power industry. Both small and large hydroelectric power developments were instrumental in the early expansion of the electric power industry.

Hydroelectric power comes from flowing water—winter and spring runoff from mountain streams and clear lakes. Water, when it is falling by the force of gravity, can be used to turn turbines and generators that produce electricity.

Hydroelectric power is important to our Nation. Growing populations and modern technologies require vast amounts of electricity for creating, building, and expanding. In the 1920s, hydroelectric plants supplied as much as 40 per cent of the electric energy produced. Although the amount of energy produced by this means has steadily increased, the amount produced by other types of power plants has increased at a faster rate and hydroelectric power presently supplies about 10 per cent of the electrical generating capacity of the United States. Hydropower is an essential contributor in the national power grid because of its ability to respond quickly to rapidly varying loads or system disturbances, which base load plants with steam systems powered by combustion or nuclear processes cannot accommodate.

Reclamation's 58 power plants throughout the Western United States produce an average of 42 billion kWh (kilowatt-hours) per year, enough to meet the residential needs of more than 14 million people. This is the electrical energy equivalent of about 72 million barrels of oil. Hydroelectric power plants are the most efficient means of producing electric energy. The efficiency of today's hydroelectric plant is about 90 per cent. Hydroelectric plants do not create air pollution, the fuel—falling water—is not consumed, projects have long lives relative to other forms of energy generation, and hydroelectric generators respond quickly to changing system conditions. These favourable characteristics continue to make hydroelectric projects attractive sources of electric power.

10.2 Hydropower

The power of water was exploited in the ancient world for irrigation, grinding corn, metal forging, and mining. Waterwheels were common in Western Europe by the end of the first millennium; over 5000 waterwheels were recorded in the Domesday book of 1086 shortly after the Norman conquest of England. The early waterwheels were of the undershot design (Fig. 10.1a) and very inefficient. The development of overshot waterwheels (Fig. 10.1b), and improvements in the shape of the blades to capture more of the incident kinetic energy of the stream, led to higher efficiencies.

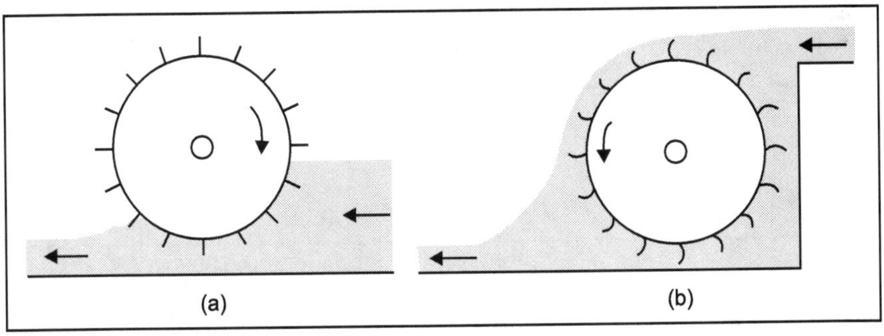

Figure 10.1 (a) Undershot, and (b) overshot waterwheels

A breakthrough occurred in 1832 with the invention of the Fourneyron turbine, a fully submerged vertical axis device that achieved efficiencies of over 80 per cent. Fourneyron's novel idea was to employ fixed guide vanes that directed water outwards through the gaps between moving runner blades as shown in Fig. 10.2. Many designs of water turbines incorporating fixed guide vanes and runners have been developed since. Modern water turbines are typically over 90 per cent efficient.

The main economic advantages of hydropower are low operating costs, minimal impact on the atmosphere, quick response to sudden changes in electricity demand, and long plant life—typically 40 years or more before major refurbishment. However, the capital cost of construction of dams is high and the payback period is very long. There are also serious social and environmental issues to be considered when deciding about a new hydroelectric scheme, including the displacement of population, sedimentation, changes in water quality, impact on fish, and flooding. Mountainous countries like Norway and Iceland are virtually self-sufficient in hydropower but, in countries where the resource is less abundant, hydropower is mainly used to satisfy peak-load demand. The hydroelectric capacity by country and the largest sites are shown in Table 10.1.

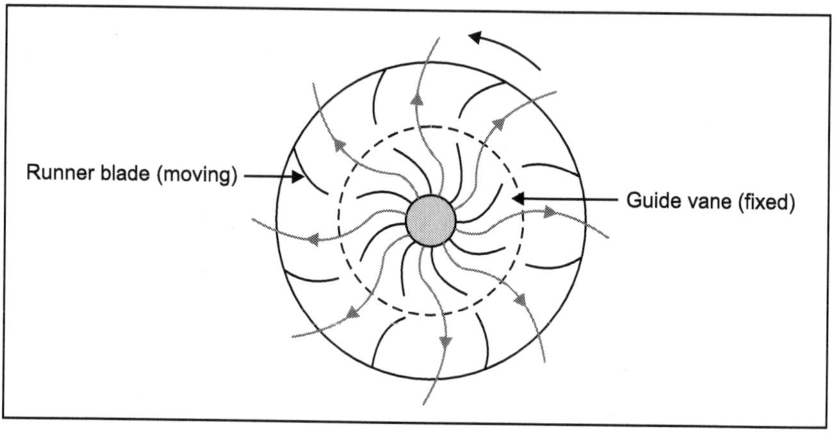

Figure 10.2 Fourneyron water turbine

Table 10.1 Installed hydropower.

Country		Hydroelectric capacity in 2005 (GW)
USA		120
Canada		98
China		95
Brazil		85
Norway		54
Japan		52
Largest sites for hydropower		
Country	Site	Hydroelectric capacity (GW)
China	Three Gorges	18.2
Brazil/Paraguay	Itaipu	12.6
Venezuela	Guri	10.3
USA	Grand Coulee	6.9
Russia	Sayano–Shushenk	6.4
Russia	Krasnoyarsk	6

10.3 Power output from a dam

Consider a turbine situated at a vertical distance h (called the head) below the surface of the water in a reservoir (Fig. 10.3). The power output P is the product of the efficiency η, the potential energy per unit volume ρgh, and the volume of water flowing per second Q, i.e.

$$P = \eta\rho g h Q \qquad \text{... (10.1)}$$

Note that the power output depends on the product hQ. Thus a high dam with a large h and a small Q can have the same power output as a run-of-river installation with a small h and large Q. The choice of which design of water turbine is suitable for a particular location depends on the relative magnitudes of h and Q.

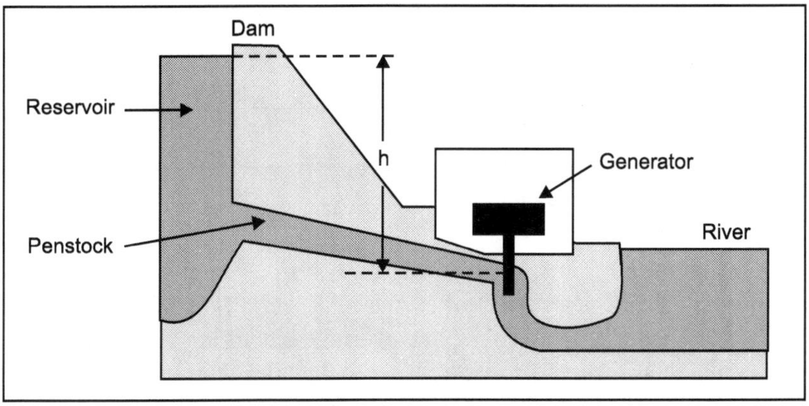

Figure 10.3 Hydroelectric plant

10.4 Measurement of volume flow rate using a weir

For power extraction from a stream it is important to be able to measure the volume flow rate of water. One particular method diverts the stream through a straight-sided channel containing an artificial barrier called a weir (Fig. 10.4). The presence of the weir forces the level of the fluid upstream of the weir to rise. The volume flow rate per unit width is related to the height of the undisturbed level of water gmin above the top of the weir by the formula:

$$Q = g^{1/2}(2/3\ \gamma_{min})^{3/2} \qquad \text{... (10.2)}$$

10.5 Water turbines

When water flows through a waterwheel, the water between the blades is almost stationary. Hence the force exerted on a blade is essentially due to the difference in pressure across the blade. In a water turbine, however, the water is fast moving and the turbine extracts kinetic energy from the water. There are two basic designs of water turbines: impulse turbines and reaction turbines.

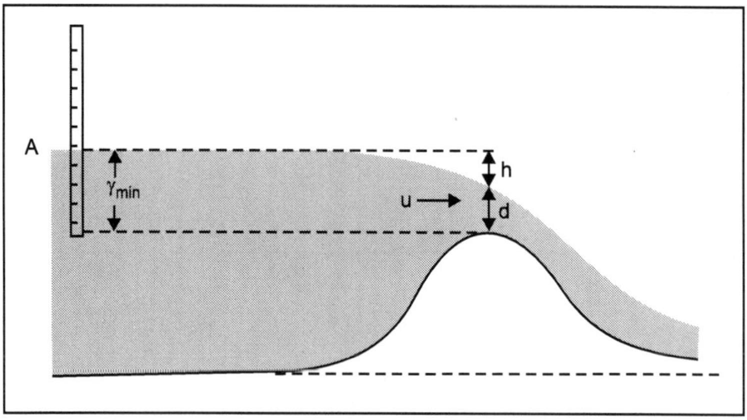

Figure 10.4 Flow over broad-crested weir

In an impulse turbine, the blades are fixed to a rotating wheel and each blade rotates in air, apart from when the blade is in line with a high speed jet of water. In a reaction turbine, however, the blades are fully immersed in water and the thrust on the moving blades is due to a combination of reaction and impulse forces. An impulse turbine called a Pelton wheel is shown in Fig. 10.5. In this example there are two symmetrical jets, and each jet imparts an impulse to the blade equal to the rate of change of momentum of the jet. The speed of the jet is controlled by varying the area of the nozzle using a spear valve. Thomas Pelton went to seek his fortune in the Californian Gold Rush during the nineteenth century. By the time he arrived on the scene the easy pickings had already been taken and the remaining gold had to be extracted from rocks that needed to be crushed.

Impulse turbines were being used to drive the mills to grind the rocks into small lumps. Pelton observed the motion of the turbine blades and deduced that not all the momentum of the jets was being utilised. He realised that some momentum was being lost because the water splashed in all directions on striking the blades.

He redesigned the cups so that the direction of the splash was opposite to that of the incident jet. This produced a marked improvement in efficiency and Pelton thereby made his fortune.

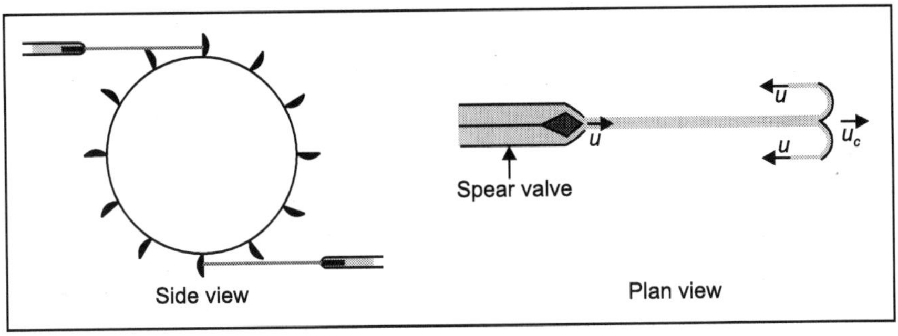

Figure 10.5 Impulse turbine (Pelton wheel)

To calculate the maximum power output from a Pelton wheel, we consider a jet moving with velocity u and the cup moving with velocity uc. Relative to the cup, the velocity of the incident jet is $(u - u_c)$ and the velocity of the reflected jet is $-(u - u_c)$. Hence the total change in the velocity of the jet is $-2(u - u_c)$. The mass of water striking the cup per second is ρQ, so the force on the cup is given by:

$$F = 2\rho Q(u - u_c) \qquad \text{... (10.3)}$$

The power output P of the turbine is the rate at which the force F does work on the cup in the direction of motion of the cup, i.e.

$$P = Fu_c = 2\rho Q\,(u - u_c)u_c \qquad \text{... (10.4)}$$

To derive the maximum power output we put $dP/du_c = 0$, yielding $u_c = ½u$.

Substituting in Eq. 10.4 then yields the maximum power as:

$$P_{max} = ½\rho Qu^2 \qquad \text{... (10.5)}$$

Thus the maximum power output is equal to the kinetic energy incident per second.

As in the Fourneyron turbine, modern reaction turbines use fixed guide vanes to direct water into the channels between the blades of a runner mounted on a rotating wheel (Fig. 10.6). However, the direction of radial flow is inward (In the Fourneyron turbine, the outward flow caused problems when the flow rate was either increased or decreased).

The most common designs of reaction turbines are the Francis turbine and the Kaplan turbine. In a Francis turbine the runner is a spiral annulus, whereas in the Kaplan turbine it is propeller-shaped. In both designs the kinetic energy of the water leaving the runner is small compared with the incident kinetic energy.

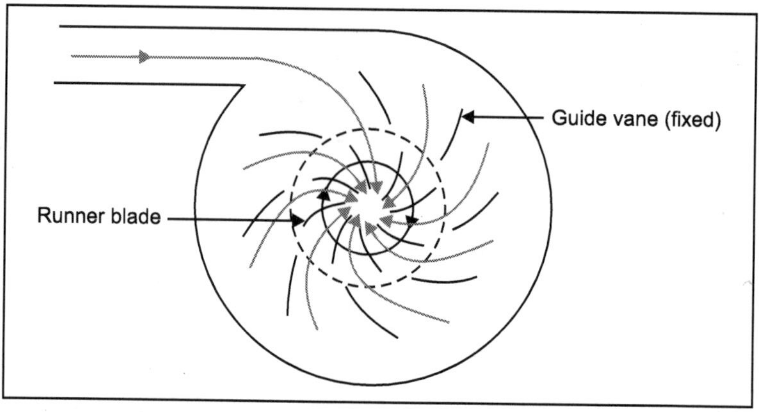

Figure 10.6 Reaction turbine (plan view)

The term 'reaction turbine' is somewhat misleading in that it does not completely describe the nature of the thrust on the runner. The magnitude of the reaction can be quantified by applying Bernoulli's equation to the water entering and leaving the runner, i.e.

$$\frac{p_1}{\rho} + \frac{1}{2}q_1^2 = \frac{p_2}{\rho} + \frac{1}{2}q_2^2 + E \qquad \text{... (10.6)}$$

where, E is the energy per unit mass of water transferred to the runner. Consider two cases: (i) $q_1 = q_2$, and (ii) $p_1 = p_2$. In case (i), Eq. 10.6 reduces to:

$$E = \frac{p_1 - p_2}{\rho} \qquad \text{... (10.7)}$$

i.e. the energy transferred arises from the difference in pressure between inlet and outlet. In case (ii), E is given by:

$$E = \frac{1}{2}(q_1^2 - q_2^2) \qquad \text{... (10.8)}$$

i.e. the energy transferred is equal to the difference in the kinetic energy between inlet and outlet. In general we define the degree of reaction R as:

$$R = \frac{p_1 - p_2}{\rho E} = 1 - \frac{(q_1^2 - q_2^2)}{2E} \qquad \text{... (10.9)}$$

The velocity diagrams in the laboratory frame of reference for an impulse turbine and a reaction turbine are shown in Fig. 10.7(a) and (b), respectively. The symbols u, q and w denote the velocity of the runner blade, the absolute velocity of the fluid, and the velocity of the fluid relative to the blade. Figure 10.7 shows the velocity triangles on the outer radius of the runner $r = r_1$ and the inner radius $r = r_2$. The runner rotates with angular velocity ω, so that the velocity of the blade is $u_1 = r_1\omega$ on the outer radius and $u_2 = r_2\omega$ on the inner radius. The torque on the runner is:

$$T = \rho Q(r_1 q_1 \cos \beta_1 - r_2 u_2 \cos \beta_2).$$

Putting $r_1 = u_1/\omega$ and $r_2 = u_2/\omega$, the work done per second is given by:

$$P = T\omega = \rho Q(u_1 q_1 \cos \beta_1 - u_2 q_2 \cos \beta_2).$$

The term in brackets represents the energy per unit mass:

$$E = u_1 q_1 \cos \beta_1 - u_2 q_2 \cos \beta_2. \quad \text{... (10.10)}$$

Equating the incident power due to the head of water h from Eq. 10.1 to the power output of the turbine, given by Euler's turbine equation, we have

$$\eta\rho g h Q = \rho Q(u_1 q_1 \cos \beta_1 - u_2 q_2 \cos \beta_2).$$

The term $\rho Q u_2 q_2 \cos \beta_2$ represents the rate at which kinetic energy is removed by the water leaving the runner. We define the hydraulic efficiency as:

$$\eta = \frac{u_1 q_1 \cos \beta_1 - u_2 q_2 \cos \beta_2}{gh} \quad \text{... (10.11)}$$

The maximum efficiency is achieved when the fluid leaves the runner at right angles to the direction of motion of the blades, i.e. when $\beta_2 = \pi/2$ so that $\cos \beta_2 = 0$. Equation 10.11 then reduces to

$$\eta_{max} = \frac{u_1 q_1 \cos \beta_1}{gh} \quad \text{... (10.12)}$$

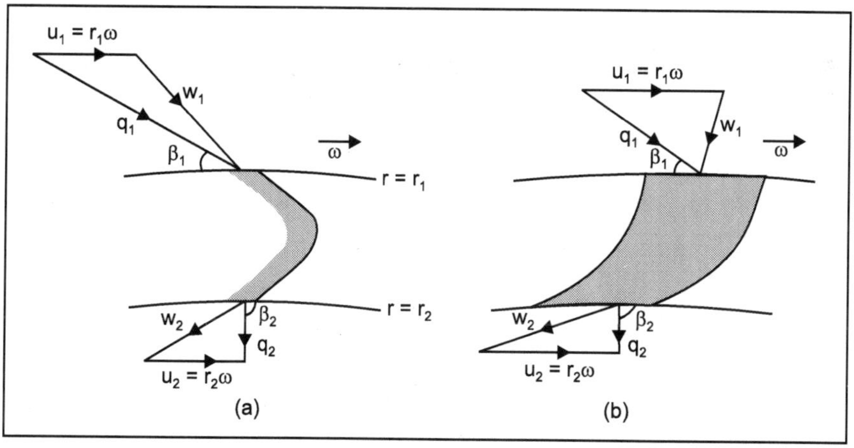

Figure 10.7 Velocity diagrams for: (a) an impulse turbine, and (b) a reaction turbine

10.5.1 Choice of water turbine

The choice of water turbine depends on the site conditions, notably on the head of water h and the water flow rate Q. Figure 10.8 indicates which turbine is most suitable for any particular combination of head and flow rate. Impulse turbines are suited for large h and a low Q, e.g. fast moving mountain streams. Kaplan turbines are suited for low h and large Q (e.g. run-of-river sites) and Francis turbines are usually preferred for large Q and large h, e.g. dams. A useful parameter for choosing the most suitable turbine is the shape (or type) number S.

10.6 Impact, economics and prospects of hydropower

Hydropower sites tend to have a large impact on the local population. Over 1.1 million people were displaced by the Three Gorges dam in China, and it has been estimated that 30–60 million people worldwide have had to be relocated due to hydropower. Proposed hydropower plants often provoke controversy and in some countries public opposition to hydropower has stopped all construction except on small-scale projects. Also, dams sometimes collapse for various reasons, e.g. overspilling of water, inadequate spillways, foundation defects, settlement, slope instability, cracks, erosion, and freak waves from landslides in steep-sided valleys around the reservoir. As with nuclear plants, the risk of major accidents is small but the consequences can be catastrophic. Given the long lifetime of dams, even a typical failure rate as low as one per 6000 dam years means that any given dam has a probability of about 1 per cent that it will collapse at sometime in its life. In order to reduce the environmental impact and the consequences of dam failure, the question arises as to whether it is better to build a small number of large reservoirs or a large number of small ones. Though small reservoirs tend to be more acceptable to the public than large ones, they need a much larger total reservoir area than a single large reservoir providing the same volume of stored water.

An argument in favour of hydropower is that it does not produce greenhouse gases or acid rain gases. However, water quality may be affected both upstream and downstream of a dam due to increases in the concentrations of dissolved gases and heavy metals. These effects can be mitigated by inducing mixing at different levels and oxygenating the water by auto-venting turbines. The installation of a hydropower installation can also have a major impact on fish due to changes in the habitat, water temperature, flow regime, and the loss of marine life around the turbines.

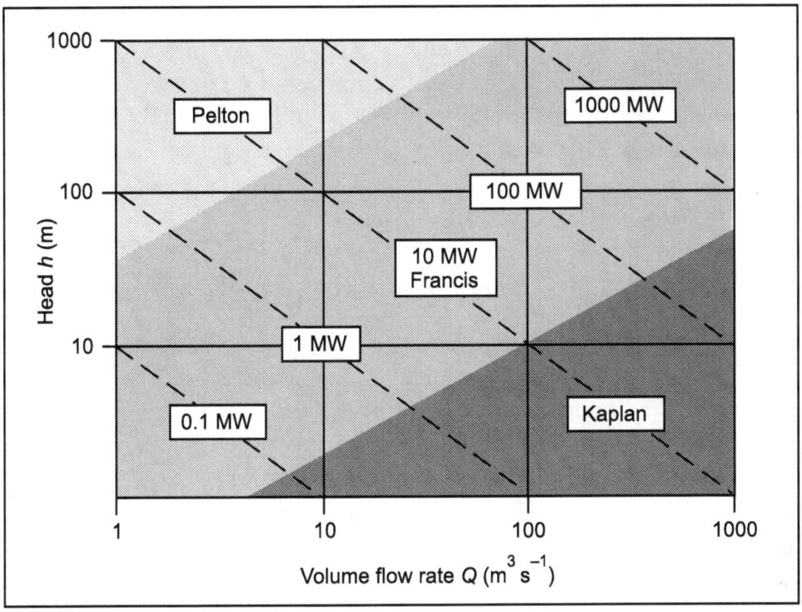

Figure 10.8 Choice of turbine in terms of head h and volume flow rate Q

The capital cost of construction of hydropower plants is typically much larger than that for fossil fuel plants. Another cost arises at the end of the effective life of a dam, when it needs to be decommissioned. The issue as to who should pay for the cost involved in decommissioning is similar to that for nuclear plants: the plant owners, the electricity consumers or the general public? On the positive side, production costs for hydropower are low because the resource (rainfall) is free. Also, operation and maintenance costs are minimal and lifetimes are long: typically 40–100 years. The efficiency of a hydroelectric plant tends to decrease with age due to the build-up of sedimentation trapped in the reservoir. This can be a life-limiting factor because the cost of flushing and dredging is usually prohibitive.

The economic case for any hydropower scheme depends critically on how future costs are discounted. Discounting reduces the benefit of long-term income, disadvantaging hydropower compared with quick payback schemes such as CCGT generation. Hydropower schemes therefore tend to be funded by governmental bodies seeking to improve the long-term economic infrastructure of a region rather than by private capital. Despite the strong upward trend in global energy demand, the prospects for hydropower are patchy. In the developed world the competitive power market has tilted the balance away from capitally intensive projects towards plants with rapid payback of capital. As long as relatively cheap fossil fuels are available, the growth of hydropower is likely to be limited to parts of the world where water is abundant and labour costs for construction are low. However, it is a source of carbon-free energy and its importance would be enhanced by restrictions on carbon emissions aimed at tackling global warming.

10.7 How hydropower works

Hydroelectric power comes from water at work, water in motion. It can be seen as a form of solar energy, as the sun powers the hydrologic cycle which gives the earth its water. In the hydrologic cycle (Fig. 10.9), atmospheric water reaches the earth's surface as precipitation. Some of this water evaporates, but much of it either percolates into the soil or becomes surface runoff. Water from rain and melting snow eventually reaches ponds, lakes, reservoirs or oceans where evaporation is constantly occurring.

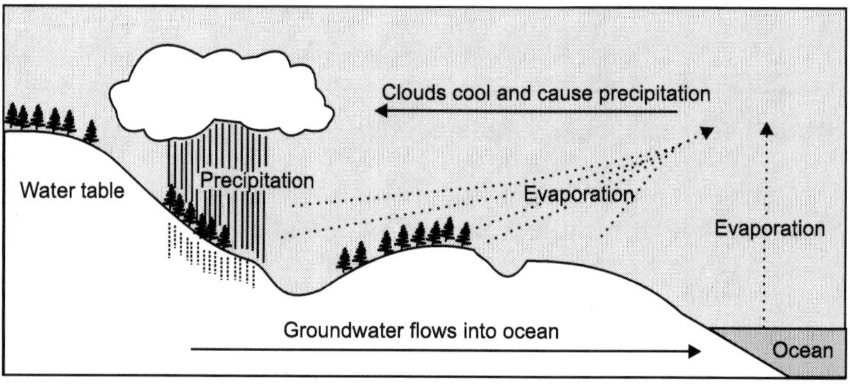

Figure 10.9 Hydrological cycle

Moisture percolating into the soil may become groundwater (subsurface water), some of which also enters water bodies through springs or underground streams. Groundwater may move upward through soil during dry periods and may return to the atmosphere by evaporation.

Water vapour passes into the atmosphere by evaporation then circulates, condenses into clouds, and some returns to earth as precipitation. Thus, the water cycle is complete. Nature ensures that water is a renewable resource.

10.7.1 Generating power

In nature, energy cannot be created or destroyed, but its form can change. In generating electricity, no new energy is created. Actually one form of energy is converted to another form. To generate electricity, water must be in motion. This is kinetic (moving) energy. When flowing water turns blades in a turbine, the form is changed to mechanical (machine) energy. The turbine turns the generator rotor which then converts this mechanical energy into another energy form—electricity. Since water is the initial source of energy, we call this hydroelectric power or hydropower for short. At facilities called hydroelectric power plants, hydropower is generated (Fig. 10.10). Some power plants are located on rivers, streams, and canals, but for a reliable water supply, dams are needed. Dams store water for later release for such purposes as irrigation, domestic and industrial use, and power generation. The reservoir acts much like a battery, storing water to be released as needed to generate power. The dam creates a 'head' or height from which water flows. A pipe (penstock) carries the water from the reservoir to the turbine. The fast-moving water pushes the turbine blades, something like a pinwheel in the wind. The waters force on the turbine blades turns the rotor, the moving part of the electric generator. When coils of wire on the rotor sweep past the generator's stationary coil (stator), electricity is produced.

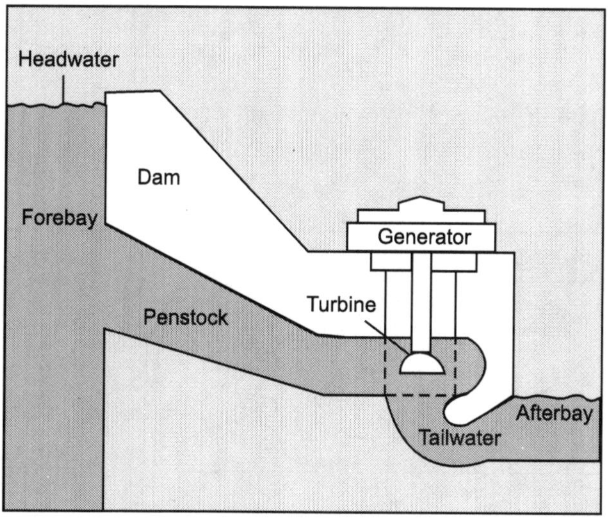

Figure 10.10 Hydroelectric power plants

This concept was discovered by Michael Faraday in 1831 when he found that electricity could be generated by rotating magnets within copper coils. When the water has completed its task, it flows on unchanged to serve other needs.

10.7.2 Transmitting power

Once the electricity is produced, it must be delivered to where it is needed—our homes, schools, offices, factories, etc. Dams are often in remote locations and power must be transmitted over some distance to its users. Vast networks of transmission lines and facilities are used to bring electricity to us in a form we can use. All the electricity made at a power plant comes first through transformers which raise the voltage so it can travel long distances through power lines. (Voltage is the pressure that forces an electric current through a wire.) At local substations, transformers reduce the voltage so electricity can be divided up and directed throughout an area (Fig. 10.11). Transformers on poles (or buried underground, in some neighbourhoods) further reduce the electric power to the right voltage for appliances and use in the home. When electricity gets to our homes, we buy it by the kilowatt-hour, and a meter measures how much we use. While hydroelectric power plants are one source of electricity, other sources include power plants that burn fossil fuels or split atoms to create steam which in turn is used to generate power. Gas-turbine, solar, geothermal, and wind-powered systems are other sources. All these power plants may use the same system of transmission lines and stations in an area to bring power to you. By use of this 'power grid', electricity can be interchanged among several utility systems to meet varying demands. So the electricity lighting your reading lamp now may be from a hydroelectric power plant, a wind generator, a nuclear facility, or a coal, gas or oil-fired power plant or a combination of these (Fig. 10.12).

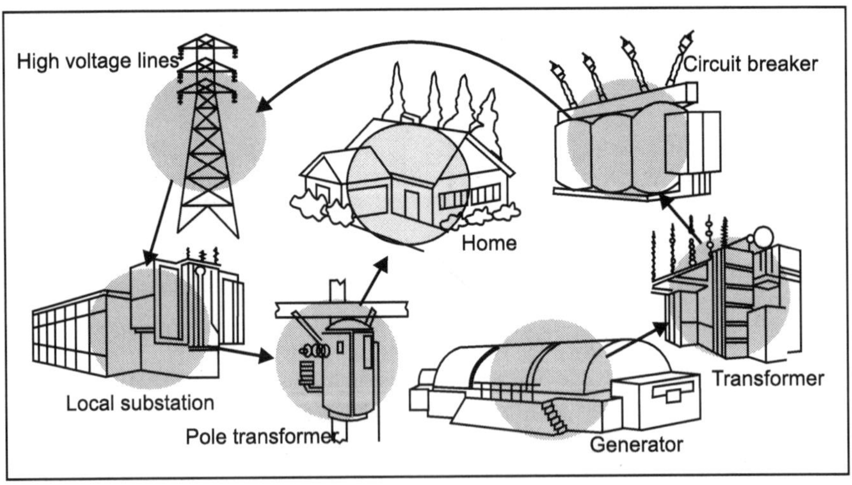

Figure 10.11 Transmitting power

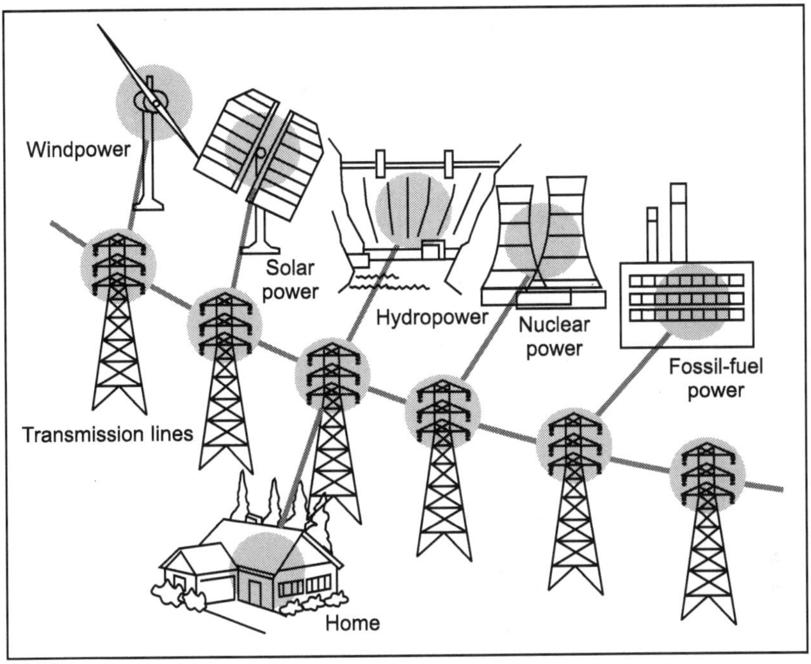

Figure 10.12 Power plants

The area where you live and its energy resources are prime factors in determining what kind of power you use. For example, in Washington State hydroelectric power plants provided approximately 80 per cent of the electrical power during 2007. In contrast, in Ohio during the same year, almost 87 per cent of the electrical power came from coal-fired power plants due to the area's ample supply of coal.

Electrical utilities range from large systems serving broad regional areas to small power companies serving individual communities. Most electric utilities are investor-owned (private) power companies. Others are owned by towns, cities, and rural electric associations. Surplus power produced at facilities owned by the Federal Government is marketed to preference power customers (A customer given preference by law in the purchase of federally generated electrical energy which is generally an entity which is nonprofit and publicly financed.) by the Department of Energy through its power marketing administrations.

10.7.3 How power is computed

Before a hydroelectric power site is developed, engineers compute how much power can be produced when the facility is complete. The actual output of energy at a dam is determined by the volume of water released (discharge) and the vertical distance the water falls (head). So, a given amount of water falling a given distance will produce a certain amount of energy. The head and the discharge at the power site and the desired rotational speed of the generator determine the type of turbine to be used. The head produces a pressure (water pressure), and the greater the head, the greater the pressure to drive turbines. This pressure is measured in pounds of force (pounds per square inch). More head or faster flowing water means more power.

To find the theoretical horsepower (the measure of mechanical energy) from a specific site, this formula is used:

$$THP = (Q \times H)/8.8$$

where, THP = theoretical horsepower
Q = flow rate in cubic feet per second (cfs)
H = head in feet
8.8 = a constant

A more complicated formula is used to refine the calculations of this available power. The formula takes into account losses in the amount of head due to friction in the penstock and other variations due to the efficiency levels of mechanical devices used to harness the power.

To find how much electrical power we can expect, we must convert the mechanical measure (horsepower) into electrical terms (watts). One horsepower is equal to 746 watts.

10.7.4 Turbines

While there are only two basic types of turbines (impulse and reaction), there are many variations. The specific type of turbine to be used in a power plant is not selected until all operational studies and cost estimates are complete. The turbine selected depends largely on the site conditions.

A reaction turbine is a horizontal or vertical wheel that operates with the wheel completely submerged, a feature which reduces turbulence. In theory, the reaction turbine works like a rotating lawn sprinkler where water at a central point is under pressure and escapes from the ends of the blades, causing rotation. Reaction turbines are the type most widely used. Various types of turbine are shown in Fig. 10.13.

An impulse turbine is a horizontal or vertical wheel that uses the kinetic energy of water striking its buckets or blades to cause rotation. The wheel is covered by a housing and the buckets or blades are shaped so they turn the flow of water about 170 degrees inside the housing. After turning the blades or buckets, the water falls to the bottom of the wheel housing and flows out.

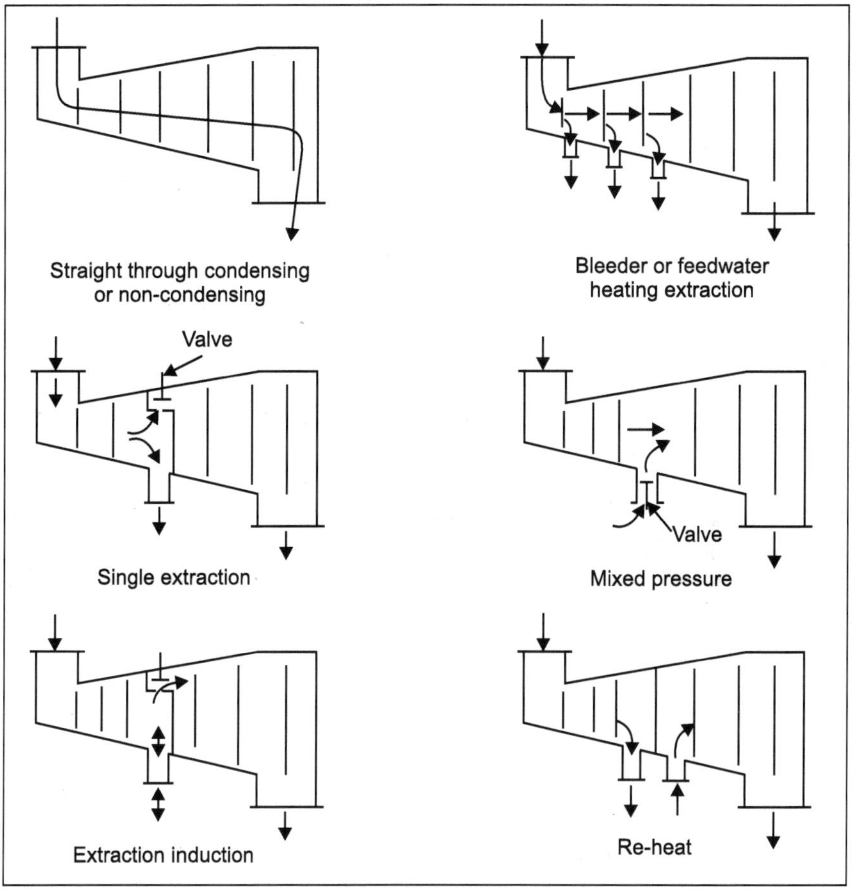

Figure 10.13 Various types of turbine

10.8 Modern concepts and future role

Hydropower does not discharge pollutants into the environment; however, it is not free from adverse environmental effects. Considerable efforts have been made to reduce environmental problems associated with hydropower operations, such as providing safe fish passage and improved water quality in the past decade at both Federal facilities and non-Federal facilities licensed by the Federal Energy Regulatory Commission. Efforts to ensure the safety of dams and the use of newly available computer technologies to optimise operations have provided additional opportunities to improve the environment. Yet, many unanswered questions remain about how best to maintain the economic viability of hydropower in the face of increased demands to protect fish and other environmental resources.

Reclamation actively pursues research and development (R&D) programs to improve the operating efficiency and the environmental performance of hydropower facilities.

Hydropower research and development today is primarily being conducted in the following areas:

1. Fish passage, behaviour, and response
2. Turbine-related projects
3. Monitoring tool development
4. Hydrology
5. Water quality
6. Dam safety
7. Operations and maintenance
8. Water resources management

Reclamation continues to work to improve the reliability and efficiency of generating hydropower. Today, engineers want to make the most of new and existing facilities to increase production and efficiency. Existing hydropower concepts and approaches include:

1. Uprating existing power plants
2. Developing small plants (low-head hydropower)
3. Peaking with hydropower
4. Pumped storage
5. Tying hydropower to other forms of energy

10.8.1 Uprating

The uprating of existing hydroelectric generator and turbine units at power plants is one of the most immediate, cost-effective, and environmentally acceptable means of developing additional electric power. Since 1978, Reclamation has pursued an aggressive uprating program which has added more than 16,00,000 kW to Reclamation's capacity. Reclamation's uprating program has essentially provided the equivalent of another major hydroelectric facility of the approximate magnitude of Hoover Dam and Power plant at a fraction of the cost and impact on the environment when compared to any other means of providing new generation capacity.

10.8.2 Low-head hydropower

A low-head dam is one with a water drop of less than 65 feet and a generating capacity less than 15,000 kW. Large, high-head dams can produce more power at lower costs than low-head dams, but construction of large dams may be limited by lack of suitable sites, by environmental considerations or by economic conditions. In contrast, there are many existing small dams and drops in elevation along canals where small generating plants could be installed. New low-head dams could be built to increase output as well. The key to the usefulness of such units is their ability to generate power near where it is needed, reducing the power inevitably lost during transmission.

10.8.3 Peaking with hydropower

Demands for power vary greatly during the day and night. These demands vary considerably from season to season, as well. For example, the highest peaks are usually found during summer daylight hours when air conditioners are running. Nuclear and fossil fuel plants are not efficient for producing power for the short periods of increased demand during peak periods. Their operational requirements and their long start-up times make them more efficient for meeting base load needs.

Since hydroelectric generators can be started or stopped almost instantly, hydropower is more responsive than most other energy sources for meeting peak demands. Water can be stored overnight in a reservoir until needed during the day, and then released through turbines to generate power to help supply the peakload demand. This mixing of power sources offers a utility company the flexibility to operate steam plants most efficiently as base plants while meeting peak needs with the help of hydropower. This technique can help ensure reliable supplies and may help eliminate brownouts and blackouts caused by partial or total power failures.

Today, many of reclamation's 58 power plants are used to meet peak electrical energy demands, rather than operating around the clock to meet the total daily demand. Increasing use of other energy-producing power plants in the future will not make hydroelectric power plants obsolete or unnecessary. On the contrary, hydropower can be even more important. While nuclear or fossil-fuel power plants can provide base loads, hydroelectric power plants can deal more economically with varying peak load demands. This is a job they are well suited for.

10.8.4 Pumped storage

Like peaking, pumped storage is a method of keeping water in reserve for peak period power demands. Pumped storage is water pumped to a storage pool above the power plant at a time when customer demand for energy is low, such as during the middle of the night. The water is then allowed to flow back through the turbine-generators at times when demand is high and a heavy load is place on the system.

The reservoir acts much like a battery, storing power in the form of water when demands are low and producing maximum power during daily and seasonal peak periods (Fig. 10.14). An advantage of pumped storage is that hydroelectric generating units are able to start up quickly and make rapid adjustments in output. They operate efficiently when used for one hour or several hours.

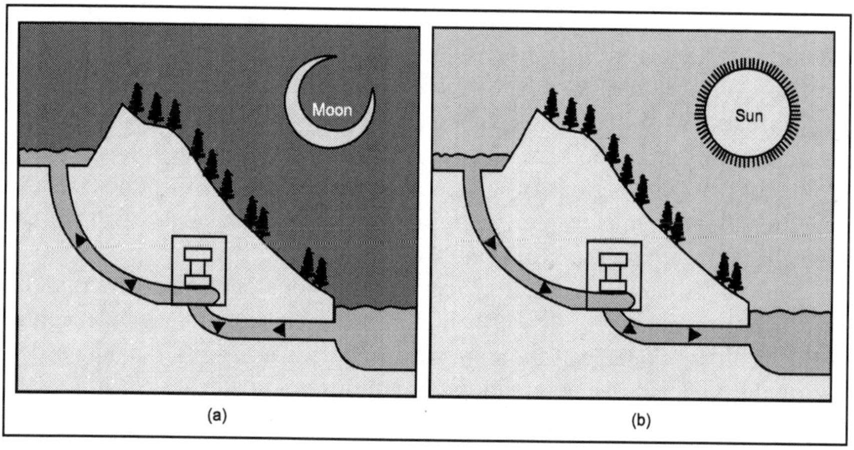

Figure 10.14 Reservoir acts much like a battery: (a) at night when customer demand for energy is low, water is pumped to a storage pool above the dam, and (b) when demand is high and a heavy load is placed on the system, water is allowed to flow back through the turbine-generators.

Because pumped storage reservoirs are relatively small, construction costs are generally low compared with conventional hydropower facilities.

10.8.5 Comparing hydropower to other energy forms

When we hear the term 'solar energy', we usually think of heat from the sun's rays which can be put to work. But there are other forms of solar energy. Just as hydropower is a form of solar energy, so too is windpower. In effect, the sun causes the wind to blow by heating air masses that rise, cool, and sink to earth again. Solar energy in some form is always at work—in rays of sunlight, in air currents, and in the water cycle.

Solar energy, in its various forms, has the potential of adding significant amounts of power for our use. The solar energy that reaches our planet in a single week is greater than that contained in all of the earth's remaining coal, oil, and gas resources. However, the best sites for collecting solar energy in various forms are often far removed from people, their homes, and work places. Building thousands of miles of new transmission lines would make development of the power too costly.

Because of the seasonal, daily, and even hourly changes in the weather, energy flow from the wind and sun is neither constant nor reliable. Peak production times do not always coincide with high power demand times. To depend on the variable wind and sun as main power sources would not be acceptable to most American lifestyles. Imagine having to wait for the wind to blow to cook a meal or for the sun to come out from behind a cloud to watch television!

As intermittent energy sources, solar power and wind power must be tied to major hydroelectric power systems to be both economical and feasible. Hydropower can serve as an instant backup and to meet peak demands.

Linking wind power and hydropower can add to the nation's supply of electrical energy. Large wind machines can be tied to existing hydroelectric power plants. Wind power can be used, when the wind is blowing, to reduce demands on hydropower. That would allow dams to save their water for later release to generate power in peak periods.

The benefits of solar power and wind power are many. The most valuable feature of all is the replenishing supply of these types of energy. As long as the sun shines and the wind blows, these resources are truly renewable.

The inherent technical, economic and environmental benefits of hydroelectric power make it an important contributor to the future world energy mix, particularly in the developing countries. These countries have a great and ever-intensifying need for power and water supplies and they also have greatest remaining hydro potential. Development is a basic human right, as few would deny. Energy policy makers must meet their responsibility in exploring the most rational options for meeting the energy needs of the developing world, while protecting the environment to the maximum possible extent, for example by limiting greenhouse gas emissions.

10.9 Benefits of hydropower

Hydropower provides unique benefits, rarely found in other sources of energy. These benefits can be attributed to the electricity itself, or to side-benefits, often associated with reservoir development. Despite the recent debates, few would disclaim that the net environmental benefits of hydropower are far superior to fossil-based generation. In 2006, for example, it has been calculated that hydropower saved GHG emissions equivalent to all the cars on the planet (in terms of avoided fossil fuel generation).

While development of all the remaining hydroelectric potential could not hope to cover total future world demand for electricity, implementation of even half of this potential could thus have enormous environmental benefits in terms of avoided generation by fossil fuels.

Carefully planned hydropower development can also make a vast contribution to improving living stands in the developing world (Asia, Africa, Latin America), where the greatest potential still exists. Approximately 2 billion people in rural areas of developing countries are still without an electricity supply. As the most important of the clean, renewable energy options, hydropower is often one of a number of benefits of a multipurpose water resources development project. As hydro schemes are generally integrated within multipurpose development schemes, they can often help to subsidise other vital functions of a project. Typically, construction of a dam and its associated reservoir results in a number of benefits associated with human well-being, such as secure water supply, irrigation for food production and flood control, and societal benefits such as increased recreational opportunities, improved navigation, the development of fisheries, cottage industries, etc. This is not the case for any other source of energy.

10.10 Characteristics of hydropower

1. Its resources are widely spread around the world. Potential exists in about 150 countries, and about 70 per cent of the economically feasible potential remains to be developed. This is mostly in developing countries.
2. It is a proven and well advanced technology (more than a century of experience), with modern power plants providing the most efficient energy conversion process (>90 per cent), which is also an important environmental benefit.
3. The production of peak load energy from hydropower allows for the best use to be made of base load power from other less flexible electricity sources, notably wind and solar power. Its fast response time enables it to meet sudden fluctuations in demand.
4. It has the lowest operating costs and longest plant life, compared with other large scale generating options. Once the initial investment has been made in the necessary civil works, the plant life can be extended economically by relatively cheap maintenance and the periodic replacement of electromechanical equipment (replacement of turbine runners, rewinding of generators, etc. in some cases the addition of new generating units). Typically a hydro plant in service for 40–50 years can have its operating life doubled.

5. The ‘fuel’ (water) is renewable, and is not subject to fluctuations in market. Countries with ample reserves of fossil fuels, such as Iran and Venezuela, have opted for a large scale program of hydro development, recognising environmental benefits. Hydro also represents energy independence for many countries.

10.11 Electrical system benefits

Hydropower, as an energy supply, also provides unique benefits to an electrical system. First, when stored in large quantities in the reservoir behind a dam, it is immediately available for use when required. Second, the energy source can be rapidly adjusted to meet demand instantaneously. These benefits are part of a large family of benefits, known as ancillary services. They include:

1. Spinning reserve—the ability to run at a zero load while synchronised to the electric system. When loads increase, additional power can be loaded rapidly into the system to meet demand. Hydropower can provide this service while not consuming additional fuel, thereby assuring minimal emissions.
2. Non-spinning reserve—the ability to enter load into an electrical system from a source not on line. While other energy sources can also provide non-spinning reserve, hydropower's quick start capability is unparalleled, taking just a few minutes, compared with as much as 30 minutes for other turbines and hours for steam generation.
3. Regulation and frequency response—the ability to meet moment-to-moment fluctuations in system power requirements. When a system is unable to respond properly to load changes its frequency changes, resulting not just in a loss of power, but potential damage to electrical equipment connected to the system, especially computer systems. Hydropower's fast response characteristic makes it especially valuable in providing regulation and frequency response.
4. Voltage support—the ability to control reactive power, thereby assuring that power will flow from generation to load.
5. Black start capability—the ability to start generation without an outside source of power. This service allows system operators to provide auxiliary power to more complex generation sources that could take hours or even days to restart. Systems having available hydroelectric generation are able to restore service more rapidly than those dependent solely on thermal generation.

10.12 Environmental issues and management for hydropower peaking operations

Hydropower schemes are built on many scales, involve different project types, and play different roles in an integrated energy system. Operating patterns vary, and discharge patterns to the downstream river environment reflect whether the station is operating in base load, peak load or frequency mode.

When meeting base load requirements, a power station usually discharges a constant flow all day and can maintain this for days, weeks and even months depending on the scale of the scheme and the generation needs. When meeting peak load requirements, a power station is turned on at a particular time during the day, generates power at a constant load for a certain number of hours, and is then turned off or set to a different load for another time period, resulting in a high variability in flow discharges. Hydropower stations may also operate in frequency mode, where generators are brought on or off depending on the changing electricity demand throughout the day, essentially 'following' the load.

Many hydropower plants are operated as peak generators or frequency controllers, because they can change their output quickly to follow the fluctuating power demand. A hydropower generating unit can start up or stop within tens of seconds, which provides an important role in an integrated energy system where each type of energy source can be used to its best advantages. Because of its flexibility in generating patterns, hydropower can optimise the efficiency of less flexible fossil or nuclear generation options, and also offers a backup for other more intermittent renewable energy sources such as wind and solar.

Reservoir hydro schemes in particular provide considerable flexibility in energy provision, because with their larger storage they can be operated to provide either base load or peak load services. Run-of-river hydropower schemes provide base load options, but with less flexibility in their ability to provide peak power because of their smaller storage capacities. Pumped-storage plants are particularly well suited to meeting the peaks in electricity demand; they essentially work as a huge storage battery by charging or discharging power according to the system's demand.

10.12.1 Potential environmental issues with hydropeaking

Where reservoir hydro schemes are operated primarily to provide peak load services, there are particular environmental risks that should be considered in any environmental impact assessment. With a sound understanding of the potential environmental issues, there are strategies that can be employed at the siting and design stage to minimise or mitigate these risks.

With any hydro scheme, the downstream river environment has an altered hydrograph due to the curtailing of major floods and a flow range restricted to the turbine discharge capacity. For a peaking station, a typical hydrograph shows twice-daily fluctuations from off to full capacity discharges often with weekend shutdowns. A peaking station may show consistent daily to weekly patterns of discharge throughout the year rather than the strong seasonal pattern that might be shown for base load providers, and depending on the scale of the scheme inter-annual variability may be low.

Downstream effects on water quality depend on the storage configuration and offtake depth. If the storage is deep and stratifies and the offtake is low, the downstream environment may experience frequent temperature and dissolved oxygen fluctuations particularly during the summer period, with the power station injecting cold water from deep in the reservoir into the warmer waters of the receiving environment. In situations where there are downstream pollution sources draining into the river system, peaking power station discharges can cause pulses of polluted water downstream rather than a general dilution effect when operating to meet baseload demand.

Downstream effects on the fluvial geomorphology and dominant geomorphic processes differ if a power station is operating to meet base versus peak demand. Major issues with any hydro operations are with the reduced sediment supply to the downstream environment and the erosive capabilities of continuous larger than natural baseflows for baseload operations. With peaking operations, the significant flow discharge patterns affecting the channel form are the rate and frequency of water level rise, the time the station is at its maximum discharge level, and the rate and frequency of water level drop. A rapid increase in water level has considerable bank scouring capabilities. The amount of time the power station discharges at its full capacity influences the degree of saturation of the river banks, which in turn influences the degree of seepage-induced erosion that may occur when the power station turns off. Frequent and rapid drawdowns in water level result in considerable pore water pressures as the water drains out of the banks, so with peaking operations the frequency of seepage-induced erosion events increases, however the severity of any one event may be less than otherwise if the power station has not been on long because the banks will not be as saturated.

Downstream effects on riparian vegetation with any hydro operations can be a loss of species cover and diversity in the riparian zone due to waterlogging and inundation, lack of regeneration and recruitment, and habitat alteration due to bank erosion. Inundation is the submergence of vegetation that prevents gas exchange, and prevents plants carrying out photosynthesis and respiration through their leaves. Waterlogging is the submergence of the root zone, which causes depletion of oxygen in the soil and prevents respiration by plant roots. Light limitation is also a stress because plants require adequate daylight hours without inundation or waterlogging to acquire carbon through photosynthesis. With peaking operations there are reduced risks of impact due to waterlogging and inundation, as the riparian zone is drained and exposed to sunlight on a daily basis. However there is still the case of limited regeneration and recruitment, because seedlings cannot establish on the banks where water levels rise rapidly several times per day and wash them away. Banks close to the power station may eventually end up with a high percentage cover of mineral substrates, and riparian tall woody shrub species may be replaced with ephemerals such as grasses, gramonoids (grass-like plants) and tolerant semi-aquatic herb which may provide some structural stability to river banks.

Downstream effects on macroinvertebrates (e.g. aquatic insects and micro-crustaceans) for any hydro operations are often a reduction in species diversity and abundance, as well as loss of edge and snag habitat. The significant drivers of these impacts are that the water levels with the power station off are often lower than mean summer baseflows resulting in less habitat availability for colonisation, and water levels with the power station on are often higher than mean winter baseflows resulting in greater depths and current velocities than optimal for some species. With peaking operations, water levels change across this range several times per day, and frequent water level as well as temperature changes would cause high stresses on the instream biota. For example, high shear stresses, that is the force applied to the river bed from rapidly rising water levels, is associated with faunal displacement and possibly bed movement under rapidly varying flows.

Hydropower schemes can affect migratory aquatic species due to the physical barrier to upstream migration presented by the dam itself. Baseload discharge patterns can reduce fish populations in the downstream environment due to reduced macroinvertebrate food supplies, loss of snag habitat, and impacts on spawning and migration cues due to changes in the seasonality of flows (and in cases temperature). Where there are natural downstream obstacles to fish migration such as river gorges, baseload discharge patterns may cause fish migration difficulties due to sustained high flow releases, whereas peaking discharge patterns may provide more frequent opportunities for migration through these gorges. Social issues can also arise due to peaking operations, for example human safety issues with rapidly changing water levels, risks of stock strandings, and issues with pump setups for landowners.

10.12.2 Investigating environmental impacts due to hydropeaking

Investigations of environmental impacts of hydropeaking operations must be underpinned by a sound analysis of operating patterns and downstream hydrology. The drivers of impact in the downstream environment are the rapid changes in flow, and at any given point in the system this will result in different water heights, current velocities, degree of backwater inundation or channel dewatering, ramp-up and drawdown rates. A detailed set of gauging sites and water level recorders will provide basic data that can then be fed into a hydrological model of the river system to assist in modelling of environmental impacts. Because of the rapid changes in water level with peaking operations, water quality impacts are best assessed using sites that continuously record temperature, dissolved oxygen and conductivity, at several sites downstream of the power station. Sites should be in conjunction with a gauging or water level station so that readings can be related to flow/level changes, and ideally upstream and downstream of significant tributaries. Investigations of the impacts of peaking operations on fluvial geomorphology require particular attention because the dominant geomorphic processes can vary considerably depending on the type of discharge regime from the power station. These investigations can employ a number of different approaches. Development of a sediment budget for the downstream environment provides a valuable framework for more detailed reach-specific assessments, as does mapping of riverbank and bed attributes using a hand-held GPS to identify those zones most susceptible to bank erosion. For existing schemes, broadscale geomorphic change in over time can be assessed using comparative aerial photography. Changes to channel profile, depth and channel geometry can be assessed using repeat survey cross-sections. Changes to bank stability and profile can be assessed using erosion pins and photo monitoring. Scour chains in association with erosion pins offer the ability to show maximum scour that occurred during the period between erosion pin measurements. Water samples analysed for suspended sediment concentrations at different points in the river, particularly with continuous autosamplers, are valuable where rivers transport fine-grained sediments.

Further techniques can be employed for predicting the fluvial geomorphic impacts of peaking operations for proposed hydropower projects. Sediment transport capacity of the flow regime can be assessed with dedicated experiments and hydraulic calculations requiring good particle size data and hydraulic characteristics at a cross-section. Penetrometer readings of sediment banks can provide an indication of bank cohesion and strength. Piezometers are a valuable tool to determine groundwater changes in the near river sediment banks in response to changes in river level, so that degree of bank saturation, pore water pressures and degree of risk of draw-down seepage induced erosion can be assessed, often with the aid of computer models. Close investigation of the inter-relationships of riparian vegetation composition and cover with bank stability processes is essential in any investigations.

Investigations of the impacts of peaking operations on riparian vegetation require a basic broadscale mapping exercise of riparian plant communities, and more site-specific surveys of cover and abundance of plant species using a quadrant based approach. Quadrants need to be located in a profile up the bank so that they can be related to different water levels and inundation times. Additional riparian vegetation investigations should encompass assessments of recruitment within each survey quadrant, sampling and analysis of root mat densities in different bank sediment types, and assessment of the contributions of mosses or ground cover species to stream bank stability.

For assessment of hydropeaking impacts on macroinvertebrate populations, investigations should consider species presence/absence, species abundance, habitat availability and shear stress. Several techniques exist to obtain samples for identification. A rapid presence/absence assessment can be undertaken using kick-net sampling in riffles, with samples identified to the family level, and ideally this data would be fed into a predictive bioassessment model which can compare observed taxa to expected taxa and thus provide a rating of degree of impact. Considerable work must go into the development of a bioassessment model if no appropriate model exists. Macroinvertebrate abundance can be assessed using quantitative (surber) samples that are typically identified to genus level, and in cases to species level to identify any threatened species.

Habitat availability analyses requires two data sets—hydraulic data (e.g. velocity, depth, substrate characteristics) collected in field surveys from representative transects across the river under power station on and off conditions, and habitat preference data for key aquatic taxa derived from either the literature and/or from field sampling—to derive plots of 'weighted usable area' for different flow levels. Finally, shear stress analyses can be undertaken in the field by placing hemispheres of known densities on the stream bed and observing their movement under changing flow conditions, an exercise that requires diving and so safety is a prime consideration when considering this technique.

Assessment of the impacts of peaking operations on fish species are largely based on field fish surveys. Backpack electrofishing is often chosen as a standard sampling method. It allows for a repeatable approach with minimal mortality rates, has known biases, and offers a method of comparison between sites. Results can be standardised into a comparative Catch Per Unit Effort figure for each visit to each site, and site ordination. Habitat availability and preference should also be undertaken as per the macroinvertebrate assessments. Site selection should include significant tributary streams as well as the mainstem river, and particular attention paid in sampling program design to upstream and downstream of natural flow obstacles to migration such as gorges. It can be valuable to dissect some fish to analyse their food sources, to better enable linkages in fish condition to be made with macroinvertebrate populations. More specific assessments are likely to be required for significant species such as waterbirds or aquatic mammals, which would vary considerably in different parts of the world.

10.12.3 Environmental management and mitigation measures for hydropeaking

A range of management approaches and mitigation measures can be employed to address the potential environmental issues with peaking operations, and investigations of impacts should be directed at identifying and evaluating management options as much as assessing impacts.

Discipline-specific options and objectives

(a) Water quality

1. Storage siting and design should consider whether reservoir stratification is likely to occur, and whether a multi-level offtake may be required to ensure release of oxygenated and ambient temperature water.
2. Air injection in the turbines can ensure sufficient oxygenation of water releases.
3. Siting upstream of a significant tributary can ensure mixing of power station discharges with water of ambient temperatures to the further downstream environment.

(b) Geomorphology

1. Physical buttressing of riverbanks.
2. Reduction of the maximum power station discharge to reduce the phreatic surface gradient in the banks.
3. Minimising the duration of maximum discharges to reduce the extent of bank saturation.
4. Maintenance of a minimum environmental flow to lessen scour of the bank toes and reduce phreatic surface gradient.
5. Measures that would increase the viability of riverbank vegetation.

(c) *Riparian vegetation*

1. Instigating low flow rates for three summer months every year to allow riparian plants to grow and reproduce and for recruitment to occur during the season of greatest metabolic activity.
2. Ensure 24–48 hour shutdowns on approximately a weekly basis to reduce stresses of waterlogging and inundation (note this would likely occur anyway with peaking operations).
3. Facilitate regeneration by direct-seeding of the river banks with local riparian species.
4. Measures that would improve physical stability of the riverbanks.

(d) *Macroinvertebrates*

1. A minimum flow to ensure that a proportion of the channel is permanently inundated, that snag habitats on the channel margins are inundated, and that the channel can maintain a constant macroinvertebrate community when the power station is not discharging.
2. Management of rates of increase and decrease of power station discharge can slow the rates of downstream river level changes, and thus reduce shear stresses on the bed particularly on the rising limb of the hydrograph by reducing water surface slopes, and reduce incidences of stranding of fish and macroinvertebrates.

(e) *Fish*

1. A minimum environmental flows would benefit macroinvertebrate communities and so would indirectly benefit fish that feed on the macroinvertebrates.
2. A partial or stepped ramp-down would provide cues to the fish of dropping flows before full dewatering of habitats occurred, hence reducing the potential for fish stranding under peaking operations.
3. Restocking with native species.

Options assessment

A number of the management and mitigation options to address environmental impacts of peaking operations involve dedicated water release patterns. Water management options include minimum environmental flows, power station ramp-downs, power station rates of flow increase, reducing maximum discharges, and minimising durations of full gate discharges.

Controls on patterns of release to the downstream environment to address potential impacts to the downstream ecosystem can be provided either through the power station discharges, through a dedicated release valve or by a re-regulating structure downstream of the power station.

In general, water management options constitute significant constraints on power station operations and can incur considerable losses in generating potential. Large generating turbines may not be able to generate small discharges required for minimum environmental flows, and many have rough running bands that should be largely avoided. For delivery of an environmental flow, an upfront major capital cost may be preferable to ongoing constraints on discharge patterns. If the siting allows it and the cost-benefit analysis supports it, construction of a re-regulation storage to allow downstream release patterns to be dedicated to environmental management outcomes can be very successful. A re-regulating structure is a dam or weir that impounds the regulated flows from the power station, and allows control over release patterns to downstream of the structure. Mini-hydro turbines can also/alternatively be employed to recover generating capability with minimum flow releases.

A ramp-down or step-down rule for the power station could be compatible with peaking operations. This would be particularly of benefit in reducing seepage-induced erosion in riverbanks, and would also offer benefits to fish to reduce risks of fish strandings.

Ramping up constraints would not be desirable with peaking operations, as they would reduce the rapid start advantage that the hydropower station offers with the provision of peak load. The provision of a minimum environmental flow can lessen the need for a ramp-up rate, as it would ensure that shear stresses are reduced as water levels rise in the downstream river system.

Depending on the circumstances, it is likely that several of the options are utilised as a package. Water management options in combination with localised treatment works can be successful. Localised treatment options include bank protection works, bank revegetation works, local willow control, and fencing/stock exclusion. There may also be site-specific opportunities, such as diversion of part of the downstream flow to lessen the degree of water level rise and fall.

To sum up, there are a number of potential environmental issues associated with hydropeaking operations, as well as a range of mitigation measures available to substantially address these issues. Potential environmental issues should be thoroughly assessed prior to finalising the siting and design for a proposed scheme, so that potential impacts can be minimised, and that the scheme can include any mitigation or management measures as part of its design. Many of these measures can also be employed to minimise impacts arising from peaking operations with an existing scheme. Assessment of management measures should consider any possibilities of trade-offs that might arise amongst the different ecosystem components, and should be subject to a thorough cost-benefit analysis.

11
Small hydroelectric plants

11.1 Introduction

Small hydro is the development of hydroelectric power on a scale serving a small community or industrial plant. The definition of a small hydro project varies but a generating capacity of up to 10 megawatts (MW) is generally accepted as the upper limit of what can be termed small hydro. For generations water has been used as a source of energy by industry and by a limited number of utility companies. In the continental United States, most rivers and streams capable of producing huge amounts of hydroelectric power have been harnessed; however, this does not preclude the possibility of using mini-hydroelectric power as a source of energy supply for home or farm. Harnessing a stream for hydroelectric power is a major undertaking. Careful planning is necessary if a successful and economical power plant is to result. State water laws and environmental concerns must be determined.

Precise field data must be gathered to compare the amount of power that can be expected from a hydroelectric installation to the electrical requirements of the home or farm. Then detailed plans that consider both construction and maintenance can be drawn up. Perhaps the greatest mistake made when considering small hydroelectric installations is the overestimation of a proposed plant's capability. This chapter bulletin will help to start the planning of a small power plant on a given stream of water. One of the first steps in planning is to measure the power potential of the stream. The amount of power that can be obtained from a stream depends on:

1. The amount of water flow.
2. The height which the water falls (head).
3. The efficiency of the plant to convert mechanical energy to electrical energy.

11.2 Stream flow

Stream flow varies greatly from season to season and depending on the nature of the terrain. A typical discharge from a 22 square mile hilly to mountainous drainage area in the Northeast during a year of normal precipitation is summarised in Fig. 11.1. The smallest commercially manufactured hydroelectric power plant at one-half kilowatt (kW) or 500 watts needs 1.1 cfs with a 12′ height of water fall or head. Therefore no power can be generated during low-flow periods without reservoir storage. It is also interesting to look at the peak flow. A ten-kilowatt hydroelectric plant needs 16.3 cfs with a 12′ head, leaving a substantial part of the peak flow to be contended with—not a simple task. This extreme variation illustrates the value of a reservoir to regulate and even the flow.

11.3 Measuring flow

Two methods are commonly suggested for measuring the flow in small or medium sized streams. Large discharges are best determined by a hydraulic engineer. The float method of testing stream flow is the easiest test to conduct and will yield satisfactory data, except in cases where a stream is shallow or rocky and thus impedes the movements of a weighted float.

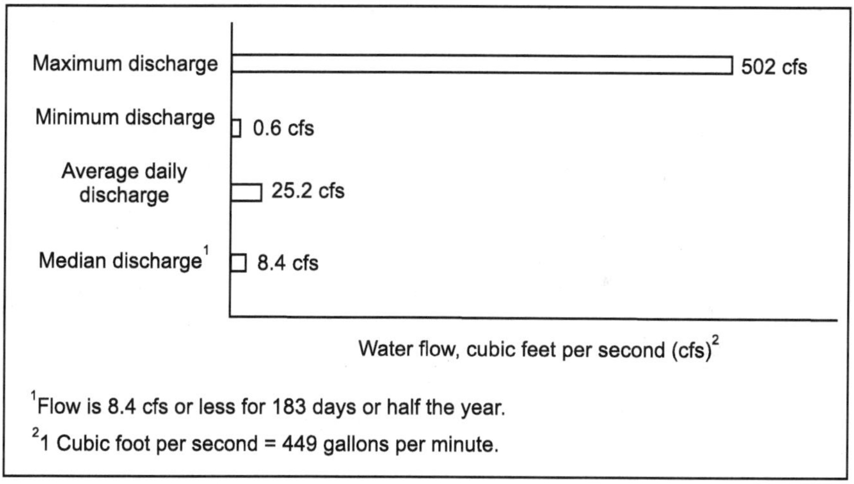

Figure 11.1 Stream flow

Basically the cross section of an unobstructed area of the stream is measured and a weighted float such as a bottle weighted with pebbles is timed as it floats down a 100 foot course.

The weir method is more time consuming but may be the most satisfactory test if the stream is very small, shallow, rocky, obstructed, or if there is an existing dam (Fig. 11.2). A weir is a dam with an opening or notch through which the entire stream flows. The flow may be calculated by precisely measuring the depth of water flowing over the crest of the weir. Tongue and groove planking makes a good temporary weir for streams not more than 1 or 2 ft deep and 6–10 ft wide. The selected references give more detail on how to measure stream flow by the weir or float method. No matter what method of flow measurement is used, it is very important to measure stream flow many times over a year or more (Fig. 11.3).

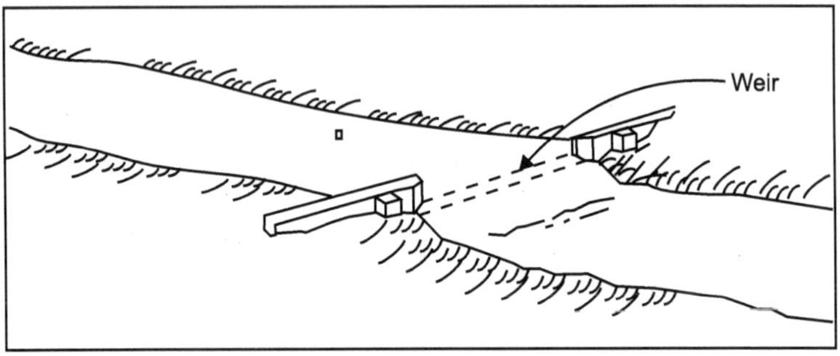

Figure 11.2 Measuring flow with a weir

11.4 Dams

Water must fall in order to generate power from a stream. In most cases the fall is enhanced and increased by constructing a dam which creates a head (Head is the vertical distance from the surface of the water at the dam down to the water in the stream below where the turbine is located). The higher the dam or head, the greater the power a given amount of water will produce. A dam also provides a storage basin to regulate stream flow and thereby increases power potential.

Before pursuing dam construction further, you should consider these points:

1. The construction of a dam is a highly technical undertaking. You need to collect extensive field data to choose the best site and design a safe dam. A professional engineer can best advise you on construction of a safe dam for your particular site.
2. Find out what permits are required if a stream is to be impounded. Laws vary from state to state: some are lenient; some are very stringent. Your state environmental conservation office should be able to supply you with the necessary information.
3. Be aware that you must control all land to be flooded.

Step 1: Measuring cross-section area

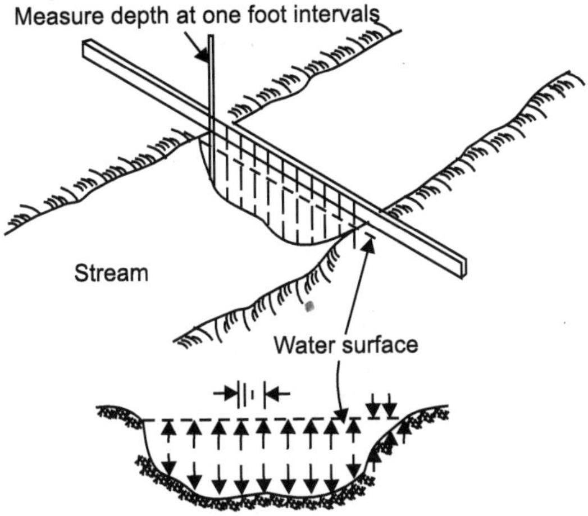

$$\text{Average depth} = \frac{\text{Sum of measured depths}}{\text{\# Intervals measured}}$$

Step 2: Measuring water velocity

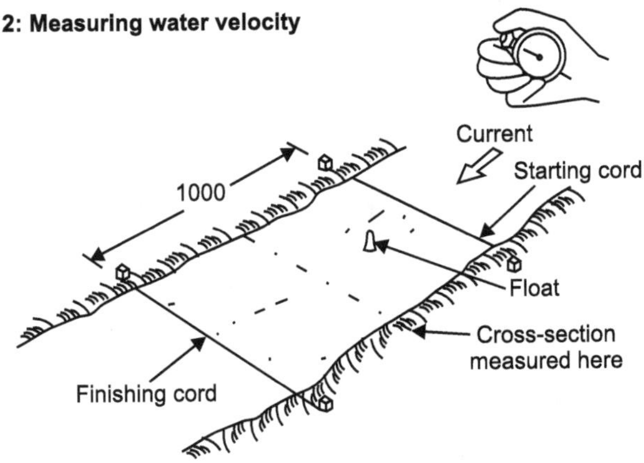

Step 3: Calculating flow rate

The flow rate is calculated by the formula:

$$Q = 0.83\left(\frac{bd}{144}\right)\left(\frac{100}{t}\right)$$

Q = flow rate, cubic feet per second
b = stream width, inches
d = average stream depth, inches
t = time for float to drift 100 feet, seconds

Figure 11.3 Measuring flow with a float

4. In the event of dam failure, you will be responsible for all downstream damage. Your insurance underwriter may be able to advise you on the availability of liability insurance.
5. You must respect the rights of others to stream use outside of your property.
6. You may be faced with resistance from nearby landowners and conservation-minded groups or individuals.
7. Obtain solid cost estimates for the dam construction. Costs vary widely, depending on the particular site, but it is reasonable to assume that the construction of any worthwhile dam will run to several thousand dollars. Will the capital outlay for construction be justified by the long-term benefits of an installation?

Figure 11.4 shows the hydroelectric installation with reservoir.

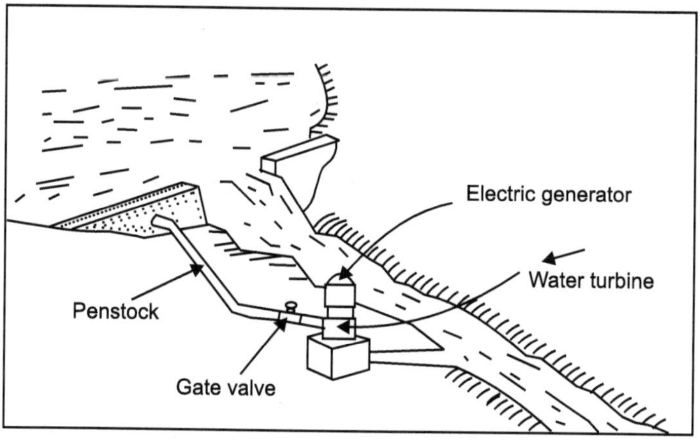

Figure 11.4 Hydroelectric installation with reservoir

11.5 Diversions

Because legal requirements and high costs restrict new dam construction, other methods of water development may be more suitable. In some places only a round concrete box structure is needed to divert part of the flow of a stream through a pipe to a downstream turbine (Fig. 11.5).

On any given stream, the best site for a diversion is usually quite obvious: a natural waterfall, a swift current or a steep slope. The terrain should be steep enough (10 per cent slope or more) so the diversion pipe doesn't have to be too long to obtain sufficient head. Because a diversion has no backup storage for periods of low flow, potential power is sometimes calculated by determining the runoff in a year of normal rainfall which is exceeded on half the days of the year. The example watershed has a stream flow of 8.4 cubic feet per second or less for half the year. With a 12" head, this stream can operate a 5 kW generator at full capacity for half the year and at about 50 per cent capacity for the remainder. Some people install batteries to provide electrical storage for periods of low flow.

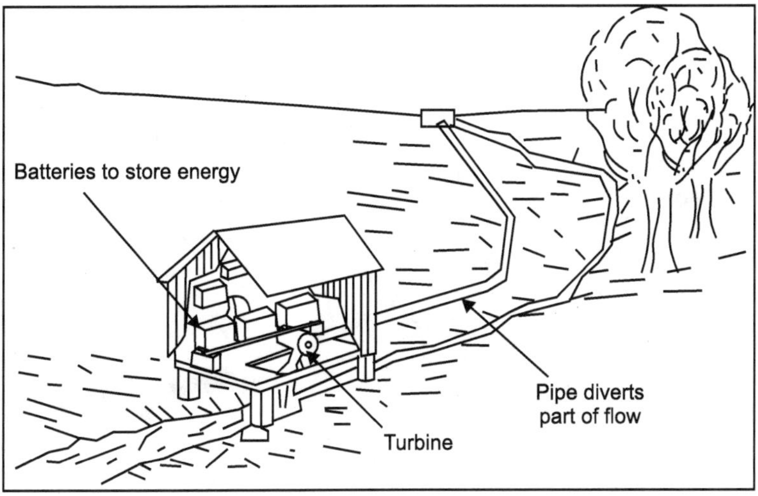

Figure 11.5 Diversion

11.6 Measuring head

After the height of the water behind the proposed dam or diversion has been decided, it is necessary to measure the head of water that will result. To determine the difference in level between two points, set a surveyor's level about midway between the points. Have an assistant hold a surveyor's rod at one point, sight through the level and record the height reading on the rod. Move the rod to the second point and read. The difference of the readings is the difference in elevation of the two points (Fig. 11.6).

Often it is impossible to see the two points from a single setting of the level so rods must be read at intermediate or turning points. The differences in readings between each pair of points can be added together to calculate the total elevation drop from the dam or diversion.

11.7 Calculating power

Power output, in kilowatts is calculated by the formula:

$$kW = 0.0846 \times E \times Q \times H$$

where, Q = Water flow, in cubic feet per second
H = Head, in ft.
E = Efficiency of hydroelectric plant, per cent divided by 100.

Friction, generator losses and turbine losses reduce the efficiency of a power plant. Small plants are about 40 per cent efficient. Five to ten kilowatt plants may be 60–70 per cent efficient when operating at full capacity. For initial estimates, calculate power based on 50 per cent efficiency.

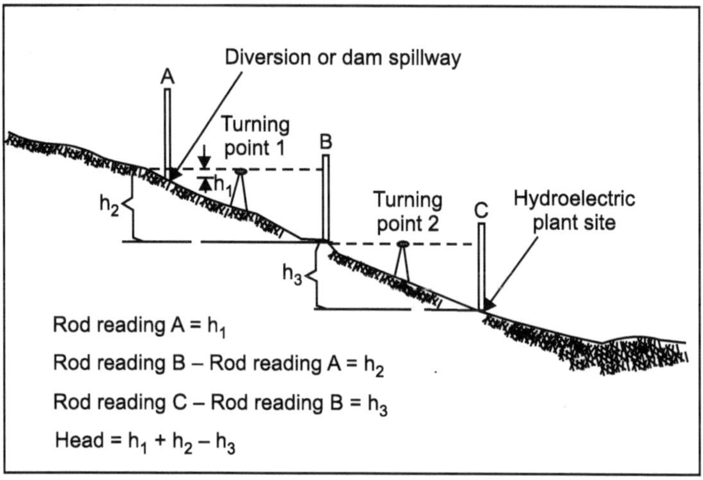

Figure 11.6 Measuring head

Table 11.1 lists the water flow and head needed to run various small hydroelectric plants operating at 50 per cent efficiency. Manufacturers of hydroelectric plants also list power outputs of their units at various flows and heads.

Table 11.1 Required water flow and head for small hydroelectric plants.

Generator output, kilowatts[1]	Head (ft)				
	10′	20′	50′	100′	200′
	Water flow, cubic feet per second				
0.5	1.2	0.6	0.24	0.12	0.06
1	2.4	1.2	0.5	0.24	0.12
2	4.7	2.4	0.9	0.5	0.24
5	11.8	5.9	2.4	1.2	0.6
10	23.6	11.8	4.7	2.4	1.2

[1] overall hydroelectric plant efficiency is 50 per cent.

11.8 Components

11.8.1 Penstocks

Friction in the pipe (penstock) or open channel that carries water to the generator is another cause of power loss. Most small hydroelectric sites have a small or moderate head, so it is very important to use large penstocks to reduce losses.

If you are diverting a water source far up the hill, plastic or aluminium irrigation pipe and the heavier walled, pressure rated PVC plastic pipe make good penstocks. Table 11.2 shows head losses for typical plastic pipe. The flow rates to the right of the dashed lines have velocities that exceed 5–7 ft/s and are not recommended. Use a larger pipe instead.

For example, Table 11.1 shows a 2 kW generator requires 0.6 cfs with a 100-foot head. Three hundred feet of 3" PVC pipe carrying 0.6 cfs from an upstream diversion has a head loss of $3 \times 10.2 = 30.6$ ft. Thus the diversion must be $100 + 30.6 = 130.6$ feet above the generator to compensate for the friction or head loss in the pipe. A 6" pipe has only a $0.54 \times 3 = 1.62$ ft head loss.

Table 11.2 Head loss for plastic pipe.

Nominal pipe diameter	Inside diameter	Flow rate, cubic feet per second[1]											
		0.1	0.2	0.3	0.4	0.6	0.8	1.0	2.0	4.0	6.0	8.0	10.0
		Head loss, feet per 100 feet of pipe											
160# PVC plastic pipe													
2"	2.193"	2.44	8.22	18.7									
3"	3.230"	0.37	1.34	2.84	4.83	10.2							
4"	4.154"	0.11	0.39	0.83	1.42	3.01	5.12	7.74					
80 psi plastic irrigation pipe													
6"	5.900"	–	0.07	0.15	0.26	0.54	0.93	1.40	5.06				
8"	7.840"	–	–	–	0.06	0.14	0.23	0.35	1.27	4.58	9.70		
10"	9.800"	–	–	–	–	–	0.08	0.12	0.43	1.54	3.27	5.58	8.42
12"	11.760"	–	–	–	–	–	–	0.05	0.18	0.64	1.35	2.29	3.47

[1]One cubic foot per second = 449 gallons per minute

11.8.2 Trash racks and head gates

Even small streams can become torrents carrying large trees and other debris. Plan to protect the generator and water passages from debris by installing a trash rack at the head of the penstock. Set steel bars on edge to the flow of water and space about 1" apart. Normally trash racks are set on an incline to increase area so water velocity is less than 1.5 ft/s through the rack. An inclined rack is easier to clean with a rake. This feature is particularly important in the fall because leaves may blanket a rack in an hour or two. A head gate or valve should be installed below the trash rack to control flow and to allow the turbine to be inspected and repaired.

11.8.3 Turbines

The towering water wheel driving the old mill's grinding stones creates a romantic image, but it is too slow and ponderous to efficiently convert water power to electric power. For example, a 5 ft diameter wheel that is 16" wide will generate only 300 W or less. A compact turbine and generator is a better choice unless you are renovating an old mill site. Hydroelectric plants are available in capacities ranging from 0.5 to 12 kW. A reaction turbine, either the Francis type or propeller wheel type, is turned by a mass of water falling through a duct encasing a wheel. Reaction-type generators are good choices if you have ample water supply but a low head. A reaction wheel is subject to greater friction losses than an impulse wheel; however, it has greater flexibility in installation (Fig. 11.7).

An impulse (Pelton) turbine turns by the velocity of a jet of water striking the turbine's wheel cups and can operate on as little as 1.5 cfm of water. In order to be most effective, a head of at least 50" is required (Fig. 11.8). The type of facility you wish to provide with electrical service will largely determine whether you use an alternating or direct current generator. Lights and the universal motors that operate small appliances and tools will operate on DC. Large motors, TVs and many appliances require AC to operate. Alternating current may be transmitted greater distances and on smaller wires than is possible with direct current; however, an AC installation does require an extra investment in governing equipment. Direct current generators are usually less expensive than AC generators but they do require expensive inverters to convert to AC. The potential of storing DC in batteries during low-usage periods and at times of uneven water flow is a compensation of such a system.

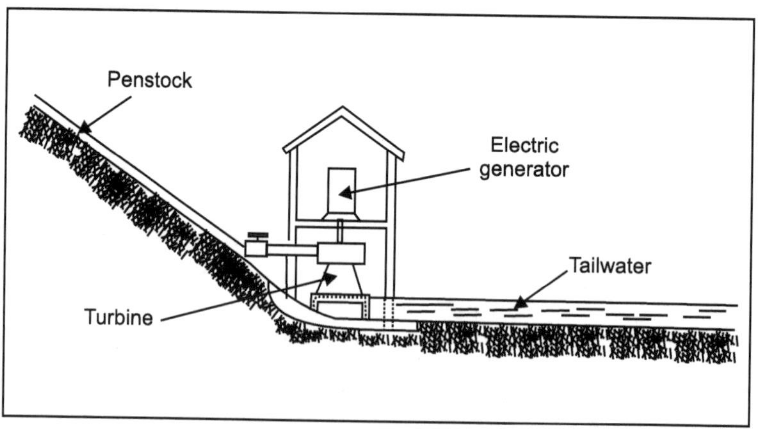

Figure 11.7 Reaction turbine

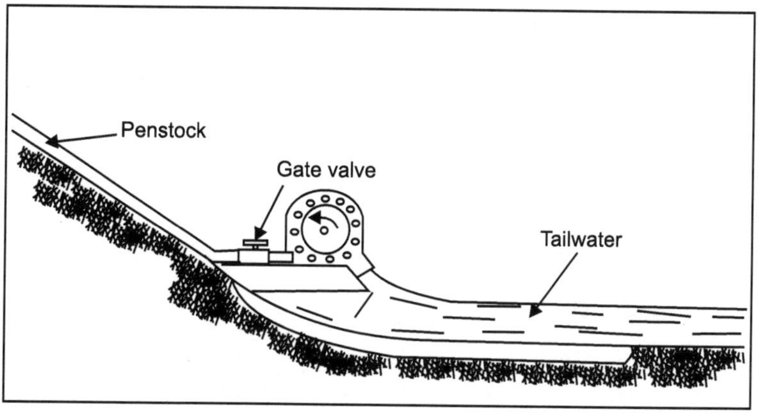

Figure 11.8 Impulse turbine

11.9 Causes of failure

The main reasons for lack of success with small water power developments are:

1. Failure to realise how important full field data is for proper design.
2. Failure of home-made equipment made with junked parts.
3. Overestimating the amount and constancy of the stream flow.
4. Penstocks or flumes, which are too small to allow the plant to operate at full capacity.
5. Failure to anticipate the expense of keeping trash racks clear and machinery in good repair.
6. Failure to design and plan for winter ice buildup.
7. Overestimation of a proposed plant's capability. The average home has demand peaks varying from 4 to 12 kW.

12

Hydropower development in India

12.1 Introduction

India has immense economically exploitable hydropower potential of over 84,000 MW at 60 per cent load factor (1,48,700 MW installed capacity), with Brahmaputra, Indus and Ganges basins contributing about 80 per cent of it. In addition to this, small, mini and micro hydropower schemes (with capacity less than 3 MW) have been assessed to have 6781.81 MW of installed capacity. Of this enormous hydro potential, India has harnessed only about 15 per cent so far, with another 7 per cent under various stages of development. The remaining 78 per cent remains unharnessed due to many issues and barriers to the large scale development of hydropower in the subcontinent.

Various studies have established the ideal hydro:thermal power mix for India at to be at 60:40. The present mix of 75:45 is creating much problem in the Indian power system with country facing energy shortage of 9.3 per cent and peaking shortage of 12.8 per cent. The total requirement ending XI plan is set to be 2,06,000 MW. The current installed thermal and hydropower capacity stands at 66 and 26 per cent of the total power generated with 83,272 and 32,726 MW respectively. Remaining 8 per cent of 10,091 MW is achieved from other forms including wind and nuclear. The current captive generation amounts to 14,636 MW.

India's power system is divided into five major region namely, the Northern region, Western region, Southern region, Eastern region and North–Eastern region, with each region facing separate issues. While the Eastern and North–Eastern regions are power abundant, the Northern and Western regions have greater power demands. The hydropower potential is largest in NE region with 98 per cent of it still untapped. Northern, Eastern, Western and Southern regions have 79, 77, 23 and 33 per cent untapped hydropower potential respectively (Table 12.1).

The Central Electricity Authority (CEA) and Ministry of Power (MoP) are the nodal agencies involved in power sector planning and development at the central level. Being a concurrent subject under the Indian Constitution, electricity is generated, transmitted, maintained and developed both by central and state authorities, with the primary role with the states. With the central policy providing the overall direction for development, state determines the power generation, distribution and management systems. The development of water resources lies with the state government. Since hydropower development involves water resources, the responsibility of its development stays primarily with the state agencies.

Table 12.1 List of hydro electric stations with capacity above 3 MW.

Region	No. of stations	No. of units	Capacity (MW)
Northern	78	234	11,070.30
Western	45	117	6588.80
Southern	92	386	11,004.35
Eastern	26	82	2424.10
Northeastern	15	42	1094.70
Total	256	761	32,182.25

The power sector in India is still largely public with 89 per cent share in the total installed capacity.

12.2 Reforms in the electricity sector in India

With the liberalisation of the economy, the government of India has been encouraging and invited private sector for investment in the power sector. Accordingly, a conducive policy environment has been created by modifying the Electricity Act. The new Electricity Act–2003 deals with the laws relating to generation, transmission, distribution, trading and use of electricity. The Act has specific provisions for the promotion of renewable energy including hydropower and cogeneration. It has been made mandatory that every state regulatory commission would specify a percentage of electricity to be purchased from renewable by a distribution licensee. The National Electricity Policy announced in 2005 aims at access of electricity by all households and per capita availability of electricity to be increased to 1000 units by 2012.

The policy underlines that renewable energy potential needs to be exploited and private sector would be encouraged through suitable promotional measures. Regarding fixing of tariff, the government has announced Tariff Policy in 2006 wherein the State Regulatory Commissions are required to fix tariff in their respective state and also decide about the renewable purchase obligation. The Electricity Act and Tariff Policy are favourably tilted towards increasing power generation from renewable. Now, Central Electricity Regulatory Commission has also announced the tariff calculation guidelines for renewable technologies including for small hydro projects.

The existing power deficit and a rapid growing demand coupled with government commitment to provide access to electricity for all has necessitated a large scale capacity addition program. A capacity addition of 78,000 MW in the 11th Plan (2007 to 2012) and approximately one 1,00,000 MW in the 12th plan (2012–2017) is planned. Concurrent investments in transmission and distribution are also going on.

Such a gigantic task is strongly supported and complemented by the private sector. These changes facilitated the removals of barriers to investment, improved the functioning of the system and resulted in additional generation of power much in excess of that achieved in the earlier plans. Ministry of new and renewable energy (MNRE) government of India is the nodal ministry for small hydropower development in India.

12.3 Government of India (GOI) policy on hydropower development

Despite hydroelectric projects being recognised as the most economic and preferred source of electricity, the share of hydropower in our country continued declining since 1963. The hydro share declined from 50 per cent in 1963 to about 25 per cent in 2010. For grid stability the ideal hydrothermal mix ratio for Indian condition is 40:60. In order to correct the hydrothermal mix to meet the grid requirements and peak power shortage, in August, 1998 and thereafter in Nov. 2008, the government of India announced a Policy on 'hydropower development'. Project affected people have been made long term beneficiary stakeholders in the hydro projects by way of 1 per cent of free power with a matching 1 per cent support from state government for local area development thus ensuring a regular stream of benefits. An initiative of installing 50,000 MW large hydro projects in the country was announced by the government. By 1998 small hydropower projects established themselves as a techno economically viable option for generating power with some preferential treatments.

Encouraged by the growing private sector participation in the sector and the potential of SHP projects to meet power requirements of remote and isolated areas, where grid extension is relatively expensive, small scale hydro was identified as an area to provide thrust in the overall hydropower development of the country. This led to transfer of the subject of hydro up to 25 MW from Ministry of Power (MOP) to MNRE in December 1999.

The process of reforms is an ongoing one and government of India has been vigorously pursuing this path for the past five to six years. Hydropower is a renewable source of clean energy and is used to supplement the base load provided by thermal power plants and storage for wind energy through pumping. To enable the project developer in the hydro sector a reasonable and quick return on investment, merchant sale of up to a maximum of 40 per cent of the saleable energy has been allowed.

Central Electricity Authority (CEA) has issued various hydroelectric related reports and guides are available through web. Some of them are the best practices in hydroelectric generation; preliminary ranking study of hydroelectric scheme; guidelines for accord of concurrence of HE scheme; guidelines for formulation of DPRs for HE scheme; draft model contract document for hydro projects; project monitoring status reports; project clearance status reports and Status of 50,000 MW hydroelectric Initiative reports. The 12th five-year plan also suggests that for projects held up for environment and forest problems, efforts may be made by the concerned state government/developer to get the timely E&F clearances. Problems such as local agitation (law and order), land acquisition, etc. need to be resolved by the concerned state government. Tendency of converting storage projects (as identified by CEA) to run-of-river projects should be discouraged. Project developer should seek long-term open access by indicating at least the region(s) in which they intend to supply their power to enable development of transmission system.

12.4 Hydropower class

There is no worldwide consensus on definitions regarding SHP, mainly because of different development policies in different countries. Based on installed capacity of hydropower projects, classification of hydropower varies differently in various countries. A general classification may be taken as:

Pico	5 kW and below
Micro	100 kW and below
Mini	2000 kW and below
Small	25,000 kW and below
Medium	1,00,000 kW and below
Large	Above 1,00,000 kW

India has a history of about 110 years of hydropower. The first small hydro project of 130 kW commissioned in the hills of Darjeeling in 1897 to mark the development of hydropower in India. At Present the biggest capacity plant is a run of river Naptha Jhakri Hydro project of 1500 MW in Himachal Pradesh. Development of hydropower resources is important for energy security of the country. It takes about 10 years for developing a large-scale hydro project, from planning to commissioning. The hydropower development is greatly boosted with the hydropower policy announcement in 1998 and again in 2008. Advance action on the identified hydro electric schemes need is being taken during 11th Plan (2007–12) period itself as long period is required for development of detailed project reports (DPRs), obtaining various clearances like environment and forest clearances and CEA clearance, investment decision and achieving financial closure. It has been proposed to maximise hydro capacity addition during 12th plan (2012–17) for reducing CO_2 emissions and energy security of the country. A shelf of 109 hydro electric schemes aggregating to 30,920 MW has been identified.

12.5 Small hydropower development

12.5.1 Grid-based SHP

Beginning of the 21st century saw near commercialisation in the small scale hydro sector. There are 760 small hydro projects in India with total installed capacity of 2820 MW. The MNRE decided that out of the total grid interactive power generation capacity that is being installed, 2 per cent should come from small hydro. This translates to about 1400 MW capacity addition during 2007–2012. A target of 3000 MW for the 12th from small hydro has been fixed and to increase capacity addition of about 500 MW per year. The present focus of the SHP program is to lower the cost of equipment, increase its reliability and set up projects in areas that give the maximum advantage in terms of capacity utilisation. SHP projects are being set up both in public and private sector.

Today the SHP program in India is essentially private investment driven. 228 private sector SHP projects of about 1230 MW capacity have been setup. Private sector entrepreneurs are finding attractive business opportunities in small scale hydro.

The MNRE is giving financial subsidy, both in public and private sector to set up SHP projects. In order to improve quality and reliability of projects, it has been made mandatory to get the project tested for its performance by an independent agency and achieving 80 per cent of the envisaged energy generation before the subsidy is released. In order to ensure project quality/performance, the ministry has been insisting to adhere to IEC/International standards for equipment and civil works. The subsidy available from the Ministry is linked to use of equipment manufactured to IEC or other prescribed international standards.

12.5.2 Decentralised SHP

The rural energy scenario in India is characterised by inadequate, poor and unreliable supply of energy services. Realising the fact that small scale hydropower projects can provide a solution for the energy problem in rural, remote and hilly areas where extension of grid system is comparatively uneconomical, promoting small scale hydro projects is one of the objectives of the small hydropower program in India. A number of mini/micro hydro projects have been set up in remote and isolated areas, mainly in Himalayan region. While these projects are developed by various state agencies responsible for renewable energy, the projects are normally maintained with local community participation. A number of tea garden owners have also set up such micro hydro projects to meet their captive requirement of power. Isolated grid often faces the problem of poor plant load factor and making financial return difficult for the plant. But this provides opportunities for the area to have industry expansion, cottage or small industry, irrigation pumping, drinking water, agro and other application, education and entertainment activity for the overall development of the area.

Water wheels have traditionally been used in the Himalayan regions for rice hulling, milling of grain and other mechanical applications. With the R&D efforts, new and improved designs of water mills have been developed for mechanical as well as electricity generation of 3 to 5 kW. These designs were tested at AHEC, IIT Roorkee and have been replicated by about 12 very small scale manufacturers. Local organisations such as the Water Mill Associations, cooperative societies, registered NGOs, local bodies, and State Nodal Agencies are being encouraged to install watermills in their areas. The state of Uttarakhand has taken a lead in setting up electricity generation watermills and over 500 such watermills were installed in remote and isolated areas of the state.

12.6 Equipment manufacturing status

India has a wide base of manufacturers of equipment for hydropower projects. State-of-the-art equipments are available indigenously. Fifteen manufacturers produce almost the entire range and type of hydropower equipment. Most of the world's leading equipment manufacturers have their factory and/or offices in India. In addition, there are about 5 manufactures that are producing micro hydro and watermill equipment.

12.7 Performance testing of SHP stations

In the hydropower the energy transformation process is highly efficient, usually with well over 90 per cent mechanical efficiency in turbines and about 98 per cent in the generator. The inefficiency is due to hydraulic loss in the water circuit (intake, turbine, tailrace), mechanical loss in the turbogenerator group and electrical loss in the generator. Old turbines can have lower efficiency, and it can also be reduced due to wear and abrasion caused by sediments in the water. The rest of the potential energy (100 per cent – η) is lost as heat in the water and in the generator. Thus energy efficiency measurement is one of the key issues.

The efficiency in electromechanical equipment, especially in turbines, can be improved by better design and also by selecting a turbine type with an efficiency profile that is best adapted to the duration curve of the inflow.

As per MNRE directive the small hydro project when commissioned is required to be tested for its performance by Alternate Hydro Energy Centre (AHEC), Indian Institute of Technology (IIT) Roorkee.

The subsidy is released after project attaining the following:

1. Overall performance of the station should be satisfactory.
2. Plant equipment should conform to Indian/International standards.
3. Weighted average efficiency of generating units should, with certain exceptions, be at least 75 per cent.
4. Project should have attained 80 per cent of projected generation for a minimum of 3 months at a stretch.

The biggest challenge for the performance testing was observed the absence of availability of provisions required for such tests in majority of the commissioned SHP plants. With such initiative SHP developers have started taken keen interest by way providing the necessary provisions in the civil structure/equipment for facilitating the testing. Needless to say that such evaluation shall help SHP plant owners regular monitoring of performance of their plants.

However there is no mandatory condition for performance testing for large hydropower projects until the owner wishes to do so as per contractual conditions. Electricity regulators and financial institutions are being followed up to adopt conditions for remunerative tariff from hydropower.

12.8 Benefits of hydropower and reasons for its slow development

Hydropower has immense benefits and has been brought forward as a preferred option for power generation over the last decade. The reasons for these can be summed as follows:

1. Abundant potential of hydropower development in India as discussed above.
2. With relative independence from international market like oil prices, hydropower involves no extra foreign exchange outgo.
3. Hydropower is a no-inflation power as water—the 'raw material' for power generation is free of inflation.

4. Environment friendly.
5. Hydropower projects support socio-economic development of remote areas as the project site is developed.
6. Hydropower is cost effective and renewable form of energy.
7. It has additional benefits like irrigation, flood control, tourism, etc.

Even with these benefits, hydropower has had slow development in India especially in last few decades. This has primarily been due to:

1. Long gestation period.
2. Time consuming process for project clearances.
3. Until recently, the national focus has been on thermal generation.
4. Highly capital intensive and absence of committed funds.
5. Poor financial health of State Electricity Boards (SEBs).
6. Technical constraints due to complex geological nature of the projects.
7. Interstate disputes as water is a state subject.
8. Absence of long tenure loans makes it difficult for private investors.
9. Advance against depreciation is disallowed.
10. 14 per cent return on equity (ROE) is not attractive enough for investors.
11. Dearth of competent contracting agencies to construct the project site.

The national focus on thermal generation has been shifting towards hydropower development. To this effect, Government of India introduced a national policy on hydropower development in 1998, in which hydropower has been accepted a national priority with emphasis on unharnessed potential especially in the NE region. Key policy initiatives to accelerate hydropower have been introduced in the policy document, which are discussed in detail in the following section.

12.9 National policy on hydropower in India

With the aim to accelerate the development of hydropower, the Ministry of Power (MOP), Government of India (GOI) introduced the National Policy on Hydropower Development in 2004. The policy document has identified and responded to the major issues and barriers. The objectives of the National Policy document on hydropower development, 2004 are (as stated in the chapter):

1. To ensure targeted capacity addition during 9th plan (and the subsequent plans).

2. With Central, State and Private hydropower projects contributing 3455, 5810 and 550 MW respectively, the GOI aims to reach the total capacity of 9815 MW during the ninth plan. (The XIth plan aims capacity addition of 18781 MW in the hydropower sector.)
3. Exploitation of vast hydro electric potential at faster pace.
4. The government would take steps like execution of all CEA cleared projects, update and clear pending DPRs, survey new green field sites and resolve interstate disputes.
5. Promotion of small and mini hydro projects.
6. Small and mini hydro projects are especially viable for remote and hilly areas where extension of grid system is comparatively uneconomical.
7. Strengthening the role of PSUs and SEBs in taking up new hydro projects.
8. The government aims at enlarging Public sector's involvement in mega hydro projects and multipurpose projects involving interstate issues, projects for peaking power and those with rehabilitation and resettlement issues.
9. Increasing private investments for development of hydropower in India.
10. The public sector would be supported by greater private investment through IPPs and joint ventures. Private sector participation is considered vital for large scale development of hydropower.

Through these measures, GOI aims to realise 100 per cent hydropower potential of the country by year 2025–26. These objectives have been developed in response to the following constraints:

1. Technical, including difficult investigation, inadequacies in tunnelling methods.
2. Financial (deficiencies in providing long-term finance).
3. Tariff related issues.
4. Managerial weakness (poor contract management).
5. Geological surprises (especially in the Himalayan region where underground tunnelling is required).
6. Inaccessibility of the area.
7. Problems due to delay in land acquisition and resettlement of project affected families.
8. Law and order problem in militant infested areas.

12.10 Development since national hydropower policy 2004

In 2006, Central Electricity Authority (CEA) introduced a ranking study which prioritised and ranked the future executable projects. As per the study, 399 hydro schemes with an aggregate installed capacity of 10,6910 MW were ranked in A, B and C categories depending upon their interse attractiveness. This was followed by a 50,000 MW Hydro Initiative in which preparation of pre-feasibility reports of 162 projects was taken up by CEA through various agencies. The project-feasibility reports (PFRs) for all these projects have been prepared and projects with first year tariff less than INR 2.50/kWh have been identified for preparation of detailed project report (DPR).

12.10.1 Renovation, modernisation and up-rating

In order to augment the hydro generation and improve the availability of existing hydropower projects, Government of India has put emphasis on renovation, modernisation and up-rating (RM&U) of various existing hydro electric (HE) power projects in the country. RM&U of the existing/old hydro electric power projects is considered the best option, as this is cost effective and quicker to achieve than setting up of green field power projects. The cost per MW of a new HE power project works out to about INR 4 to 5 crores whereas the cost per MW of capacity addition through up-rating and life extension of old HE power project works out to about 20 per cent. Further, the RM&U of a hydro project can be completed in 1 to 3 years depending upon scope of works as compared to gestation period of 5 to 6 years for new hydro projects.

Under the hydro RM&U program, 33 hydro electric projects (13 up to the VIIIth plan and 20 in the IXth plan) with an installed capacity of 6174.10 MW have been completed by the end of the IXth plan. During the Xth plan (2002–07), 47 HE power projects with an installed capacity of 7449.20 MW are programmed for completion of RM&U works. For the XIth plan (2007–12) a total of 59 hydro electric power projects having an installed capacity of 10325.40 MW are programmed for completion of RM&U works to accrue a benefit of 5461.18 MW.

12.10.2 Capacity addition scenario XIth plan (2007–12)

In support to the 2004 national hydropower policy, GOI plans to wipe out all energy shortage by the end of 2011–12, i.e. by end of XIth Plan and also to provide spinning reserve and ensure uninterrupted quality power at affordable cost. With coal based power plant as the backbone of Indian Power sector, during the XIth plan capacity addition of about 40200 MW coal based plant is planned with introduction of supercritical technology (Table 12.2).

Table 12.2 Sector-wise plan of capacity addition in XIth plan.

Prime mover	Hydro (MW)	Thermal (MW)	Nuclear (MW)	Total (MW)
State sector	3957	15,538	–	19,495
Central sector	11,080	19,880	3160	34,120
Private sector	3744	11,145	–	14,889
Total	18,781	46,563	3160	68,504

12.10.3 Three-stage clearance system

A three-stage clearance system has been set up to enable relatively faster and hindrance free clearance of suitable projects and includes survey, investigation and pre-construction activities. The three-stage clearance system works as under:

Stage I: Survey and preparation of pre-feasibility reports.

Stage II: Detailed investigation, preparation of DPR and pre-construction activity including land acquisition.

Stage III: Execution of the project after investment decision through PIB/CCEA.

Small hydropower projects up to 25 MW can be set up in private sector without Central government's involvement. Techno-economic clearance needs to be obtained from CEA if the estimated cost of the project exceeds INR 2500 million and/or there are interstate issues involved.

12.11 Current issues/problems with hydropower in India

The Government of India set up a National Committee in 1987 and a Standing Committee in 1998 to oversee the progress on hydropower development.

This section derives largely from the report submitted by the Standing Committee on Energy (2005–2006)—Hydropower: a Critique which discusses the actions taken by the Government on the recommendations made by the Committee in the forty-second report on hydropower in India.

12.11.1 Technical issues

1. To expedite early execution of hydro projects, bankable detailed project report (DPR) based on detailed survey should be prepared to avoid geological uncertainties. Survey and investigation and analysis of geological, geo-morphological, geo-electrical, hydrological data, etc. should be done at the time of preparation of a DPR itself in order to minimise the impact of risks. It is, therefore, necessary to expedite survey and investigations with the latest state of the art technology and prepare a shelf of projects for execution. The quality of DPRs should be of high standard which should infuse confidence in the national/ international developers to take up the execution of projects without losing time in rechecks, etc. at the same time, contract monitoring as distinct from project monitoring should be emphasised and land acquisition and infrastructure development be settled and completed before the start of the project.
2. Renovation and modernisation (R&M) has been recognised world over as a well proven cost effective technique for improving the performance/efficiency of older power plants. The useful life of the plants can be increased by R&M and the plants yield benefits in the shortest possible time at a reasonable cost. GOI in its policy on hydropower development, 1998 has laid stress on need for renovation and modernisation of hydropower plants. Accordingly, Government of India set up a Standing Committee to identify new hydro R&M schemes to be undertaken for implementation under Phase-II.
3. The pump storage potential should be harnessed as these are essential in optimising energy generation from base load thermal stations and in meeting peak load and system contingencies. Only 2.45 per cent of total identified potential of 94,000 MW pump storage schemes had been harnessed and another 2.5 per cent are under construction. New exclusive program/action plan for PSS should be launched, to tap the vast potential.
4. Contingency plan for hydro projects affected by natural calamities need to be prepared and made public.
5. National policy on rehabilitation and resettlement (R&R) should be finalised and made public.

12.11.2 Infrastructural issues

1. There is a need to setup single window clearance for hydro projects. Various authorities such as the Central Electricity Authority (CEA), the Ministry of Finance, Ministry of Environment and Forests, etc.

are involved in the appraisal of a hydropower project before it is certified for development. It will be desirable to have a single window dispensation/authority so that a project is cleared without many hassles. Any hydro project submitted for clearance should receive all the statutory/non-statutory clearances/approvals within six months of submission of the proposal. The certification of commercial viability should be given within 15 days, especially to private developers. The techno-economic, MoEF and CCEA clearances should be given within 1 and 2 months respectively. The Ministry of Power should have a set of hydro projects cleared from all the angles. MoEF should also be involved in the appraisal process.

2. There are long delays on account of land acquisition for the project. The process of land (both private and Govt.) acquisition for a project differed from state to state as per Land Acquisition Act. The Government should amend Land Acquisition Act and include hydropower projects in the priority list and the State Governments should be persuaded to provide land to the project authority in agreed time frame to facilitate shifting of project affected persons (PAPs).
3. The hydro projects which involve lesser risk element and entail lesser capital investment can be considered for development in the private sector. Public sector can take up: (i) multipurpose projects, (ii) projects involving interstate issues and in interstate river systems, (iii) projects involving cooperation with neighbouring countries, (iv) projects for complementary peaking with regional benefits, and (v) projects in the North–Eastern region, etc.

12.11.3 Financial issues

There is also a need to off-load indirect cost components on hydro project. Many hydro projects are located in troubled areas and infested by militancy and terrorist activities. There is an urgent need to amend the present policy of the Government in regard to charging the entire security expenditure from concept and until commissioning—on the project cost. However, the recurring expenditure incurred on security, once a project goes on stream could to be charged on the project developer.

The cost of access roads should not be included in the project cost, as development of hydro projects triggers economic and commercial activities around the project site and results in economic benefit to the state. Inclusion of R&R, flood moderation costs, along with the provision of 12 per cent free power to the state in the capital cost of the project needed reconsideration as the provision did not apply to thermal power projects.

Although the government has planned to achieve 50,000 MW of additional power by the end of 11th plan period, the incentives such as benefits/concession in custom duties and local levies/taxes on project components are being denied for projects even up to 250 MW resulting in low investments in new power schemes.

A premium as well as lease rent @10 per cent is charged where forest land is diverted for a hydropower project. This needs to be discussed with the state governments as land is a subject matter.

12.12 State-specific hydropower issues

12.12.1 Issues related to hydropower development in Uttarakhand

Uttaranchal has an estimated hydropower potential of 20,236 MW against which only about 1407 MW has been harnessed so far. The Government of Uttarakhand (GOU) has introduced separate policy guidelines for development of hydropower projects of capacity up to 25 MW and 25–100 MW. The policy promotes small hydropower (SHP) projects with installed capacity of up to 25 MW through IPP. The main issues from the policy document are discussed as follows.12.12

Lack of detailed project reports (DPRs) and technical data

Lack of bankable DPRs of the projects for development in the private sector before their allotment makes it very difficult for the private investors/developers to take investment decisions. Similarly technical data on availability of water at the diversion site of the project should be made available to the developer before project allocation. Currently, river discharge observations are not made available to the developers on pretext of confidentiality and are made available to the concerned government department only after the approval of the Ministry of Water Resources, GOI. Considerable time is lost in getting the approval. It is suggested that the states should organise measurement of hydrological and meteorological data of all identified sites of hydroelectric projects through Central Water Commission or other dependable agencies. The data collected should be made available to the developers allowing for better assessment of river hydrology and reduced hydrological risk.

Project allotment on the basis of premium

Allotment for projects above 100 MW is made on the basis of the highest bid received from the developers after short listing. The premium paid by the developers increases the tariff, making hydropower expensive. It is suggested that the promoters should be selected on the basis of their competence and experience and not alone on the basis of premium.

Upfront fee

The developers are charged an upfront fee of INR 5 lacs per MW by the state. This increases the cost of electricity generated by hydropower making implementation of Electricity Act 2003 difficult.

Land procurement and development cost

The policy on land provision for hydropower differs from state to state. While in Uttarakhand the land is available at market price for project development, Uttar Pradesh government provides land at INR 100 per acre. This difference in land cost makes hydropower development very difficult in few states as compared to others. It is suggested that the land for project development should be acquired by the state and handed over to the project developer.

The state should provide for the cost of approach roads, resettlement and rehabilitation of the project affected people instead of loading the cost on the project.

Payment security

The states with hydro potential are generally power surplus making hydropower power sale financially less lucrative. It is suggested that the State should guarantee to buy power from developers at the rates fixed by the Regulatory Commission. They should also be allowed to sell the power independently, if so desired by them.

Back loading the free power

Being capital intensive, the tariff in hydroelectric projects is high during initial years of project operation. If free power is back loaded, the tariff during the initial years will be reduced, thus making hydropower more competitive compared to other sources of power.

12.12.2 Issues related to hydropower development in Himachal Pradesh

The total identified potential in HP is 20386.07 MW. Out of this 6045.07 MW has been harnessed so far and 2720.5 MW is under execution. For which DPR is ready for 3011.50 MW, projects for 3671.50 MW are Under Investigation and 4187.50 MW of schemes are yet to be investigated. The main issues related to Hydropower development in the state are as follows.

Purchase rate of power by state

In Himachal Pradesh the purchase rate of power was fixed as INR 2.50 in 1997–98 for small hydro. Since then, the cost of steel, cement and labour has increased considerably. However, the purchase rate for hydropower has not been revised by the state. It should be revised at the earliest.

Local area development

In HP 1.5 per cent of the project cost is realised from the developers for local area development. This unnecessarily increases the tariff. As stated above, the local area development should be paid from the revenue of free power received by the State.

Fishery department compensation

Compensation at the rate of INR 5 lac per MW and 1 lac per km of breeding area lost due to project is payable by the developer. It is suggested that the developer should not be made liable for this.

Charges for augmentation of transmission system of HPSEB

It is understood that Himachal Pradesh State Electricity Board is going to charge INR 40 lac per MW from developers for augmentation of transmission system beyond interconnection point. This should be reconsidered.

12.12.3 Recommendations for hydropower development in Himachal Pradesh

Various issues and barriers to large scale development have been identified in the above discussion. As the first step towards hydropower development, the state government guarantee has been withdrawn comforting financers as well as Independent Power Producers (IPPs). Following policy recommendations are suggested to the Government of Himachal Pradesh to further attract significant investment in the sector of Hydropower.

One-stop window for clearances

Private investors expect clear rules, a streamlined application process and a reasonable timeline for response from the HP state government.

Several clearances including IPH, Public Water Department (PWD), Gram Panchayat, Wildlife, Forest, conversion of private land under section 118 of HP. Land Reforms and Tenancy Act, No-objection certificate (NOC) from Fisheries Dept. and NOC from Pollution Control Board are to be taken within the six months time. The offices of respective departments are located in different places and the given time frame is very inadequate for the application process.

The procedure should be simplified. Once the project is conceived, the government should indirectly give clearances to set it up. Considering the small size of projects, the government should give one clearance to cover all the items. Alternatively, there should be a single point body to give the clearances.

Streamlining fees and royalties

Several added costs have burdened the project. The net present value (NPV) payment for government land to Forest Dept., fees to Pollution Control Board and compensation to Fisheries Dept. are significantly high. Recently, new norms for Fisheries have been proposed @INR 5 lacs per MW and INR 1 lac per kilometre (KM) upstream of tail race which may add up to 30 lacs for a 5 MW project.

In addition, government is proposing a development fund of 1 to 1.5 per cent of the project cost for local development works which can be up to INR 40 lacs for a 5 MW project. The NPV compensation to Forest Department also can be up to INR 40 to 50 lacs for the project land and transmission line. All these total to more than INR 100 lacs per project.

Recent news in the press suggest that 15 per cent water is to be released for downstream use. Earlier a provision of 10 per cent was made. During lean season, the plant may have to be shut down. The wheeling charges and transmission charges should be fixed for projects up to 25 MW. Currently, there is no such provision and it is to be decided by the ERC. There should be a government policy.

The IA draft has been revised in the recent past. There are some provisions which should be reviewed, e.g. Arbitration by sole Arbitrator and he is Secretary to Government of Himachal Pradesh (GOHP) other than Energy Secretary. There is a penalty or extension fee of INR 10,000 per month per MW proposed for extension beyond 6 months. Only one extension of 6 months is allowed. This time frame is inadequate. The IPP is anyway incurring cost on the project and should not be burdened with such extra fees.

Escalation inclusive tariff

A tariff of INR 2.50 per unit for 40 years is unrealistic. The prices of cement and steel have significantly increased, and an escalation provision must be added.

Transparency in land acquisition

The private land owners have started asking very high price from IPPs. The government should acquire land for the project and provide to IPPs at a fair cost.

Expansion of transmission

The transmission system as existing is inadequate to cater to large number of small projects which will need to be developed in HP. Evacuation is required at a manned substation. Substations are also limited and located very far from the projects requiring longer transmission lines. Besides high cost towards forest compensation and long transmission lines, there are transmission losses. HPSEB is implementing a master plan with multilateral funds, e.g. Asian Development Bank, etc. for improving the infrastructure in transmission. This revamping should be done keeping the requirement of small projects in view. HPSEB is undertaking a policy that IPPs shoulder the expenditure towards improving the basic transmission infrastructure beyond interconnection point. This should not be burdened on the projects.

Several projects have been awarded in different regions with limited availability of substations for interconnection. A few substations have already started facing a space and capacity crunch. As a result, the evacuation is becoming difficult as well as expensive because the capacity is sent to other substations located at longer distances.

12.13 Establishment of R&D hydraulic turbine laboratory

An R&D hydro turbine laboratory of international level being established at AHEC IIT Roorkee with the financial support from MNRE, at AHEC, IIT Roorkee for the multiple purposes, viz. turbine-model testing, research and development (R&D), human resource development (HRD) along with the other purposes of verification of designs, generation of design data, design validation through CFD analysis, witnessing tests on turbines/pumps in field. The laboratory is expected to be fully functional by the end of year 2012.

12.14 Standards for small hydropower

There is a series of standards, guidelines and manuals on hydropower issued by international standards organisations like ISO, IEC, IEEE, ASME, USBR and the national statuary bodies of several countries, including India where it has from CEA, REC, BIS and CBIP. But most of them were prepared keeping in view the large water/hydropower projects. SHP needs to be made profitable and a long-term investment opportunity, while ensuring quality and reliability of the power. To make SHP cost effective and reliable, standards, guidelines and manuals are required covering entire range of SHP activities. Necessity of the standards/guidelines and manuals has been strongly felt by developers, manufactures, consultants, regulators and others. The efforts of Government of India have taken initiatives to prepare about 30 standards for SHP and are expected available by the year 2013.

12.15 Real time simulator

Training and human resource development is given due importance by the ministry of new and renewable energy. Towards this, apart from regular training programs, a real time digital simulator (RTDS) for SHP plants has been established at alternate hydro energy centre (AHEC), IIT Roorkee, and India with the aim of providing efficient initial and advanced training to operators and engineering staff of different types of SHP plants.

Training conditions have been created very close to real operating conditions. This will meet the large requirement of trained personnel for operation and maintenance of SHP plants, reduce O&M costs, reduce damage to plant, increase plant life and reduce training time. The hydraulic part, the generator, the transformers as well as auxiliary electricity systems are taken care in the simulation. Present and future operators are taking the benefits of such training and is being utilised by national and international personnel.

12.16 Performance of Tawa hydroelectric power plant

Small hydropower project may be used as one of the option for achieving the energy targets in a developing country like India where centre or state governments have limited financial resources to put in large projects which require long gestation period. One additional advantage with the small hydropower project is that private partners may get attracted due to low investment and quicker return in comparison to large projects. The last but not least is the most ecofriendliness of small power projects which is a point of serious concern in case of thermal, or nuclear or sometimes in big Hydropower projects depending upon the location of the projects.

Small hydropower potential in India is still underutilised and there is need to tap this potential for optimum utilisation of natural resources. In Madhya Pradesh, small hydro plants are not many; however there is good scope for developing such plants. Tawa is one of such plants in MP, which has been developed as canal head powerhouse on the left bank canal (LBC) of Tawa irrigation project by a private investor. This plant is working in a very efficient manner addressing both the power and irrigation aspects successfully. This example will attract the private investments in small hydropower sector in the developing countries like India.

Hydropower is a renewable, non-polluting and environment friendly source of energy. It is perhaps the oldest energy technique known to mankind for conversion of mechanical energy into electrical energy. Hydropower represents use of water resources towards inflation free energy due to absence of fuel cost. Hydropower contributes around 22 per cent of the world electricity supply generated. The total potential of small hydropower of the whole world is 7,80,000 MW out of which 50,000 MW has already been utilised.

Small hydro is also the highest density resources in generation of electricity due to the reason of being it environment friendly, flexibility in operation and suitability in giving support in peak time to the local grid.

Due to the small gestation period, small capital investment and quicker return involved, in recent years it has become the point of attraction for private sector. Fiscal incentive announced by the central and state governments time to time for investment in this sector have further caused private investor to give attention to this sector.

Small hydropower plants (SHP) provide maximum benefits in minimum time. And offers the fastest economical means to enhance power supply, improve living standards, stimulate industrial growth and enhance agriculture with the least environmental impact and without heavy transmission losses. Due to less transmission losses there is a reduction in distribution cost as well. Its availability at the head of the irrigation canals and small streams is also a one of the added advantage of it.

12.17 Small hydro development in India

India has a rich history of small hydropower development. The first hydropower plant was set up in the year 1897 with a capacity of 130 kW in Darjeeling district. After independence due to increase in power requirement the thrust shifted from small hydropower to large size hydro and thermal project.

After opening of our economy and envisaging an aspire growth rate of 7–8 per cent, the need of energy at low cost has been felt much. Here it would be desirable to clarify that India is a developing country, which cannot afford to invest huge amount of money in only large projects with long gestation period, and hence it was felt more reasonable also to go for small hydropower project with participation of private sector. Further the small gestation period and its ability to cater local grid need, with minimum transmission losses reinforced the idea for going towards small hydropower projects. It was felt by the ministry of power that out of total power of the grid to be installed, 2 per cent should come from small hydropower and if this is translated into reality then by 2013 there will be an addition of 2000 MW capacity from small hydropower.

India has a potential of 15000 MW. To tap this potential 4404 potential sites with an aggregate capacity of 10477 MW have already been identified. The installed capacity of SHP project as on 31.03.2008 is 1826 MW from 556 projects. In addition 203 SHP projects with an aggregate capacity of 468 MW are under completion. The growth of capacity addition from small hydro projects during last few years is on an average 100 MW per year.

12.17.1 Scope for investments in small hydropower sector

There is lot of scope for investment in small hydropower sector. Very few power plants are under the private sector management. One of such hydroelectric plant, which is working successfully, is Tawa hydroelectric power plant under the efficient management of M/s HEG (Hindustan Electro Graphite Ltd.) in Madhya Pradesh (MP).

Hydroelectric power is one of the cleanest and most viable renewable sources of energy. Government has been constantly promoting setting up of non-conventional power plants, including hydro, through various policies, incentives and agencies specially created for this purpose. The policies formulated so far by the central and state governments had been all encompassing and applicable to all kinds of non-conventional power producing sources like, wind, biomass, solar, municipal solid waste and hydro.

12.17.2 Status of private participation in India in small hydropower sector

After adaptation of liberalisation policy in year 1991 it was felt necessary to open the energy sector to private partners. In order to achieve the targets set up by Government of India to cater the requirement of power in coming years state as well as central nodal agencies were set up to provide assistance for obtaining necessary clearance in allotment of land and project site to private partners. A provision was also set in the electricity act of 2003 to set up state electricity Regulatory commissions, which would promote generation of electricity from non-conventional sources by providing suitable measures for connectivity with grid. In continuation of this till now 16 states have policies (including Madhya Pradesh) for private sector participation. These states have offered sites to private sector to achieve an aggregate potential of 2500 MW. So far over 120 SHP projects with a total capacity of 500 MW have been set up in private sector in Andhra Pradesh, Karnataka, HP, Punjab, Uttarakhand and Maharashtra.

Ministry of non-conventional energy sources has been providing subsidiary for public sector as well as private sector. In case of private sector, subsidiary is released after successful commissioning of the project. Renewable Energy Development Agency (IREDA) and Rural Electrification Corporation are providing loan assistance for setting up of small hydropower projects. The financial assistance is also being provided by World Bank through IREDA.

12.17.3 Status of small hydropower sector in Madhya Pradesh

The estimated potential for generation from small hydropower projects (SHPs) is 15,000 MW in India, out of which in undivided MP, estimated potential is 410 MW. Despite various promotional schemes and measures announced by MP government, it could set up only 41 MW of SHPs which is only 2.2 per cent of national installed capacity of 1826 MW, as GOI aims to provide 10 per cent of gross power generation in the country through non-conventional energy sources by the year 2013. Therefore MP state has to travel very long distance and sincere efforts needs to be made, especially to make up the deficit in the small hydropower sector.

12.17.4 Policy on SHP in Madhya Pradesh

MP sate government like other state governments and central government had announced policy on small hydropower and on the participation of private sector to increase the small hydropower generation in the state. The main objectives of the policy are:

1. To promote generation of power through small hydropower projects with private sector participation.
2. To define the incentives and benefits available to a private sector.
3. To create a favourable environment for promoting the sector.
4. To lay down the framework for the implementation of the project.

12.18 Case study: Power plant in private sector in Madhya Pradesh

Tawa project is an irrigation project located on River Tawa a tributary of Narmada river. The river Tawa, which is one of major tributaries of River

Narmada, rises from the Satpura range of hills at an elevation of 2500 ft (762.00 m).

The head works comprises earthen dam of average height 22.528 m and masonry dam of 57.912 m heights with central masonry spillway. 13 no. radial gates each of size 15.24 m × 12.192 m are provided with spillway crest at a RL 343.205. The MWL of the dam is 356.692 m. The right bank canal has a cultural command area of 98079 ha and an annual irrigation of 75878 ha (Kharif, Rabi and Summer crops). The left bank irrigation canal has cultural command area 1,86,162 ha and an annual irrigation area of 2,56,904 ha (Kharif, Rabi, and Summer crops).

12.18.1 Hydroelectric plant

Small hydropower plant was setup on left bank to utilise the tailrace water for irrigation purpose. The tailrace channel from the powerhouse leads to LBC. The two units of 2 × 6.75 MW were setup by M/s HEG Ltd., Mandideep. The Power production in this plant was started in the year 1998. During Kharif and Rabi season the water is supplied through powerhouse tailrace channel to command area for irrigation.

12.18.2 Highlights of project

1. The construction of the project was completed in record time of 22 months and at a cost of around 65 crore against the estimated time of 40 months and Rs. 75 crore estimated cost.
2. Early and efficient completion made possible by RSWI, Canada (Joint Venture Company of India and Canada).
3. Optimal scheduling and project monitoring.
4. Low on capital cost and efficient in generation.

12.18.3 Technical highlights

1. Canal head project
2. Catchments area spreads over approximately 6000 sq. km
3. Full reservoir level (FRL) 335.397 m
4. Head range 7 to 21 m and discharge varying from 25 to 54 cumecs
5. Two turbo generators 6.75 MW rated capacity (20 per cent overload)

12.18.4 Machinery and equipment

1. Vertical shaft kaplan turbines and its auxiliaries

2. 11 kV semi umbrella synchronous generator ad auxiliaries
3. 33 kV vacuum circuit breakers
4. 1 MW 33 kV/415 V auxiliary transformer.
5. Fire protection system
6. 110 DC with battery back up

12.18.5 Performance of Tawa Hydroelectric Plant (Year 2006–07)

The generation of hydropower depends mainly on storage of dam resulting due to rainfall. The southwest monsoon arrives in Madhya Pradesh state around June 15th and withdraws by October 15. The reservoir filled up to and above FRL during the past five years.

It is a known fact that small hydropower offers maximum benefit with least environmental impact and less investment compared to other conventional source of energy. On the basis of the following data and discussion it has been tried to establish that small hydropower plays an important role in development of the country with minimum maintenance cost as far as non-conventional energy sources are concerned.

The study of the available data indicates that the plant was neither shut down nor there was any major break down observed during the period 2006–07. This also shows the efficient management, technical expertise and well maintained condition of the plant by the company with optimum technical manpower. It is also learnt that plant worked in the same efficient manner since the year when power generation was started in 1998.

Performance details of plant

Period of generation	4200 hrs
Duration of power generation in %	48%
Power production	5.720 crore units
Performance	94%
Technical manpower deployed	Two technicians per shift
	One engineer per shift
	(8–10 technicians per day)
Highest post at plant	One GM
Total manpower	26*

Year	Reservoir water level on Oct. 15th in m	FRL in m	Generated units (crores)	Approximate selling rate (Rs.)	Approximate revenue in crores of Rs.	Benefit in crores of Rs.
2004–05	354.086	335.397	3.00	2.30	6.9	5.4
2005–06	355.397	355.397	5.20	2.30	11.96	10.46
2006–07	355.427	355.397	5.70	2.30	13.11	11.61

Approximate expenditures details in plant maintenance for one year.

S. no.	Description	No.	Days/ Month	Rate (Rs.)	Month/ Day	Amount (Rs.)
1.	Water charges					
	The water charges are to be paid at the rate of Rs. 60,000/day if the unit runs completely, since there was no breakdown /shut down	–	175	60,000	Day	1,05,00,000 say 1 crore
2.	Establishment charges					
	General manager	1	12	40,000*	Month	4,80,000
	Engineer	3	12	20,000*	Month	7,20,000
	Technicians	10	12	10,000	Month	12,00,000
	Other staff	12*	12	7000*	Month	10,08,000
3.	Annual maintenance charges	–	–	–	LS	10,00,000*
*Approximate charges					Total Rs.	1.49 crores

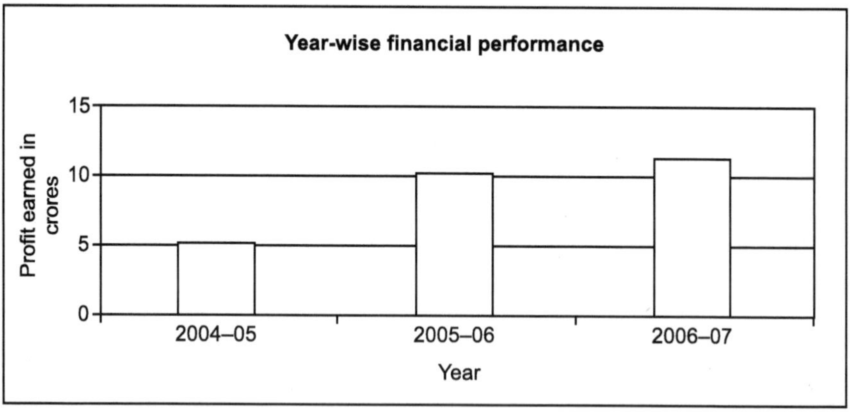

Thus, in order to achieve a growth rate of 7–8 per cent as envisaged in National policy of India, it is also required to tap all the small hydropower potential of the country. The encouraging performance shown by the Tawa project itself set an example for inviting private investment in the small hydropower project sector, especially in view of the fact that large hydropower projects involve huge capital investment and long gestation period which private partners do not afford to bear. The utilisation of small hydropower potential is especially required in all states where the utilised potential is very low like in MP and therefore optimum utilisation of the same may set up a stepping up stone for achieving self-sufficiency in power sector in country.

12.19 Way ahead

Government of India is vigorously promoting large scale development of hydropower in India through changes in policies, all driven to involve greater private entrepreneurs. The major issues plaguing the hydropower sector have been identified and policy changes to mend these barriers are now required. These issues have been discussed earlier in the chapter.

The way ahead to meet the energy demand is as follows:

1. The recommendations made by the Standing Committee for hydropower development are crucial and should be enforced for maximum benefit to the Indian hydropower sector.
2. Consistent policies and regulations should be made through the states. Any variation in policies and benefits offered by different states will cause problems in development of many project sites in different states.
3. Large scale hydro projects which involve greater risks due to geological uncertainties, etc. should be implemented by the state nodal agencies, while the relatively safer projects with reduced risks and smaller capital investments should be offered to the private entrepreneurs.
4. A single window clearance setup for hydro projects will solve most problems related to the clearances, etc.
5. Renovation and modernisation of existing/old hydropower plants should be promoted and planned for, instead of complete focus on setting up of green field power projects.
6. Financial issue like long-term debt financing/long tenure loans, differential tariff for peak and off-peak hours need serious thinking and early implementation.
7. The private sector participation in the large scale development of hydropower should be promoted. This can be achieved either through public–private partnerships or through independent private projects (IPP way). The Government of India has already recognised the need

to increased private involvement and has referred to it in the national policy document.

8. The hydro sector needs to develop a set of competent civil engineers/ contracting agencies that have the technical and the management expertise to conceptualise and develop a project of the required scale. The contract management practices with a transparent system of selection of contractors and resolve any disputes that may arise need to be reviewed
   (a) Underground works (like tunnel, powerhouse, surge tanks)
   (b) Pooling of electricity and marketing
   (c) Efficient operation of hydro reservoirs
   (d) Hydrological estimation planning
   (e) Overall planning and design consultancy

13

Tide, wave and ocean energy

13.1 Introduction

Tidal power, also called tidal energy, is a form of hydropower that converts the energy of tides into useful forms of power—mainly electricity. Although not yet widely used, tidal power has potential for future electricity generation. Tides are more predictable than wind energy and solar power. Among sources of renewable energy, tidal power has traditionally suffered from relatively high cost and limited availability of sites with sufficiently high tidal ranges or flow velocities, thus constricting its total availability. However, many recent technological developments and improvements, both in design (e.g. dynamic tidal power, tidal lagoons) and turbine technology (e.g. new axial turbines, crossflow turbines), indicate that the total availability of tidal power may be much higher than previously assumed, and that economic and environmental costs may be brought down to competitive levels.

13.2 Tides

There are two high tides and two low tides around the earth at any instant. One high tide is on the longitude closest to the Moon and the other on the longitude furthest from the moon. The low tides are on the longitudes at 90° to the longitudes where the high tides are situated. On any given longitude the interval between high tides is approximately 12 hours 25 minutes. The difference in height between a high tide and a low tide is called the tidal range. The mid-ocean tidal range is typically about 0.5–1.0 metres but is somewhat larger on the continental shelves. In the restricted passages between islands and straits the tidal range can be significantly enhanced, up to as much as 12 m in the Bristol Channel (UK) and 13 m in the Bay of Fundy (Nova Scotia). Tidal power has the advantage over other forms of alternative energy of being predictable. For conventional tidal power generation it is necessary to construct huge tidal basins in order to produce useful amounts of electricity. However, in recent years, an alternative technology for exploiting strong tidal currents has been under development using underwater rotors, analogous to wind turbines. Table 13.1 shows the potential of some large tidal range sites in various locations around the world.

13.2.1 Physical cause of tides

The main cause of tides is the effect of the moon. The effect of the sun is about half that of the moon but increases or decreases the size of the lunar tide according to the positions of the sun and the moon relative to the earth. The daily rotation of the earth about its own axis only affects the location of the high tides.

Table 13.1 Tidal potential of some large tidal range sites.

Country	Site	Mean tidal range (m)	Basin area (km^2)	Capacity (GW)
Argentina	Golfo Nuevo	3.7	2376	6.6
Canada	Cobequid	12.4	240	5.3
India	Gulf of Khambat	7.0	1970	7.0
Russia	Mezen	6.7	2640	15.0
Russia	Penzhinsk	11.4	20,530	87.4
UK	Severn	7	520	8.6

For simplicity we assume that the earth is covered by water. Consider a unit mass of water situated at some point P as shown in Fig. 13.1. The gravitational potential due to the moon is given by:

$$\phi = -\frac{Gm}{s}$$

where, G is the gravitational constant, m is the mass of the moon, and s is the distance from P to the centre of the moon. For d p r we can expand 1/s as follows:

$$\frac{1}{s} = \frac{1}{[d_2 + r_2 - 2dr\cos\theta]^{\frac{1}{2}}} = \frac{1}{2}\left[1 + \left(-\frac{2r}{d}\cos\theta + \frac{r^2}{d^2}\right)\right]^{-\frac{1}{2}}$$

$$= \frac{1}{d}\left[1 + \frac{r}{d}\cos\theta + \frac{r^2}{d^2}\left(\frac{3}{2}\cos^2\theta - \frac{1}{2}\right) + \ldots\right].$$

The first term in the expansion does not yield a force and can be ignored. The second term corresponds to a constant force, Gm/d^2, directed towards N, which acts on the earth as a whole and is balanced by the centrifugal force due to the rotation of the earth–moon system. The third term describes the variation of the moon's potential around the earth. The surface profile of the water is an equipotential surface due to the combined effects of the moon and the earth. The potential of unit mass of water due to the earth's gravitation is gh, where, h is the height of the water above its equilibrium level and $g = GM/r^2$ is the acceleration due to gravity at the earth's surface, where, M is the mass of the earth. Hence, the height of the tide $h(\theta)$ is given by:

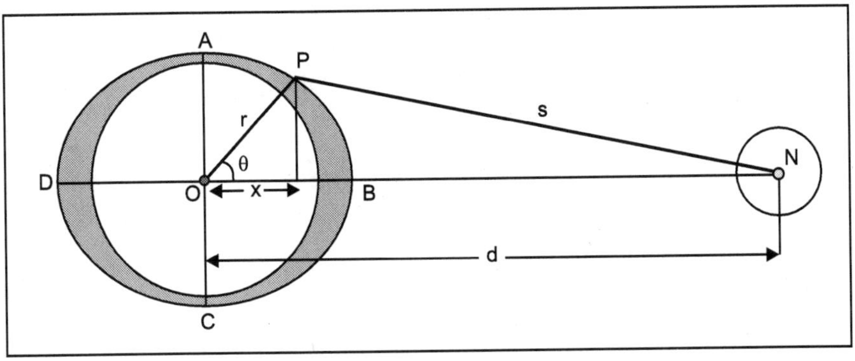

Figure 13.1 Tidal effects due to the moon (not to scale)

$$gh(\theta) - \frac{Gmr^2}{d^3}\left(\frac{3}{2}\cos^2\theta - \frac{1}{2}\right) = 0$$

or

$$h(\theta) = h_{max}\left(\frac{3}{2}\cos^2\theta - \frac{1}{2}\right) \quad \text{... (13.1)}$$

where,

$$h_{max} = \frac{mr^4}{Md^3} \quad \text{... (13.2)}$$

is the maximum height of the tide, which occurs at points B and D ($\theta = 0$ and $\theta = \pi$). Putting $m/M = 0.0123$, $d = 3{,}84{,}400$ km, and $r = 6378$ km we obtain $h_{max} \approx 0.36$ m, which is roughly in line with the observed mean tidal height.

13.2.2 Tidal waves

There are two tidal bulges around the earth at any instant. A formula for the speed of a tidal wave in a sea of uniform depth h_0 is obtained from shallow water theory as:

$$c = \sqrt{gh_0}. \quad \text{... (13.3)}$$

The tidal bulges cannot keep up with the rotation of the earth, so the tides lag behind the position of the moon, the amount dependent on latitude. The presence of continents and bays significantly disturbs the tides and can enhance their range.

13.2.3 Shallow water theory

We consider a wave such that the wavelength λ is much greater than the mean depth of the sea h_0. We also assume that the amplitude of the wave is small compared with the depth, in which case the vertical acceleration is small compared with the acceleration due to gravity, g.

Hence the pressure below the surface is roughly hydrostatic and given by:

$$p = p_0 + \rho g(h - y) \quad \text{... (13.4)}$$

where, p_0 is atmospheric pressure and $y = h(x, t)$ is the wave profile on the free surface (Fig. 13.2). Differentiating Eq. 13.4 with respect to x we have $\partial p/\partial x = -\rho g \partial h/\partial x$. Neglecting second-order terms, the equation of motion in the x-direction is of the form:

$$\partial u/\partial t = -g\partial h/\partial x \quad \text{... (13.5)}$$

Since h(x, t) is independent of y it follows from Eq. 13.5 that u is also independent of y. This allows us to derive an equation of mass conservation in terms of u and h. Consider a slice of fluid between the planes x and $x + \delta x$. The volumc flowing per second is uh across x and $(u + \delta u)(h + \delta h)$ across $x + \delta x$. By mass conservation, the difference in the volume flowing per second from x to $x + \delta x$ is equal to the volume displaced per second vdx in the vertical direction. Hence:

$$uh = (u + \delta u)(h + \delta h) + v\delta x$$

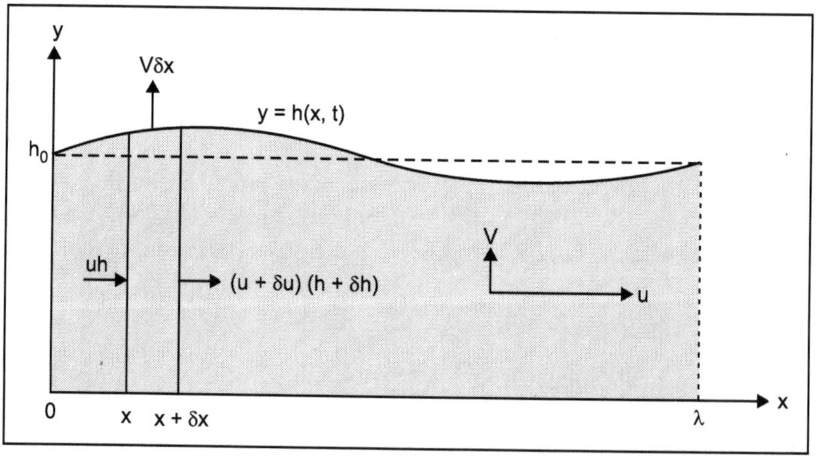

Figure 13.2 Shallow water wave

Putting $v = \partial h/\partial t$, $\delta u \approx (\partial u/\partial x)\delta x$, $\delta h \approx (\partial h/\partial x)\delta x$, and noting that $h\partial u/\partial x$ p $u\partial h/\partial x$ and $h \approx h_0$, yields the mass continuity equation as:

$$-\frac{\partial u}{\partial x} = \frac{1}{h_0}\frac{\partial h}{\partial t} \quad \text{... (13.6)}$$

Eliminating u between Eqs. 13.5 and 13.6, we obtain the wave equation:

$$\frac{\partial^2 h}{\partial x^2} = \frac{1}{c^2}\frac{\partial^2 h}{\partial t^2} \qquad \text{... (13.7)}$$

for the height profile h(x, t) of the wave, where $c = \sqrt{gh_0}$ is the wave speed.

13.3 Tidal power

The earliest exploitation of tidal power was in tidal mills, created by building a barrage across the mouth of a river estuary. Sea water was trapped in a tidal basin on the rising tide and released at low tide through a waterwheel, providing power to turn a stone mill to grind corn. Tidal barrages for electricity generation use large low-head turbines and can operate for a greater fraction of the day. An important issue is whether it is better to use conventional turbines that are efficient but operate only when the water is flowing in one particular direction or less efficient turbines that can operate in both directions (i.e. for the incoming and the outgoing tides). The first large-scale tidal power plant in the world was built in 1966 at La Rance in France. It generates 240 MW using 24 low-head Kaplan turbines. A number of small tidal power plants have also been built more recently in order to gain operational experience and to investigate the long-term ecological and environmental effects of particular locations. Various proposals during the last century to build a large-scale tidal barrage scheme for the River Severn in the UK have been turned down due to the large cost of construction, public opposition and the availability of cheaper alternatives.

13.4 Power from a tidal barrage

A rough estimate of the average power output from a tidal barrage can be obtained from a simple energy balance model by considering the average change of potential energy during the draining process. Consider a tidal basin of area A as shown in Fig. 13.3. The total mass of water in the tidal basin above the low water level is $m = \rho Ah$, where, h is the tidal range. The height of the centre of gravity is ½h, so the work done in raising the water is mg(½h) = ½ρgAh^2. Hence the average power output is:

$$P_{ave} = \frac{\rho gAh^2}{2T} \qquad \text{... (13.8)}$$

where, T is the time interval between tides, i.e. the tidal period. In practice, the power varies with time according to the difference in water levels across the barrage and the volume of water allowed to flow through the turbines. Also, the operating company would seek to optimise revenue by generating electricity during periods of peak-load demand when electricity prices are highest.

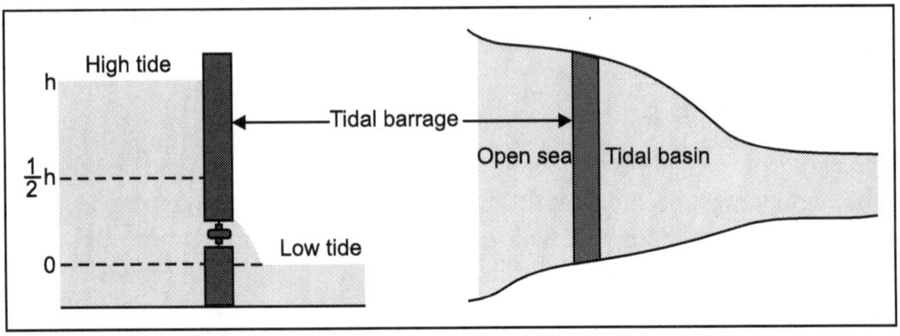

Figure 13.3 Tidal barrage

13.5 Tidal resonance

The tidal range varies in different oceans of the world due to an effect known as tidal resonance. For example, the Atlantic Ocean has a width of about 4000 km and an average depth of about 4000 m, so the speed of a shallow water wave Eq. 13.3 is about $c = \sqrt{gh_0} \approx \sqrt{10 \times 4000} \approx 200 \text{ m s}^{-1}$. The tidal frequency is about $2 \times 10^{-5} \text{ s}^{-1}$, so the wavelength is $\lambda = c/f \approx 200/(2 \times 10^{-5})$ m $\approx 10^4$ km. This is about twice the width of the Atlantic and so resonance occurs; the time taken for the shallow water wave to make the round trip, reflecting off both shores, is about the same as the tidal period, so the amplitude builds up.

The wave amplitude also increases on the continental shelf and can reach about 3 m at the shores. River estuaries can also exhibit large tidal resonance if the length and depth of the estuary are favourable. From Eq. 13.3 the time taken for a wave to propagate the length of the channel and back to the inlet is given by $t = 2L/c = 2L/\sqrt{gh_0}$. If this time is equal to half the time between successive tides then the tidal range is doubled.

13.5.1 Tidal resonance in a uniform channel

For simplicity, consider a uniform channel of length L such that the end at x = 0 is open to the sea and the other end of the channel at x = L is a vertical wall. Suppose that the height of the incident tidal wave varies with time as hi (t) = a cos (ωt). We consider a travelling wave of the form:

$$h_i(x, t) = a \cos(kx - \omega t)$$

From the mass continuity Eq. 13.6, we have:

$$-\frac{\partial u_i}{\partial x} = \frac{1}{h_0}\frac{\partial h_i}{\partial t} = \frac{\omega a}{h_0}\sin(kx - \omega t)$$

Integrating with respect to x yields the velocity in the horizontal direction as:

$$u_i(x, t) = \frac{\omega a}{h_0 k}\cos(kx - \omega t)$$

In order to satisfy the boundary condition u = 0 at x = L (since there cannot be any flow across the barrier) we superimpose a reflected wave of the form:

$$u_r(x, t) = \frac{\omega a}{h_0 k}\cos(kx - \omega t)$$

The total velocity at x = L is given by $u(L, t) = u_i(L, t) + u_r(L, t) = \omega a/h_0k\ [\cos(kL - \omega t) + \cos(kL + \omega t)] = 2\omega a/h_0k \cos(kL)\cos(\omega t) = 0$. Hence $kL = \pi/2\ (2n + 1)$, and the lowest mode of oscillation (n = 0) is given by $kL = \pi/2$. Putting $k = 2\pi/\lambda$ we then obtain the minimum length of the channel as $L = \lambda/4$. The total height of the incident and reflected waves is:

$h(x, t) = h_i(x, t) + h_r(x, t) = a\cos(kx - \omega t) - a\cos(kx + \omega t) = 2a\sin(kx)\sin(\omega t)$.

At the end of the channel, x = L, the height is $h(L, t) = 2a \sin(wt)$, i.e. double that due to the incident wave. This causes the amplitude to build up with the result that the tidal range can be very large—in the River Severn estuary between England and Wales a range of 10–14 m is observed.

13.6 Kinetic energy of tidal currents

In particular locations (e.g. between islands) there may be strong tidal currents that transport large amounts of kinetic energy. In recent years various devices for extracting the kinetic energy have been proposed. These devices are essentially underwater versions of wind turbines and obey the same physical principles. In the majority of designs the axis of rotation of the turbine is horizontal and the device is mounted on the seabed or suspended from a floating platform. Before installation, the tidal currents for any particular location need to be measured to depths of 20 m or more in order to determine the suitability of the site. The first generation of prototype kinetic energy absorbers have been operated in shallow water (i.e. 20–30 m) using conventional engineering components. Later generations are likely to be larger, more efficient, and use specially designed low-speed electrical generators and hydraulic transmission systems.

13.7 Generation of tidal energy

Tidal power is extracted from the earth's oceanic tides; tidal forces are periodic variations in gravitational attraction exerted by celestial bodies. These forces create corresponding motions or currents in the world's oceans. The magnitude and character of this motion reflects the changing positions of the moon and sun relative to the earth, the effects of earth's rotation, and local geography of the sea floor and coastlines.

Tidal power is the only technology that draws on energy inherent in the orbital characteristics of the earth–moon system, and to a lesser extent in the earth–sun system. Other natural energies exploited by human technology originate directly or indirectly with the sun, including fossil fuel, conventional hydroelectric, wind, biofuel, wave and solar energy. Nuclear energy makes use of earth's mineral deposits of fissionable elements, while geothermal power taps the earth's internal heat, which comes from a combination of residual heat from planetary accretion (about 20 per cent) and heat produced through radioactive decay (80 per cent).

A tidal generator converts the energy of tidal flows into electricity. Greater tidal variation and higher tidal current velocities can dramatically increase the potential of a site for tidal electricity generation.

Because the earth's tides are ultimately due to gravitational interaction with the moon and sun and the earth's rotation, tidal power is practically inexhaustible and classified as a renewable energy resource. Movement of tides causes a loss of mechanical energy in the earth–moon system: this is a result of pumping of water through natural restrictions around coastlines and consequent viscous dissipation at the seabed and in turbulence. This loss of energy has caused the rotation of the earth to slow in the 4.5 billion years since its formation. During the last 620 million years the period of rotation of the earth (length of a day) has increased from 21.9 hours to 24 hours; in this period the earth has lost 17 per cent of its rotational energy. While tidal power may take additional energy from the system, the effect is negligible and would only be noticed over millions of years.

13.7.1 Categories of tidal power

Tidal power can be classified into three main types:

1. Tidal stream systems make use of the kinetic energy of moving water to power turbines, in a similar way to windmills that use moving air. This method is gaining in popularity because of the lower cost and lower ecological impact compared to barrages.

2. Barrages make use of the potential energy in the difference in height (or head) between high and low tides. Barrages are essentially dams across the full width of a tidal estuary, and suffer from very high civil infrastructure costs, a worldwide shortage of viable sites, and environmental issues.
3. Dynamic tidal power exploits a combination of potential and kinetic energy: by constructing long dams of 30–50 km in length from the coast straight out into the sea or ocean, without enclosing an area. Both the obstruction of the tidal flow by the dam—as well as the tidal phase differences introduced by the presence of the dam (which is not negligible in length as compared to the tidal wavelength)—lead to hydraulic head differences along the dam. Turbines in the dam are used to convert power (6–15 GW per dam). In shallow coastal seas featuring strong coast-parallel oscillating tidal currents (common in the UK, China and Korea), a significant water level differential (2–3 metres) will appear between both sides of the dam.

Modern advances in turbine technology may eventually see large amounts of power generated from the ocean, especially tidal currents using the tidal stream designs but also from the major thermal current systems such as the Gulf Stream, which is covered by the more general term marine current power. Tidal stream turbines may be arrayed in high-velocity areas where natural tidal current flows are concentrated such as the west and east coasts of Canada, the Strait of Gibraltar, the Bosporus, and numerous sites in Southeast Asia and Australia. Such flows occur almost anywhere where there are entrances to bays and rivers or between land masses where water currents are concentrated.

13.7.2 Tidal stream generators

Tidal versus wind turbines

Tidal stream generators draw energy from currents in much the same way as wind turbines. As a relatively new technology, though first conceived in the 1970s during the oil crisis, the potential for power generation by an individual tidal turbine can be greater than that of similarly rated wind energy turbine. The higher density of water relative to air (water is about 800 times the density of air) means that a single generator can provide significant power at low tidal flow velocities compared with similar wind speed. Given that power varies with the density of medium and the cube of velocity, it is simple to see that water speeds of nearly one-tenth of the speed of wind provide the same power for the same size of turbine system; however this limits the application in practice to places where the tide moves at speeds of at least 2 knots (1 m/s) even close to neap tides. Furthermore, at higher speeds in a flow between 2 to 3 metres per second in seawater a tidal turbine can typically access four times as much energy per rotor swept area as a similarly rated power wind turbine.

Types of tidal stream generators

Since tidal stream generators are an immature technology, no standard technology has yet emerged as the clear winner, but a large variety of designs are being experimented with, some very close to large scale deployment. Several prototypes have shown promise with many companies making bold claims, some of which are yet to be independently verified, but they have not operated commercially for extended periods to establish performances and rates of return on investments.

The European Marine Energy Centre categorises them under four heads although a number of other approaches are also being tried.

Axial turbines: These are close in concept to traditional windmills operating under the sea and have the most prototypes currently operating.

Vertical and horizontal axis crossflow turbines: Invented by Georges Darreius in 1923 and patented in 1929, these turbines that can be deployed either vertically or horizontally. Neptune Renewable Energy has developed Proteus which can be used to form an array in mainly estuarine conditions.

Oscillating devices: Oscillating devices do not have a rotating component, instead making use of aerofoil sections which are pushed sideways by the flow. Oscillating stream power extraction was proven with the omni- or bidirectional Wing'd Pump windmill.

Venturi effect: This uses a shroud to increase the flow rate through the turbine. These can be mounted horizontally or vertically.

Energy calculations

(a) Turbine power

Various turbine designs have varying efficiencies and therefore varying power output. If the efficiency of the turbine 'ξ' is known the equation below can be used to determine the power output of a turbine.

The energy available from these kinetic systems can be expressed as:

$$P = \frac{\xi \rho A V^3}{2}$$

where, ξ = the turbine efficiency.
P = the power generated (in watts).

ρ = the density of the water (seawater is 1025 kg/m^3).
A = the sweep area of the turbine (in m^2).
V = the velocity of the flow.

Relative to an open turbine in free stream, depending on the geometry of the shroud shrouded turbines are capable of as much as 3 to 4 times the power of the same turbine rotor in open flow.

Resource assessment

While initial assessments of the available energy in a channel have focus on calculations using the kinetic energy flux model, the limitations of tidal power generation are significantly more complicated. For example, the maximum physical possible energy extraction from a strait is given by:

$$P = 0.221 \rho g \Delta H_{max} Q_{max}$$

where,

ρ = the density of the water (seawater is 1025 kg/m^3)
g = gravitational acceleration (9.81 m/s^2)
Δ_{Hmax} = maximum differential water surface elevation across the channel
Q_{max} = maximum volumetric flow rate through the channel

Potential sites

As with wind power, selection of location is critical for the tidal turbine. Tidal stream systems need to be located in areas with fast currents where natural flows are concentrated between obstructions, for example at the entrances to bays and rivers, around rocky points, headlands or between islands or other land masses.

13.7.3 Environmental impacts

Tidal energy is a renewable resource of electricity and does not result in the creation of any emissions. In this way it is a benefit to the environment because its use reduces the demand on fuels such as coal and oil. As for environmental impacts, some of the non-monetary costs associated with barrage systems include destruction of habitat, interruption of organisms' travel routes, potential electromagnetic interference (in the case of species that can sense electric fields), and potential acoustic pollution. In addition, hydroelectric systems are well-known for killing fish, and the waste heat that ends up in the water reduces its capacity to store dissolved oxygen, harming not only fish, but all organisms in the affected area. Despite all this, such systems are attractive to many because of their reliance on existing technology and knowledge.

13.8 Advantages and disadvantages of tidal energy

Tidal wave energy is still a very niche technology with tidal barrages generating most of the electricity in a few power stations. Most of the tidal power plants using the modern tidal turbine technology are still in the pilot phase and generate negligible power. Tidal energy has many advantages and disadvantages with the cons of massive investments and long gestation time winning over the many pros of tidal power. Note tidal energy is a favourite for scientists and has been for a long time due to the almost limitless potential.

Like geothermal energy, however tidal energy is stuck in a low growth phase and faces a real possibility that it may not be anything but an esoteric source of energy. Tidal energy shares many of the same problems as geothermal energy does despite geothermal energy having a much longer history. Here is the list of the advantages and disadvantages of tidal energy.

13.8.1 Tidal power pros

1. *Renewable, non-polluting and carbon negative:* Tidal energy is completely renewable, does not lead to any pollution of the air and does not lead to any carbon emissions like fossil fuels.
2. *Predictable:* Tidal wave energy is very predictable as the tides rise with great uniformity. Other forms of renewable energy like solar and wind energy are intermittent in nature. The electricity supply is much more uniform and reliable in case of tidal power.
3. *No fuel:* Tidal power needs water for generation of electricity in its catchment area. It does not need fuel like thermal, gas or oil powered power stations.
4. *Low costs:* Once a tidal energy power plant starts running, its costs are extremely low.
5. *Long life:* A tidal barrage has a very long life of around 100 years which is much longer than that of even nuclear power plants. The long life implies that the life-cycle cost of a tidal energy power plant becomes very low in the long-term.
6. *High energy density:* The energy density of tidal energy is much higher than that of other forms of renewable energy like wind power.
7. *High load factor:* The load factor for solar and wind energy ranges from 15–40 per cent which is quite low compared to fossil fuel energy. Tidal energy has a load factor of almost 80 per cent which is equal to that of thermal power.

13.8.2 Tidal power cons

1. *High initial capital investment:* Tidal barrages require massive investment to construct a barrage or dam across a river estuary. This is comparable to construction of a massive dam for hydro power. This is perhaps the biggest disadvantage of this technology.
2. *Effect on marine life:* The operation of commercial tidal power stations has known to moderately affect the marine life around the power plant. It leads to disruption in movement and growth of fishes and other marine life and can also lead to increase in silt. Turbines can also kill fish passing through it.
3. *Immature technology:* Except for tidal barrage, the other forms of technology generating tidal or wave power are quite immature, costly and unproven.
4. *Long gestation time:* The cost and time over runs can be huge for tidal power plants leading to their cancellation just like that of the Severn barrage in the UK. Many of the tidal power stations like the gigantic plant being planned in Russia will never come to fruition because of the very long gestation time.
5. *Difficulty in transmission of tidal electricity:* Some forms of tidal power generate power quite far away from the consumption of electricity. Transportation of tidal energy can be quite cumbersome and expensive.
6. *Weather effects:* Severe weather like storms and typhoons can be quite devastating on the tidal power equipment especially those place on the sea floor.

Tidal energy like all other forms of energy suffers from both advantages and disadvantages. The tidal energy phase of development has made the cons outweigh the pros unlike other renewable energy forms like wind and solar energy. However innovative methods are being developing to harness tidal and wave energy which might prove to be a game changer and move tidal energy into the mainstream energy industry.

13.9 Wave energy

The waves on the surface of the sea are caused mainly by the effects of wind. The streamlines of air are closer together over a crest and the air moves faster. It follows from Bernoulli's theorem that the air pressure is reduced, so the amplitude increases and waves are generated. As a wave crest collapses the neighbouring elements of fluid are displaced and forced to rise above the equilibrium level (Fig. 13.4). The motion of the fluid beneath the surface decays exponentially with depth. About 80 per cent of the energy in a surface wave is contained within a quarter of a wavelength below the surface.

Thus, for a typical ocean wavelength of 100 m, this layer is about 25 m deep. We now derive an expression for the speed of a surface wave using intuitive physical reasoning. The water particles follow circular trajectories, as shown in Fig. 13.4.

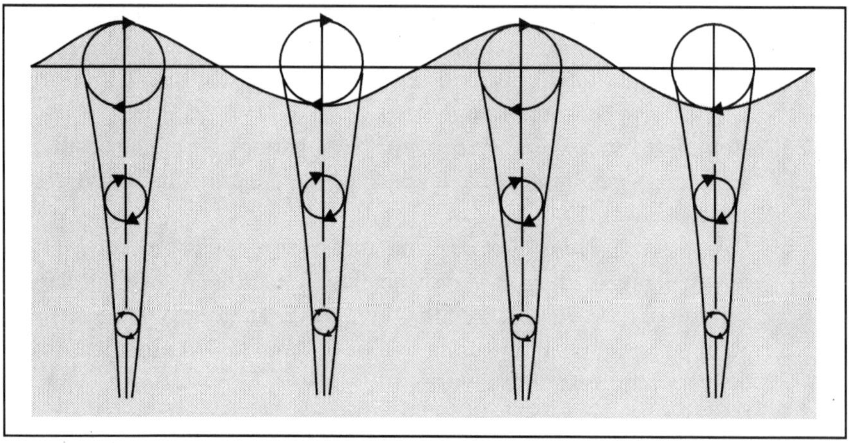

Figure 13.4 Surface wave on deep water

Consider a surface wave on deep water and choose a frame of reference that moves at the wave velocity, c, so that the wave profile remains unchanged with time. Noting that the pressure on the free surface is constant (i.e. atmospheric pressure), Bernoulli's equation yields:

$$u_c^2 - u_t^2 - 2gh = 0 \qquad \text{... (13.9)}$$

where, u_c is the velocity of a particle at a wave crest, u_t is the velocity of a particle at a wave trough, and h is the difference in height between a crest and a trough. If r is the radius of a circular orbit and s is the wave period then we can put:

$$u_c = \frac{2\pi r}{\tau} - c, \quad u_t = \frac{2\pi r}{\tau} - c, \quad h = 2r. \qquad \text{... (13.10)}$$

Substituting for u_c, u_t and h from Eq. 13.10 in Eq. 13.9, and putting $\lambda = c\tau$, we obtain the wave speed as:

$$c = \sqrt{g\lambda/(2\pi)} \qquad \text{... (13.11)}$$

It follows from Eq. 13.11 that the wave speed increases with wavelength, so that surface waves are dispersive. In practice the wave profile on the surface of the sea is a superposition of waves of various amplitudes, speeds, and wavelengths moving in different directions. The net displacement of the surface is therefore more irregular than that of a simple sine wave. Hence, in order for a wave power device to be an efficient absorber of wave energy in real sea conditions, it needs to be able to respond to random fluctuations in the wave profile. The total energy E of a surface wave per unit width of wave-front per unit length in the direction of motion is given by:

$$E = \frac{1}{2} rga^2 \qquad \text{... (13.12)}$$

The dependence of wave energy on the square of the amplitude has mixed benefits. Doubling the wave amplitude produces a fourfold increase in wave energy. However, too much wave energy poses a threat to wave power devices and measures need to be taken to ensure they are protected in severe sea conditions.

The power P per unit width in a surface wave is the product of E and the group velocity c_g, given by:

$$c_g = \frac{1}{2}\sqrt{g\lambda/(2\pi)} \qquad \text{... (13.13)}$$

Hence the incident power per unit width of wave-front is:

$$P = \frac{1}{4}\rho g a^2 \sqrt{g\lambda/(2p)} \qquad \text{... (13.14)}$$

In mid-ocean conditions the typical power per metre width of wave-front is 30–70 kW m^{-1}.

13.9.1 Wave power devices

Though the first patent for a wave power device was filed as early as 1799, wave power was effectively a dormant technology until the early 1970s, when the world economy was hit by a series of large increases in oil prices. Wave power was identified as one of a number of sources of alternative energy that could potentially reduce dependency on oil. It received financial support to assess its technical potential and commercial feasibility, resulting in hundreds of inventions for wave power devices, but most of these were dismissed as either impractical or uneconomic. The main concerns were whether wave power devices could survive storms and their capital cost. During the 1980s, publicly funded research for wave power virtually disappeared as global energy markets became more competitive. However, in the late 1990s interest in wave power technology was revived due to increasing evidence of global climate change and the volatility of oil and gas prices. A second generation of wave power devices emerged, which were better designed and had greater commercial potential. In general, the key issues affecting wave power devices are:

1. Survivability in violent storms
2. Vulnerability of moving parts to seawater
3. Capital cost of construction
4. Operational costs of maintenance and repair
5. Cost of connection to the electricity grid

We now describe various types of wave power device and examine how they operate and how they address the above challenges.

Spill-over devices

TAPCHAN (TAPered CHANnel) is a Norwegian system in which sea waves are focused in a tapered channel on the shoreline. Tapering increases the amplitude of the waves as they propagate through the channel. The water is forced to rise up a ramp and spill over a wall into a reservoir about 3–5 m above sea level (Fig. 13.5). The potential energy of the water trapped in the reservoir is then extracted by draining the water back to the sea through a low-head Kaplan turbine. Besides the turbine, there are no moving parts and there is easy access for repairs and connections to the electricity grid. Unfortunately, shore-based TAPCHAN schemes have a relatively low power output and are only suitable for sites where there is a deep water shoreline and a low tidal range of less than about a metre. To overcome these limitations, a floating offshore version of TAPCHAN called Wave Dragon is under development, with an inlet span of around 200 m, to generate about 4 MW.

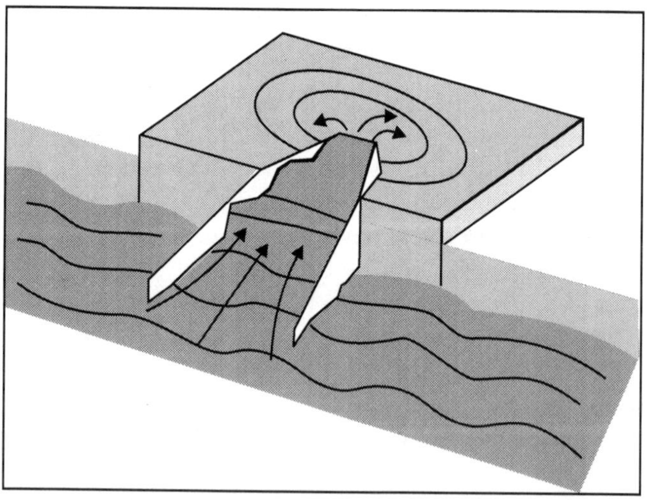

Figure 13.5 TAPCHAN

Oscillating water columns

The oscillating water column (OWC) uses an air turbine housed in a duct well above the water surface (Fig. 13.6). The base of the device is open to the sea, so that incident waves force the water inside the column to oscillate in the vertical direction. As a result the air above the surface of the water in the column moves in phase with the free surface of the water inside the column and drives the air turbine. The speed of air in the duct is enhanced by making the cross-sectional area of the duct much less than that of the column. A key feature of the OWC is the design of the air turbine, known as the Wells turbine. It has the remarkable property of spinning in the same direction irrespective of the direction of air flow in the column! Unlike conventional turbine blades, the blades in a Wells turbine are symmetrical about the direction of motion (Fig. 13.7). Relative to a blade, the direction of air flow is at a non-zero angle of attack α. The net force acting on the blade in the direction of motion is then given by:

$$F = L \sin \alpha - D \cos \alpha \qquad \text{... (13.15)}$$

where, L and D are the lift and drag forces acting on the blade. It is clear from the force diagram in Fig. 13.7(b) that the direction of the net force is the same, irrespective of whether the air is flowing upwards or downwards inside the air column.

The shape of the blade is designed such as to maximise the net force on the blade and the operational efficiency of a Wells turbine is around 80 per cent. At low air velocities the turbine absorbs power from the generator in order to maintain a steady speed of rotation, whilst for large air velocities the air flow around the blades is so turbulent that the net force in the direction of motion of the blade becomes erratic and the efficiency is reduced. Two designs of shore-based OWCs are the Limpet (UK) and the Osprey (UK); generating about 0.5 and 1.5 MW, respectively. A prototype 0.5 MW Australian OWC scheme is also being developed, which uses a 40 m wide parabolic wave reflector to focus waves onto a 10 m wide shoreline OWC; the capital cost is 30 per cent higher but the output is increased by 300 per cent. A large floating OWC known as the Mighty Whale has been developed in Japan. It generates 110 kW but its primary role is as a wave breaker to produce calm water for fisheries and other marine activities.

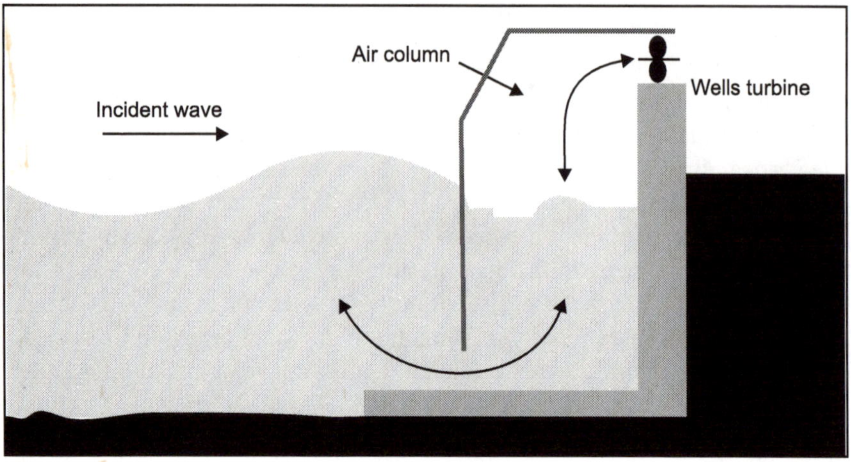

Figure 13.6 Oscillating water column (OWC)

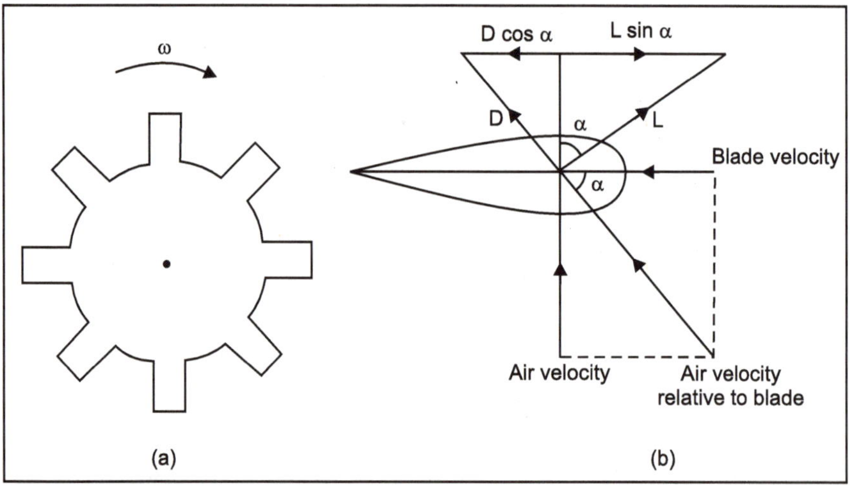

Figure 13.7 Wells turbine: (a) plan view of blades and (b) velocity and force triangles in frame of reference of a blade

Submerged devices

Submerged devices have the advantage of being able to survive despite rough sea conditions on the surface. They exploit the change in pressure below the surface when waves pass overhead: the pressure is increased for a wave crest but is decreased in the case of a wave trough. An example of this type of device is the Archimedes Wave Swing (AWS, Fig. 13.8). The AWS is a submerged air-filled chamber (the 'floater'), 9.5 m in diameter and 33 m in length, which oscillates in the vertical direction due to the action of the waves.

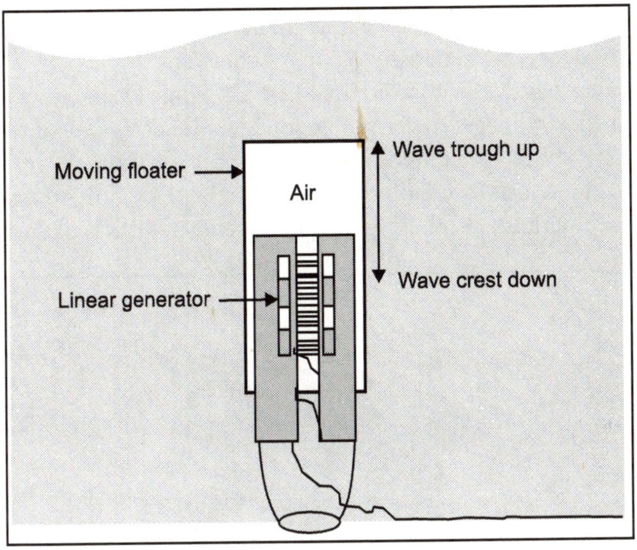

Figure 13.8 Archimedes wave swing

The motion of the floater energises a linear generator tethered to the sea bed. The AWS has the advantage of being a 'point' absorber, i.e. it absorbs power from waves travelling in all directions, and extracts about 50 per cent of the incident wave power. Also, being submerged at least 6 m below the surface it can avoid damage in violent sea conditions on the surface. The device has the advantages of simplicity, no visual impact, quick replacement and cost effectiveness in terms of the power generated per kg of steel.

A pre-commercial pilot project off the coast of Portugal has three AWS devices and produces 8 MW. A fully commercial AWS system could involve up to six devices per kilometre and it is estimated that the global potential for AWS is around 300 GW.

Floating devices

In the early 1970s public interest in wave power was stimulated by a novel device known as the Salter duck (Fig. 13.9). The device floated on water and rocked back and forth with the incident waves. The shape was carefully chosen such that the surface profile followed the circular trajectories of water particles, so that most of the incident wave energy was absorbed with only minimal reflection and transmission. Efficiencies of around 90 per cent were achieved in ideal conditions. The complete system envisaged a string of Salter ducks of several kilometres in total length parallel to a wave-front. A spinal column, of 14 m diameter, used the relative motion between each duck and the spine to provide the motive force to generate power.

The device was designed to be used in the Atlantic Ocean for wavelengths of the order of 100 m but never got beyond small-scale trials due to lack of funding in the 1980s when governmental support for wave power in the UK was dropped in favour of wind power. Nonetheless, the Salter duck provides a useful benchmark for comparing the efficiencies of all wave power devices.

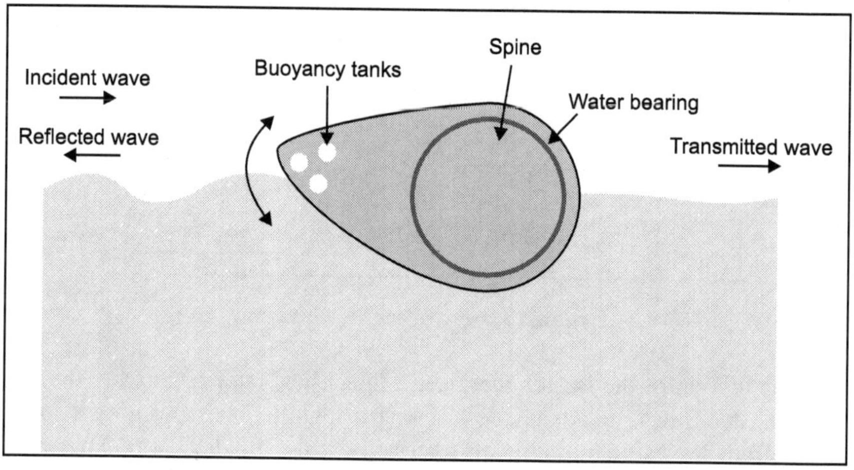

Figure 13.9 Salter duck

A much more recent type of floating device is the Pelamis (Fig. 13.10). It is a semi-submerged serpentine construction consisting of series of cylindrical hinged segments that are pointed towards the incident waves. As waves move along the device, the segments rock back and forth and the relative motion between adjacent segments activates hydraulic rams that pump high pressure oil through hydraulic motors and drive electrical generators. A three-segment version of Pelamis is 130 m long and 3.5 m in diameter and generates 750 kW. The combination of great length and small cross-section to the incident waves provides good protection to large amplitude waves. Three Pelamis devices are due to be installed 5 km off the coast of Portugal and it is estimated that about 30 machines per square kilometre would generate about 30 MW. In order to prevent unwanted interference effects, the devices are spaced apart by about 60–90 m.

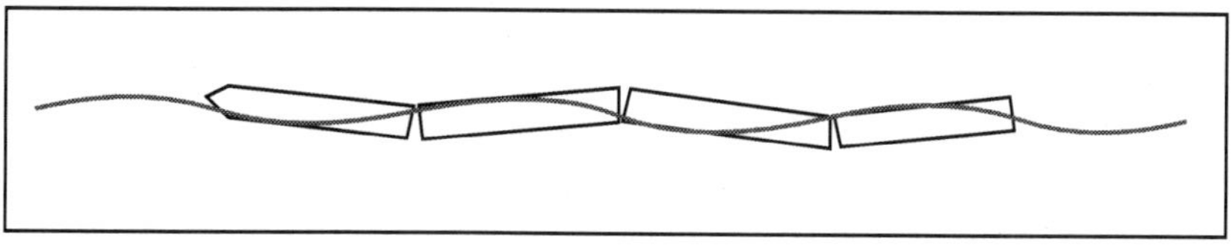

Figure 13.10 Pelamis

13.9.2 Environmental impact, economics and prospects of wave power

As with most forms of alternative energy, wave power does not generate harmful greenhouse gases. Opposition to shore-based sites could be an issue in areas of scenic beauty, on account of the visual impact (including the connections to the electricity transmission grid) and the noise generated by air turbines in the case of oscillating water columns. The visual impact is much less significant for offshore devices but providing cables for electricity transmission to the shore is an added cost.

The global potential of wave power is very large, with estimates of 1–10 TW. Around the UK the Department of Trade and Industry (DTI) estimated (2001) a potential of about 6 GW from wave power. The main challenges for the implementation of wave power are to reduce the capital costs of construction, to generate electricity at competitive prices, and to be able to withstand extreme conditions at sea. Wave power is generally regarded as a high risk technology. Moving to shore-based and near-shore devices reduces the vulnerability to storms but the power available is less than that further out at sea. Even the largest floating devices are vulnerable in freak storms: every 50 years in the Atlantic Ocean there is a wave with an amplitude about ten times the height of the average wave, so any device must be able to withstand a factor of a hundred times the wave energy. Measures to combat such conditions such as submerging the devices can provide an effective means of defence but add to the cost of the system. Another factor to consider is that the frequency of incident sea waves is only about 0.2 Hz, much lower than the frequency of 50–60 Hz for electricity transmission. Though this not a difficult electrical engineering problem, the challenge is to find cost-effective solutions. Wave power is beginning to look competitive in certain niche markets, especially in remote locations where electricity supplies are expensive. However, it is likely to take one or two decades to gather sufficient operational experience for wave power to compete with other alternative energy technologies. In the long-term as fossil fuel reserves become scarce, and concerns over global warming increase, forecasts of an eventual global potential for wave power to provide about 15 per cent of total electricity production do not seem unreasonable, as part of a diverse mix of alternative energy sources.

13.10 Ocean energy

Oceans cover more than 70 per cent of earth's surface, making them the world's largest solar collectors. The sun's heat warms the surface water a lot more than the deep ocean water, and this temperature difference creates thermal energy. Just a small portion of the heat trapped in the ocean could power the world. Ocean thermal energy is used for many applications, including electricity generation. There are three types of electricity conversion systems: closed-cycle, open-cycle, and hybrid. Closed-cycle systems use the ocean's warm surface water to vapourise a working fluid, which has a low-boiling point, such as ammonia. The vapour expands and turns a turbine. The turbine then activates a generator to produce electricity. Open-cycle systems actually boil the seawater by operating at low pressures. This produces steam that passes through a turbine/generator. And hybrid systems combine both closed-cycle and open-cycle systems.

Ocean mechanical energy is quite different from ocean thermal energy. Even though the sun affects all ocean activity, tides are driven primarily by the gravitational pull of the moon, and waves are driven primarily by the winds. As a result, tides and waves are intermittent sources of energy, while ocean thermal energy is fairly constant. Also, unlike thermal energy, the electricity conversion of both tidal and wave energy usually involves mechanical devices.

A barrage (dam) is typically used to convert tidal energy into electricity by forcing the water through turbines, activating a generator. For wave energy conversion, there are three basic systems: channel systems that funnel the waves into reservoirs; float systems that drive hydraulic pumps; and oscillating water column systems that use the waves to compress air within a container. The mechanical power created from these systems either directly activates a generator or transfers to a working fluid, water or air, which then drives a turbine/generator.

Sufficient temperature gradients are found in equatorial and tropical regions. The surface temperatures there can reach values between 25° and 30°C in a layer 100 to 200 m thick and their seasonal variations are not markedly different. The raw solar energy resource captured annually by the oceans is enormous, in the order of 400×10^{15} kWh (i.e. ~ 45,000 TW).

Considering the global output of all the deep cold currents (30 Mm^3/s) and the fact that conceivable thermodynamic machines require a very high discharge of cold water (about 2 m^3/s per MW) and weak conversion efficiency in the order of 3 per cent (principally due to the small difference in temperature between the hot and cold sources), we reach a maximum annual potential of 80,000 TWh, which is renewable.

13.10.1 Ocean thermal energy conversion

1. The thermodynamic converters may be of two types: open or closed cycles.
2. The open cycle machines use seawater directly and make it possible to avoid exchangers that are very cumbersome and risk being clogged from the effect of biofouling (growth of micro-organisms). The principle consists of having two turbines work in a belt of very low pressure (2 to 3 kPa or 20 to 30 mbar) created with vacuum pumps. The water from the hot source, at such pressures, vapourises and drives the turbine, then condenses on contact with the cold source. The freshwater obtained by condensation can then be used as a value added by-product. The major disadvantage of the open cycle resides in the large dimensions of the low pressure turbine. To produce 1 MW, a turbine diameter of 8 m is required.
3. The closed cycle, for its part, at the price of using a fluid of lower boiling point, such as ammonia, and exchangers with a large surface, allows the use of a turbine of smaller dimension (diameter 1 m for 1 MW).
4. The exchangers should have surfaces of the order of 10,000 m^2 per MW and resist corrosion and biofouling. Materials used are titanium and aluminium. The battle against biofouling requires, if we want to avoid using chemical products (chlorine, for example), almost continuous mechanical washing.
5. If these methods can be put to long-term use, ocean thermal energy conversion (OTEC) power stations could first be used in islands in tropical and equatorial regions, where basic electricity production from fossil sources is increasingly expensive.

Figure 13.11 shows the ocean thermal energy conversion: open cycle and Fig. 13.12 shows the ocean thermal energy conversion: close cycle.

13.10.2 Good areas for exploiting tidal energy

Tidal range may vary over a wide range (4.5–12.4 m) from site to site. A tidal range of at least 7 m is required for economical operation and for sufficient head of water for the turbines. Traditional tidal electricity generation involves the construction of a barrage across an estuary to block the incoming and outgoing tide. The dam includes a sluice that is opened to allow the tide to flow into the basin; the sluice is then closed, and as the sea level drops, the head of water (elevated water in the basin) using traditional hydropower technology, drives turbines to generate electricity. Barrages can be designed to generate electricity on the ebb side or flood side or both.

There are many ways to generate electricity from the huge power of the tides: building tidal barrages, exploiting naturally occurring tidal streams, and tidal fences (a scaled down version of the tidal barrage). However, all of these techniques require huge financial investment and have very long pay back periods. In addition there are well documented negative environmental effects which have to be considered.

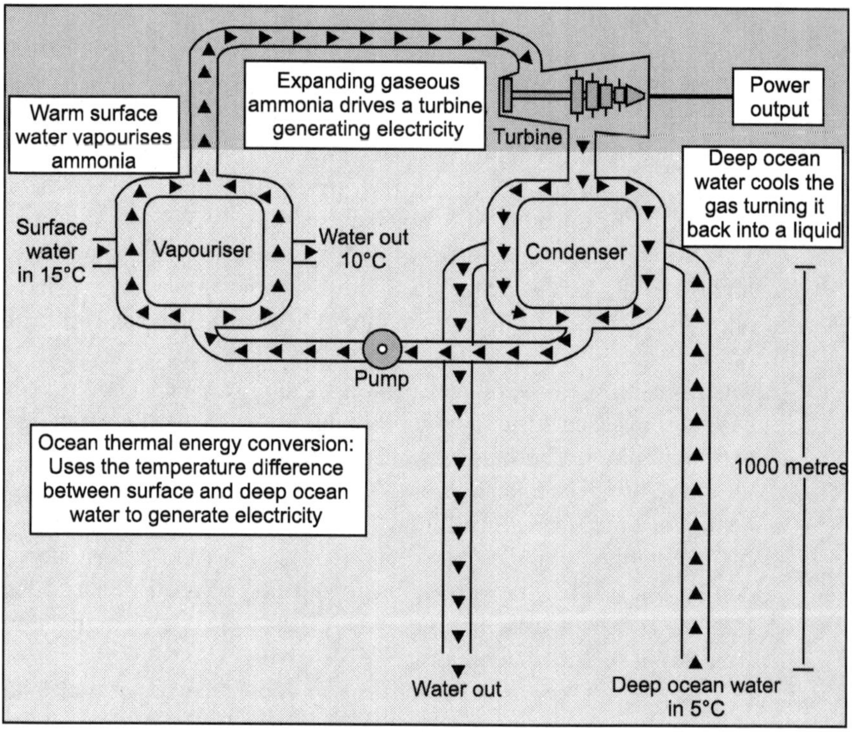

Figure 13.11 Ocean thermal energy conversion: Open cycle

The technology required to convert tidal energy into electricity is very similar to the technology used in traditional hydroelectric power plants. The first requirement is a dam or 'barrage' across a tidal bay or estuary. The best tidal sites are those where a bay has a narrow opening, thus reducing the length of dam which is required. At certain points along the dam, gates and turbines are installed. When there is an adequate difference in the elevation of the water on the different sides of the barrage, the gates are opened. This 'hydrostatic head' that is created, causes water to flow through the turbines, turning an electric generator to produce electricity. Electricity can be generated by water flowing both into and out of a bay. As there are two high and two low tides each day, electrical generation from tidal power plants is characterised by periods of maximum generation every twelve hours, with no electricity generation at the six-hour mark in between. Alternatively, the turbines can be used as pumps to pump extra water into the basin behind the barrage during periods of low electricity demand. This water can then be released when demand on the system its greatest, thus allowing the tidal plant to function with some of the characteristics of a 'pumped storage' hydroelectric facility.

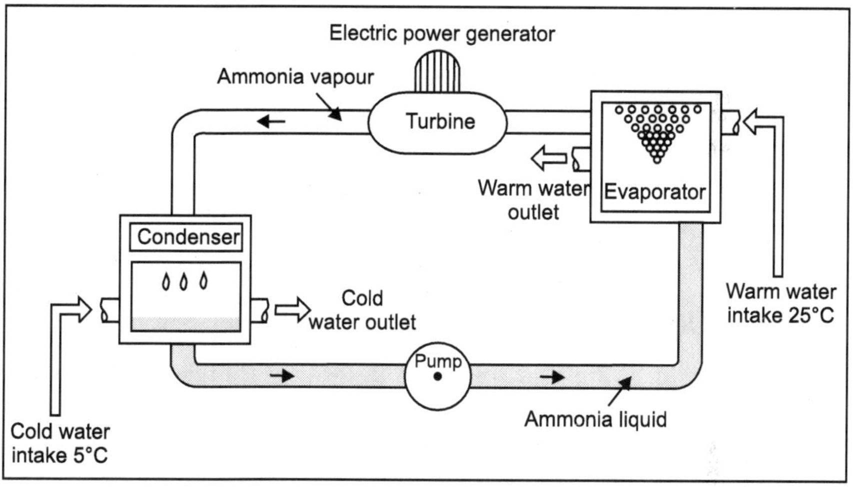

Figure 13.12 Ocean thermal energy conversion: Close-cycle

Although the technology required to harness tidal energy is well established, tidal power is expensive, and there is only one major tidal generating station in operation. This is a 240 MW station at the mouth of the La Rance river estuary in France.

Studies have been undertaken to examine the potential of several other tidal power sites worldwide. It has been estimated that a barrage across the Severn River in western England could supply as much as 10 per cent of the country's electricity needs (12 GW). Similarly, several sites in the Bay of Fundy, Cook Inlet in Alaska, and the White Sea in Russia have been found to have the potential to generate large amounts of electricity.

13.10.3 Costs of tidal energy

Tidal power is a form of low-head hydroelectricity and uses familiar low-head hydroelectric generating equipment, such as has been in use for more than 120 years. The placement of the impoundment offshore, rather than using the conventional 'barrage' approach, eliminates environmental and economic problems that have prevented the deployment of commercial-scale tidal power plants.

Offshore tidal power generators use familiar and reliable low-head hydroelectric generating equipment, conventional marine construction techniques, and standard power transmission methods.

Electricity can be generated by water flowing both into and out of a bay. As there are two high and two low tides each day, electrical generation from tidal power plants is characterised by periods of maximum generation every six hours. Alternatively, the turbines can be used as pumps to pump extra water into the basin behind the barrage during periods of low electricity demand. This water can then be released when demand on the system is at its greatest. This allows the tidal plant to function with some of the characteristics of a pumped storage hydraulic facility. In order to produce practical amounts of electricity, a difference between high and low tides of at least 5 metres is required.

13.10.4 Devices for tidal energy conversion

Tidal energy exploits the natural ebb and flow of coastal tidal waters caused principally by the interaction of the gravitational fields of the earth, moon and sun. The fast sea currents are often magnified by topographical features, such as headlands, inlets and straits, or by the shape of the seabed when water is forced through narrow channels. The tidal stream devices which utilise these currents are broadly similar to submerged wind turbines and are used to exploit the kinetic energy in tidal currents. Due to the higher density of water, this means that the blades can be smaller and turn more slowly, but they still deliver a significant amount of power. To increase the flow and power output from the turbine, concentrators (or shrouds) may be used around the blades to streamline and concentrate the flow towards the rotors.

There are four main types of tidal energy convertors (TEC):

1. *Horizontal axis turbine:* This device extracts energy from moving water in much the same way as wind turbines extract energy from moving air. Devices can be housed within ducts to create secondary flow effects by concentrating the flow and producing a pressure difference.

2. *Cross-axis turbine:*
   (a) *Vertical:* This device extracts energy from moving in a similar fashion to that above, however the turbine is mounted on a vertical axis.
   (b) *Horizontal:* This device is essentially a vertical cross axis turbine orientated horizontally. This turbine configuration allows for deployment in shallow water.
3. *Oscillating hydrofoil:* A hydrofoil attached to an oscillating arm and the motion is caused by the tidal current flowing either side of a wing, which results in lift. This motion can then drive fluid in a hydraulic system to be converted into electricity.
4. *Enclosed tips (venturi):* By housing the device in a duct, this has the effect of concentrating the flow past the turbine. The funnel-like collecting device sits submerged in the tidal current. The flow of water can drive a turbine directly or the induced pressure differential in the system can drive an air-turbine.
5. *Other designs:* This covers those devices with a unique and very different design to the more well-established types of technology or if information on the device's characteristics could not be determined.

14

Geothermal power generation

14.1 Introduction

Geothermal comes from the Greek words *'geo'* which means earth and *'therme'* which means heat. So, geothermal energy means energy or power extracted from beneath the earth. The energy inside the earth was formed by the decay of minerals and forests several years ago. Traditionally, it was used for bathing and heating purposes but today it is also used for generating electricity. Geothermal energy is called renewable source of energy because heat is continuously produced inside the earth.

This heat is brought to the near-surface by thermal conduction and by intrusion into the earth's crust of molten magma originating from great depth. As groundwater is heated, geothermal energy is produced in the form of hot water and steam.

Geothermal energy is produced inside the earth's surface. The earth's layer consists of innermost layer called iron core which itself has two layers: solid iron core and an outer core made of hard rock, called magma, mantle which surrounds the core and the outermost layer called crust which forms oceans and continents. When magma comes close to the earth's surface, i.e. crust it heats up the groundwater which gets trapped in porous rocks. They may also flow along faults and fractured rock surfaces. Now these hydrothermal resources have two ingredients: water (hydro) and heat (thermal). When these hydrothermal resources occur naturally these are called geothermal reservoirs. Various tools and techniques are used nowadays to detect geothermal reservoirs.

Geothermal energy existence: Geothermal energy exists in the form of: (i) volcanoes, (ii) hot springs, and (iii) geysers.

Geothermal energy is clean and sustainable and environment friendly. It may be noted that the so called 'ring of fire' of the earth envelopes the Pacific rim. Though there are over 300 hot springs sites in India, this form of energy is yet to be tapped. In USA, California generates highest amount of electricity through geothermal energy.

Future of geothermal energy: The future of geothermal energy depends on three factors: its demand, supply and its competitiveness among other renewable resources in terms of cost, availability, reliability etc. Demand for geothermal energy is going to increase and increase with the increase in the population and extinction of other non-renewable sources. Moreover, today government also support the resources which are cleaner and do not spoil the environment. Supply of geothermal energy is limited and confined to certain areas only. The entire resource of geothermal energy is fairly bigger than that of coal, oil and gas. Geothermal energy can be made more widely available if the methods and technologies used to extract it are improved. Geothermal energy is still not explored fully. Several miles below the earth surface is hot, dry rock being heated by the molten magma directly below it.

14.2 Geothermal fluid

Geothermal fluid—a hot, sometimes salty, mineral-rich liquid and/or vapour—is the carrier medium that brings geothermal energy up through wells from the subsurface to the surface. This hot water and/or steam is withdrawn from a deep underground reservoir and isolated during production, flowing up wells and converting into electricity at a geothermal power plant. Once used, the water and condensed steam is injected back into the geothermal reservoir to be reheated. It is separated from groundwater by thickly encased pipes, making the facility virtually free of water pollution.

A resource that uses an existing accumulation of hot water or steam is known as a 'hydrothermal' resource. While several other types of geothermal resources exist, all producing geothermal plants in the United States use hydrothermal resources.

Characteristics of the geothermal fluid, including temperature, chemistry, and non-condensable gas content (NCG), can influence power plant design.

14.2.1 Temperature

Each power plant is designed to optimise the use of the heat supplied by the geothermal fluid. Underground heat can reach thousands of degrees, as shown in Fig. 14.1.

Geothermal fluids suitable for hydrothermal electricity production generally occupy a range of 200°F, 93°C (low temperature) to 400°F, 204°C (high temperature). The type of conversion technology and size of various components, such as heat exchangers and cooling towers, is determined by the temperature of the carrier medium. As the temperature of the resource goes up, the efficiency of the power system increases.

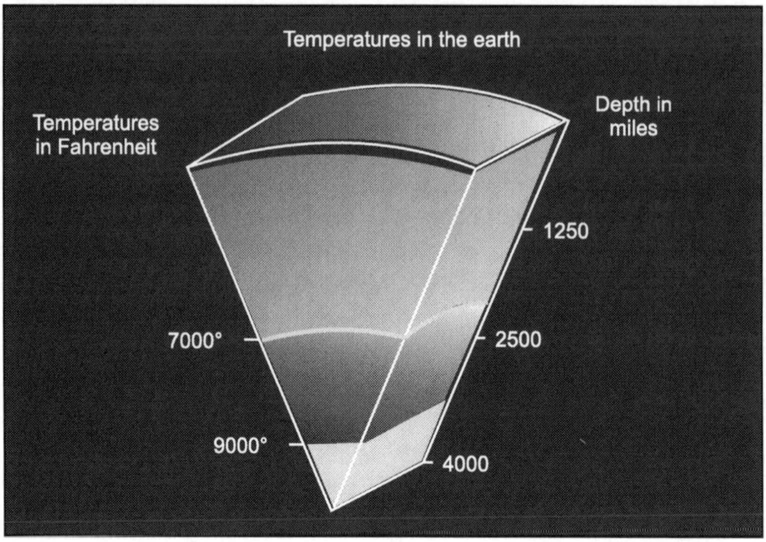

Figure 14.1 Earth's temperatures

14.2.2 Chemistry

Several chemical characteristics are addressed at the beginning of the power plant design phase, including the NCG content, corrosiveness, and geothermal liquid scaling potential, which may require additional equipment. While flash and dry steam plants may or may not produce gases as part of the conversion process, binary facilities, which function in closed loop systems, produce near zero gas emissions.

Non-condensable gases

Geothermal fluids contain entrained non-condensable gases (NCGs) that may not be easily injected back into the reservoir. These gases, which accumulate in the condenser, can decrease heat transfer and raise turbine back-pressure, thereby lowering turbine performance. Steam is sometimes used in ejectors to remove NCGs, but this reduces the amount of steam available for use in the turbines.

Typically, either steam jet ejectors, vacuum pumps or a combination are used to remove NCGs. The system's 'parasitic load'—as steam in ejectors or electricity that operates vacuum pumps—is reduced through recent improvements in vacuum systems. Reductions in parasitic load will increase the overall efficiency of the system. When parasitic load is decreased, more energy can be used to create electricity. Steam jet ejectors produce lower plant efficiencies but cost less than vacuum pumps. Because jet ejectors require steam supply, the quantity of steam available for producing electricity is reduced compared with the quantity available using vacuum pumps.

Vacuum pumps tend to be more expensive and complex, but are more energy efficient. Therefore a cost-benefit analysis will best determine how and to what extent non-condensable gases should be removed from a geothermal system.

Though some geothermal fluids contain non-condensable gases, emissions of each of these are significantly lower than those found at fossil fuel power plants. Typically less than 5 per cent of cooling tower non-condensable gases contain regulated toxic substances, such as hydrogen sulphide and mercury, discussed below. Even in those reservoirs with regulated gases, developers have no trouble meeting California's stringent standards.

Hydrogen sulphide

Hydrogen sulphide (H_2S) is a colourless gas that is harmless in small quantities, but is often regarded as an annoyance due to its distinctive rotten-egg smell. Anthropogenic (man-made) sources of hydrogen sulphide account for approximately 5 per cent of total hydrogen sulphide emissions. H_2S emissions vary with type and size of the plant and with the chemical quality of the resource.

During drilling and certain plant maintenance activities at some reservoirs, hydrogen sulphide gases can pose a worker safety issue. Appropriate plant design and drilling safety procedures developed through oil and gas and geothermal experience are therefore implemented, and standards and procedures are imposed by state and federal regulatory agencies. In addition, H_2S is abated at some geothermal power plants where necessary to meet air quality standards. The two most commonly used vent gas hydrogen sulphide abatement systems are the Stretford and LO-CAT. Both systems convert over 99.9 per cent of the hydrogen sulphide from geothermal non-condensable gases to elemental sulphur, which can then be used as a soil amendment and fertiliser feedstock. The cost to transport and sell the sulphur as a soil amendment is about equal to the revenue gained from the transaction.

Mercury

The mercury abatement measures are already in place at most geothermal facilities where mercury is present (though mercury is not present at every geothermal resource). Abatement measures that reduce hydrogen sulphide also reduce mercury: after hydrogen sulphide is removed from geothermal steam, the gas is run through a mercury filter that absorbs mercury from the gas. After removing mercury, the sulphur created from the abatement process can then be used as an agricultural product.

The rate of mercury abatement within a facility, which varies according to the efficiency of the activated carbon mercury absorber, is typically near 90 per cent, and is always efficient enough to ensure that the sulphur by-product is not hazardous. The activated carbon media is changed out periodically and is disposed of.

Corrosion

At some resource sites, geothermal liquids can gradually wear away power plant materials by chemical action, a process known as corrosion. Corrosion is particularly problematic at mineral rich resource areas, such as the Salton Sea. Much like rust corrodes a nail, geothermal liquids can corrode the metal components of a power plant (pipes, heat exchanger, tanks, etc.) if resistant materials are not used. Corrosion resistant materials such as stainless steel or titanium can be substituted for more corroded carbon steel. Protective coatings can also be applied to carbon steel at a lower cost than corrosion resistant steel alloys or titanium. Though figures vary widely, using corrosion-resistant materials can reduce costs by around 0.25 cents per kWh.

Scaling

At some resource locations, dissolved elements produce scaling. Scaling, a type of precipitation, occurs directly on a surface such as a heat transfer surface or pipe wall. Scaling results in dissolved materials that separate from solution, sometimes remaining suspended as small particles or attaching to a solid surface such as a pipe wall. Silica, a sand-like material, is the most common substance that scales out. Other common materials include metallic carbonates and sulphides.

Scaling can be induced by temperature and pH changes. When flashing a liquid to produce steam in separators, the carbon dioxide (CO_2) originally dissolved in the geothermal liquid is naturally emitted in limited amounts. This creates a positive feedback loop, because the pH increases as a result of the CO_2 emission, which results in further scaling of dissolved liquids.

Scaling can be dealt with in a variety of ways. A developer can reduce the heat captured from the geothermal liquid (thereby reducing plant efficiency), add scaling inhibitors or acidify the geothermal liquid to maintain minerals in solution. More complex equipment may be required to clean the geothermal liquid and control mineral precipitation.

Methods of scale control have improved in recent years, with technologies such as the crystallizer–reactor–clarifier and pH Mod now successfully used at geothermal facilities. One of the advantages of the binary system is avoided by scaling. By maintaining the geothermal water under pressure and injecting it at an elevated temperature (above 160°F or 71°C), the dissolved chemical constituents are maintained in solution. This mitigates/prevents scaling of heat exchangers, wells, and piping.

14.2.3 Injection

Hot water and steam gathering systems are the network of pipelines connecting the power plant with production and injection wells. The size and cost of a steam gathering system can be influenced by some or all of the following: site topography, slope stability, size and spread of the steam field, and temperature and pressure of the resource. Production wells bring the geothermal water to the surface to be used for power generation, while injection wells return the geothermal water and steam condensate back into the geothermal system to be used again. In order to maintain a geothermal system and ensure the continued availability of a resource, geothermal liquids must be injected back into the system. Benefits of injection include enhanced recovery and safe disposal of geothermal fluids, reduced possibility of subsidence, and an increased operational lifetime of the reservoir. When geothermal water is injected, it runs through pipes and cools to a typical injection temperature of 180°F (82°C). If the cooled geothermal liquid is injected too close to a production well, the resource may cool. If, however, the water is injected too far from the geothermal reservoir, it will not sufficiently replenish the system and reservoir pressure may decline.

14.3 Power plant basics

Like all conventional thermal power plants, a geothermal plant uses a heat source to expand a liquid to vapour/steam. This high pressure vapour/steam is used to mechanically turn a turbine-generator. At a geothermal plant, fuel is geothermal water heated naturally in the earth, so no burning of fuel is required.

At many power plants, a steam turbine is used to convert the thermal energy extracted from pressurised steam into useful mechanical energy. Mechanical energy is then converted into electricity by the generator. Geothermal plants rely upon one or a combination of three types of conversion technology—

binary, steam, and flash—to utilise the thermal energy from the hot subsurface fluids and produce electricity.

Each of these processes is described in greater detail in the next section of this chapter, 'conversion technologies'.

After the thermal energy has been used to turn the turbine, spent steam is condensed back to a liquid and injected into the ground where it is reused in the geothermal system, prolonging the lifetime of a geothermal plant. Electricity is then transported by transmission lines into the regional grid.

14.3.1 Design and construction

As one expert notes in his survey of geothermal power plant technology, 'power generation from geothermal resources has been around for more than 100 years; yet there continue to be advancements made that improve resource utilisation, reliability, and economics'. Although funds for research and development have tended to focus on less advanced subsurface exploration techniques, new surface developments also shape the industry. Surface technology advances improve the viability of a geothermal resource and often provide short-term payback.

A power plant typically requires 6–9 months to build once the shovel hits the ground and construction begins. However, when the time needed for exploration, discovery, permitting, and other hurdles is taken into account, the entire geothermal development process can last anywhere from 3 to 7 years or more.

A geothermal developer considers a number of factors when building a plant, particularly related to cost and the long-term viability of a project. Power plant designers must find the optimal size of power plant equipment and choose the best-suited technologies and construction materials that deal with site and resource specifics. Resource characteristics and those of the geothermal carrier medium vary in temperature, chemistry, and permeability. Site characteristics vary depending upon weather conditions, water availability, and geological factors such as ground and slope stability.

14.3.2 Recent power plant developments—turbines

Turbine efficiencies have improved in recent years, increasing as much as ten percentage points. At The Geysers, for example, turbines have been designed to more appropriately match current steam conditions and to utilise more efficient and reliable turbine blade technology. Turbines at steam plants now benefit from longer lasting stage blades and a variety of other improvements.

14.3.3 Factors affecting plant size

Economies of scale

Though the size of a power plant is determined primarily by resource characteristics, these are not the only determining factors. In some cases, a larger power plant proves more cost-effective than a smaller version due to economies of scale. A ten megawatt plant, for example, usually requires all the elements of a 50 megawatt plant. And though a few small plants are capable of running virtually by themselves with monitoring, a geothermal plant usually requires a minimum number of people to run, whether the plant is 50 or 100 MW.

Certain other construction, operation and maintenance costs must be borne independently of the project's capacity. Remote areas tend to have little existing infrastructure, so many geothermal power plants will require excavation, road building, and electric, phone and other connections. Transmission costs can also be significant regardless of power plant size.

Transmission

Because geothermal resources cannot be transported distances over more than a few miles without heat loss, geothermal plants must be built at the site of the reservoir and rely upon transmission systems. If new transmission lines are needed to access the regional grid, high costs can sometimes prohibit geothermal development. While larger transmission systems are traditionally more expensive than smaller versions, economies of scale still apply. For example, companies must negotiate and pay for easements and rights-of-way (ROW) if transmission lines cross private or public lands.

Transmission costs also depend upon topography, slope stability, site accessibility, and aesthetic sensitivities. The length of transmission lines from geothermal facilities to the grid delivery point can vary significantly.

The voltage size of transmission systems can vary considerably as well, depending upon the power involved—from a low of 60 kilovolts (kV) to a high of 230 kV. A typical transmission line averages around 100 kV.

Resource uncertainty

All aspects of the power production process must be taken into account when choosing the megawatt capacity of a power plant, and this may lead to the preference for a smaller sized plant—regardless of economies of scale. Because the ultimate reservoir capacity characteristics and response to production can be uncertain at the start of a project, developers may opt to begin small and then gradually expand the use of the resource rather than risk: (i) overusing the resource, and (ii) spending money on costly power plant infrastructure only to find resource recovery is lower than expected due to misunderstandings of subsurface resource characteristics. In the past, some oversized projects have faced serious problems due to overuse.

Other factors promoting small size

Large plants can take longer to permit than their smaller counterparts and generally require longer environmental reviews. The production tax credit (PTC), a credit awarded for renewable energy generation, offers yet another incentive for developers to construct small plants. To be eligible for the PTC, developers' plants must begin operation within a challengingly short timeframe. The PTC, which has been cited by many experts as the most important policy needed to move the geothermal industry forward, incentivises developers to create smaller plants with consequently smaller lead times.

One final factor promoting smaller units could come from a stipulation in a plant's power purchase agreement (PPA)—the contract to buy the electricity generated by a power plant. A PPA could require a company to first develop a modest number of megawatts, and then gradually work up to a larger output.

Average size

Considering these factors, some experts cite an economically viable geothermal power plant at 20 MW. In practice, plants in the states range from less than 1 MW to just over 100 MW.

14.3.4 Raw materials

Geothermal power plants require a variety of raw materials. Some can be difficult and costly to obtain due to competition for limited resources. The cost of steel—which account for 10–20 per cent of the cost of a geothermal power system—has increased substantially in recent years, particularly due to a demand from China. The steel needs of the oil and gas industry also increase worldwide demand. Other raw materials critical to geothermal development include concrete, oil, fuel (for a drill rig) and lumber. Some of these costs have doubled in recent years.

14.4 Holistic design approach for geothermal binary power plants with optimised net electricity provision

The contribution of geothermal binary power plants to the energy system is based on the provision of net electricity. That is defined by the produced gross electricity from which the auxiliary power to run power consuming components in the different parts of the plant must be deduced. In contrast to other thermal power plants the ratio of auxiliary power to gross electricity can significantly vary in geothermal binary power plants depending on site-specific conditions. It typically lies in the range of 30–50 per cent but can be also higher depending on the site-specific energetic effort to deliver the geothermal fluid from the reservoir or the effort for the recooling of the conversion cycle. In order to optimise the provision of net electricity at a specific site it is hence important to consider the different characteristics of gross electricity production and auxiliary power consumption. The chapter will therefore introduce a geothermal-specific, holistic design approach in which not only parameters, which characterise the quality of single plant components, but also site-specific reservoir and ambient conditions are considered. With the presented methodology, in contrast, it is possible to realise geothermal binary power plants with a higher net electricity output based on existing technology.

Binary power plants are used to produce power from low to medium temperature heat sources. Since the predominant part of the worldwide geothermal potential is based on a temperature level between 100° and 200°C, binary power plants will play a more and more important role for geothermal power generation in the future. The role which different power plant technologies or energy sources play in the electricity mix is typically measured by the total installed power capacity. However, the real contribution of power plants is their net electricity output which is defined by the produced gross electricity from which the auxiliary power to run the power consuming components in a plant must be deduced.

Geothermal binary power plants contain not only the power unit on the surface but also the geothermal fluid loop to deliver the fluid from the reservoir (Fig. 14.2). Therefore, the power consuming components of such plants are in the geothermal fluid loop (such as the downhole pump), in the binary power unit (such as the feed pump), and the recooling system (such as cooling pumps and fans). Existing experiences show that, in contrast to other thermal power plants, the auxiliary power demand in geothermal binary power plants can significantly vary depending on site-specific conditions.

The power demand for the geothermal fluid production, for example, is determined by the reservoir characteristics and the delivered flow rate, e.g. Heidinger, Sanyal, Legarth. When dimensioning the geothermal fluid flow from a specific reservoir, technical restrictions, such as the maximum installation depth and maximum pump capacity, must be considered.

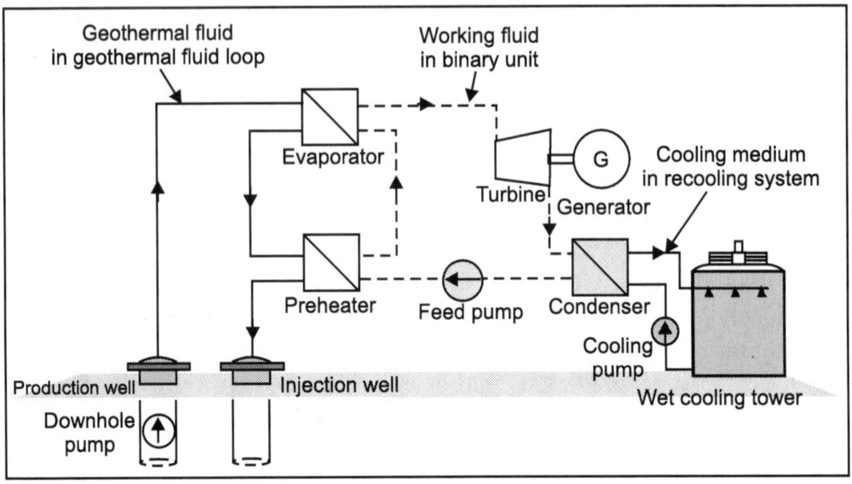

Figure 14.2 Schematic setup of a geothermal binary power plant showing the different subsystems

Another important aspect is the over-proportional increase of the pumping effort with increasing fluid flow rate. Since a higher flow rate results in a larger draw down of the fluid level in the production well, the production effort shows a quadratic dependence with respect to the flow rate. The increase of the fluid production effort with increasing geothermal fluid flow rate is stronger for lower reservoir productivities (Fig. 14.3).

The auxiliary power consumed by the recooling system—which has a considerable influence on the net power output due to relatively large waste heat amounts—is significantly influenced by site preconditions and ambient conditions, e.g. Moya and DiPippo, Kröger, IPPC, Klenke. The relative amount of waste heat thereby increases with decreasing conversion efficiencies (Fig. 14.4). For removing the waste heat from the binary cycle, a suitable heat sink, such as surface water, water from groundwater wells or ambient air is necessary. In conventional power plant engineering, water cooling is usually preferred to air cooling due to the ability to realise lower condensing temperatures and therefore a larger power output. However, the precondition of sufficient supply of cooling water can at many sites, if any, only be met with additional technical and energetic effort. The design of the recooling system in geothermal binary power plants will therefore oftentimes be a compromise between technical realisation and energetic aspects.

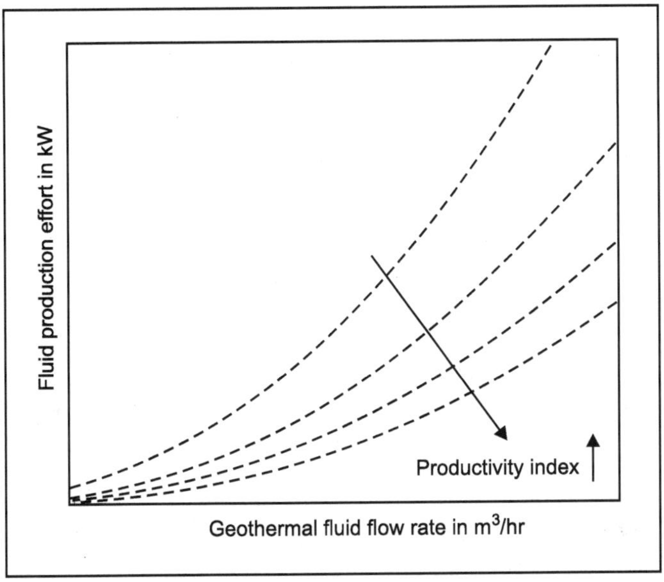

Figure 14.3 Example of fluid production effort as a function of geothermal fluid flow for different reservoir productivities

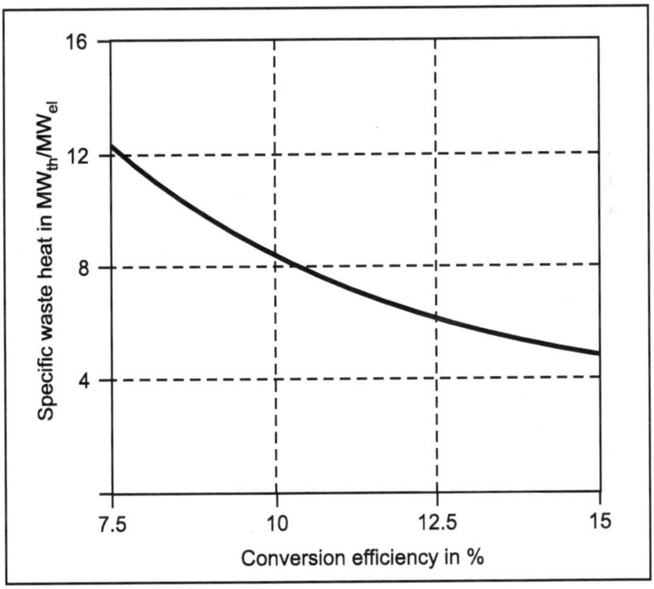

Figure 14.4 Specific waste heat as a function of the conversion efficiency

Based on the considerations above, geothermal binary power plant design must focus on net electricity provision and integrate site-specific reservoir characteristics, ambient conditions and operation characteristics in a holistic design approach. In the following, such an approach will be introduced and applied in a theoretical case study.

It will be shown that the presented methodology leads to geothermal binary power plants with a higher net electricity output based on existing and proven technology.

14.4.1 Holistic design approach—theoretical considerations

The design of the subsystems in geothermal binary power plants depends on different site-specific influences and parameters. In order to design reliable and efficient geothermal binary power plants it is therefore important to integrate these different characteristics in an overall or holistic design approach. The most important aspects or differences of a holistic design approach compared to the separate design of each subsystem are:

1. *An optimum geothermal fluid flow rate:* The gross power which can be generated from a geothermal resource linearly increases with increasing geothermal fluid flow rate. However, the considerations above have shown that increasing the flow rate at a site also results in an over-proportional increase in power consumption of the fluid production system (Fig. 14.3). This means that an optimum geothermal fluid flow rate exists for which the net power provision of a geothermal binary power plant reaches its maximum. Assuming a specific plant setup on the surface, the optimum flow rate for a site with lower reservoir productivity is therefore lower. Referring to different surface plants setups, the optimum flow rate is increasing with a more efficient and better utilisation of the geothermal heat such as with more efficient (and reliable) binary units or the supply of the residual heat in the geothermal fluid after the heat transfer to the binary unit.
2. *An optimum working fluid:* The selection of the working fluid enables the adaptation of the binary conversion cycle to the characteristics of the geothermal heat source. This is due to different shapes of the dew-point curve and different evaporation characteristics which can be realised with different media and mixtures. The choice of a suitable working fluid is therefore an important aspect in designing geothermal binary power plants. A suitable working fluid must allow reliable operation (e.g. thermally stable in the long-term, compatible with other materials used in the binary cycle), a high conversion efficiency and a good utilisation (i.e. the cooling of the geothermal

fluid) of the geothermal heat. Due to the relatively large waste heat amount in geothermal binary power plants, also a selection of the working fluid according to the heat sink and operation characteristics must be considered.

3. *An optimum evaporation temperature:* The evaporation temperature contrarily influences the conversion cycle efficiency and the utilisation of the geothermal heat. Hence, an optimum evaporation temperature exists for which the power output reaches a maximum. Regarding the design of the evaporation also annually varying ambient conditions which might influence the condensation temperature must be considered. If a geothermal binary power plant should also supply heat in serial connection to the binary power unit, the evaporation temperature also depends on the temperature which is required at the outlet of the binary unit in order to provide a certain supply temperature. The outlet temperature of the geothermal fluid can be increased with higher evaporation temperatures. Another possibility is internal heat recuperation in case a dry working fluid is used.
4. *An optimum condensation temperature:* The gross power output of a binary unit is increasing for decreasing condensation temperatures due to the increasing enthalpy difference in the expansion machine. However, also the requirements for the recooling are increasing with decreasing condensation temperatures. This is because lower temperatures must either be realised by lower cooling sink temperatures such as in case of once-through cooling systems. At many sites, where once-through cooling is not an option, lower condensation temperatures can only be realised by a larger auxiliary power input to the fans of wet cooling towers or air coolers. The correlation between condensation temperature and auxiliary power demand is determined by the recooling system, its performance and the ambient conditions. Therefore, regarding the net electricity production, also an optimum condensation temperature does exist. It must be considered that the ambient conditions can significantly vary during the year. Regarding the relatively large waste heat amounts in geothermal binary power plants, the technical use of the waste heat (e.g. conventional cogeneration) can also reduce the demand for recooling.

14.5 Conversion technologies

A conversion technology represents the entire process of turning hydrothermal resources into electricity. Of the four available to developers, one of the fastest growing is the binary cycle, which includes a Rankine cycle engine.

14.5.1 Steam

'Dry steam' plants have been operating for over 100 years—longer than any other geothermal conversion technology, though these reservoirs are rare. In a dry steam plant like those at The Geysers in California, steam produced directly from the geothermal reservoir runs the turbines that power the generator. Dry steam systems are relatively simple, requiring only steam and condensate injection piping and minimal steam cleaning devices. A dry steam system requires a rock catcher to remove large solids, a centrifugal separator to remove condensate and small solid particulates, condensate drains along the pipeline, and a final scrubber to remove small particulates and dissolved solids. Today, steam plants make up a little less than 40 per cent of US geothermal electricity production, all located at The Geysers in California. Figure 14.5 shows a dry steam plant.

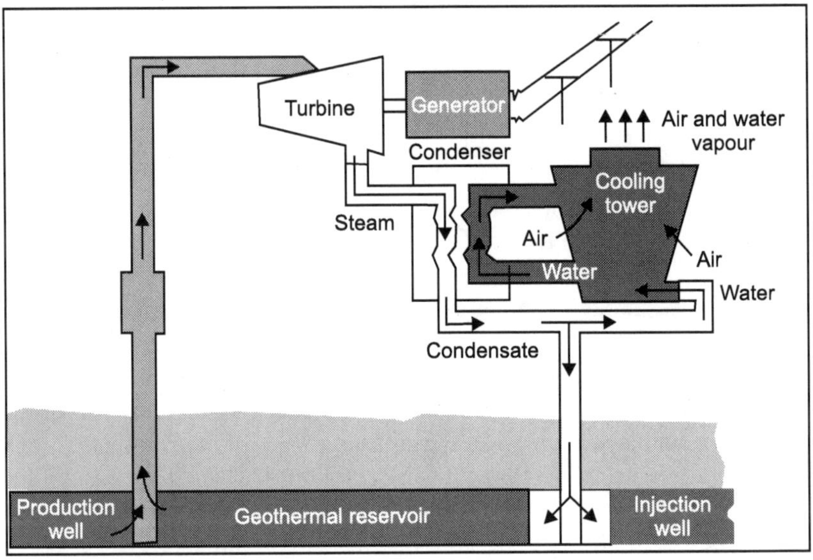

Figure 14.5 Dry steam plant

14.5.2 Flash

The most common type of power plant to date is a flash power plant, where a mixture of liquid water and steam is produced from the wells. At a flash facility, hot liquid water from deep in the earth is under pressure and thus kept from boiling. As this hot water moves from deeper in the earth to shallower levels, it quickly loses pressure, boils and 'flashes' to steam. The steam is separated from the liquid in a surface vessel (steam separator) and is used to turn the turbine, and the turbine powers a generator. Flash power plants typically require resource temperatures in the range of 350–500°F (177° to 260°C). A number of technology options can be used with a flash system. Double flashing, the most popular of these, is more expensive than a single flash, and could concentrate chemical components if they exist in the geothermal water. Even considering potential drawbacks, most geothermal developers agree that double flash is more effective than single flash because a larger portion of the resource is used. Steam processing is an integral part of the gathering system for flash and steam plants. In both cases, separators are used to isolate and purify geothermal steam before it flows to the turbine. A flash system requires three or more stages of separation, including a primary flash separator that isolates steam from geothermal liquid, drip pots along the steam line, and a final polishing separator/scrubber. A steam wash process is often employed to further enhance steam purity. All geothermal power plants require piping systems to transport water or steam to complete the cycle of power generation and injection. Figures 14.6 and 14.7 show schematics of single and double flash-type power plants.

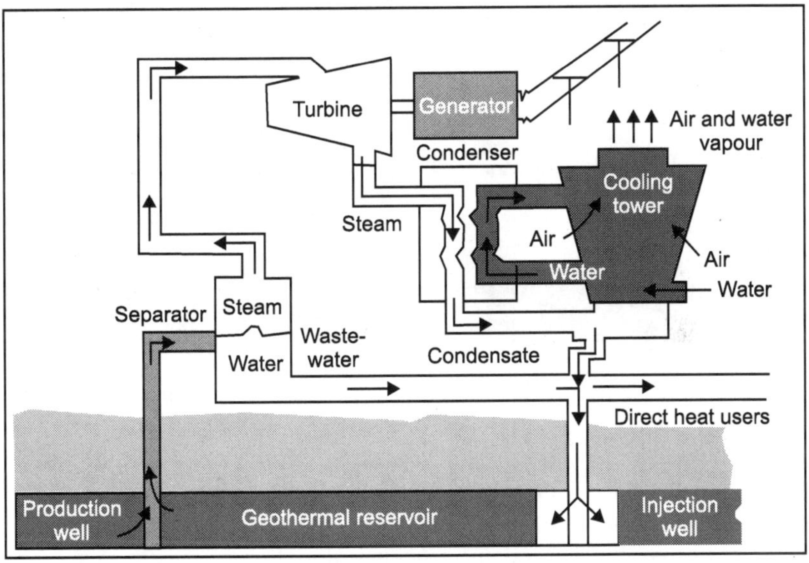

Figure 14.6 Single flash steam power plant schematic

14.5.3 Binary

Technology developments during the 1980s have advanced lower temperature geothermal electricity production. These plants, known as 'binary' geothermal plants, today make use of resource temperatures as low as 165°F or 74°C (assuming certain parameters are in place) and as high as 350°F (177°C). Approximately 15 per cent of all geothermal power plants utilise binary conversion technology. In the binary process, the geothermal fluid, which can be either hot water, steam or a mixture of the two, heats another liquid such as isopentane or isobutane (known as the 'working fluid'), that boils at a lower temperature than water. The two liquids are kept completely separate through the use of a heat exchanger used to transfer heat energy from the geothermal water to the working fluid. When heated, the working fluid vapourises into gas and (like steam) the force of the expanding gas turns the turbines that power the generators.

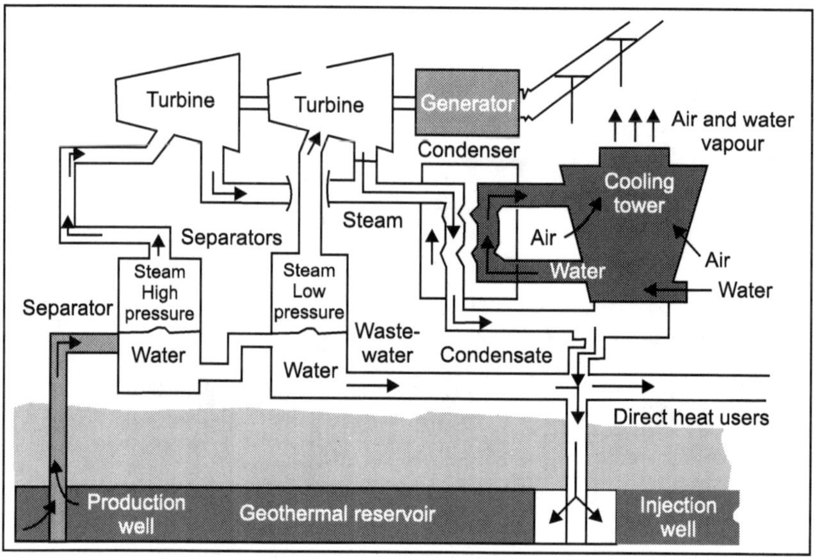

Figure 14.7 Double flash steam power plant schematic

Geothermal fluids never make contact with the atmosphere before they are pumped back into the underground geothermal reservoir. Because the geothermal water never flashes in air-cooled binary plants, 100 per cent can be injected back into the system through a closed loop. This serves the dual purpose of reducing already low emissions to near zero, and also maintaining reservoir pressure, thereby extending project lifetime. For lower pressure steam, a two phase binary cycle is sometimes used. Two-phase systems are similar to traditional binary cycles, except the steam flow enters the vapouriser/heat-exchanger, while the geothermal liquid is used to preheat the organic motive fluid. The steam condensate either flows into the preheater or is combined in the geothermal liquid after the pre-heater. Since the steam pressure in the vapouriser/heat-exchanger remains above atmospheric pressure, the non-condensable gases (NCG) can be reinjected together with cooled-geothermal fluid or simply vented without the need for a power consuming vacuum pump (Fig. 14.8).

Rankine cycle

A Rankine cycle, the commercial binary cycle in the United States, converts heat into electricity. Rankine cycles require an organic-based working fluid with a lower boiling point than water, and are thus often used with lower temperature geothermal resources. The four major pieces of the Rankine cycle include the boiler, turbine, cooling tower, and feed pump. The working fluid in a Rankine cycle follows a closed loop and is reused constantly (Fig. 14.9).

The Rankine cycle, which includes four processes that change the state of the working fluid, has been running geothermal power plants with success for over 100 years.

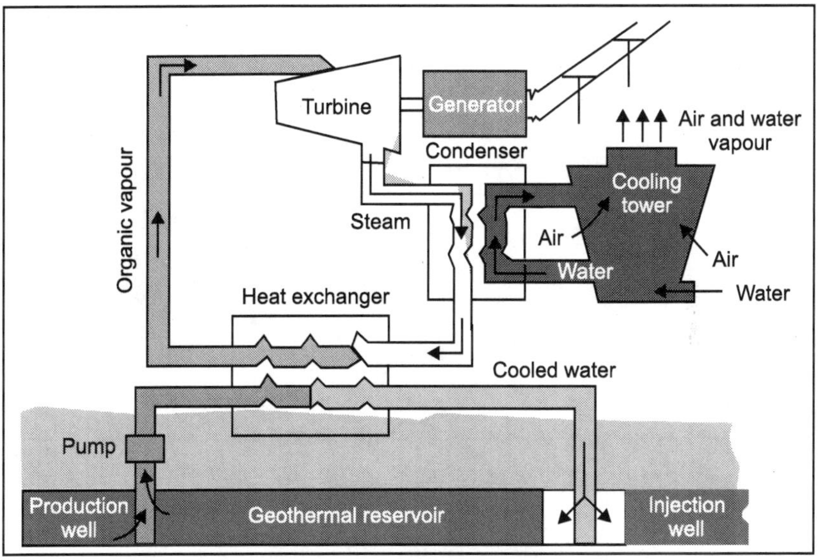

Figure 14.8 Binary power plant schematic

Outlook for binary systems

The ability to use lower temperature resources increases the number and type of geothermal reservoirs suitable for power production. According to some experts, lower temperature resources suitable for binary cycles will be the most common of all future hydrothermal resources developed. As binary use has increased, associated power technology has continued to improve. For example, advances in production pumps have allowed for sustained pump run time for years rather than months. Also, binary systems can now operate at lower temperatures than scientists previously thought possible. Improvements will likely continue as conventional energy prices increase. The post-2001 hike in energy prices has led to the replacement and/or expansion of older geothermal generation systems using newer technology, particularly incrementally improved binary technology. In addition, several geothermal resources discovered in the 1980s are today undergoing their first commercial binary development. These new developments, while substantial, represent only a small fraction of the potential for new generation using binary technology that is being promoted by developers throughout the country.

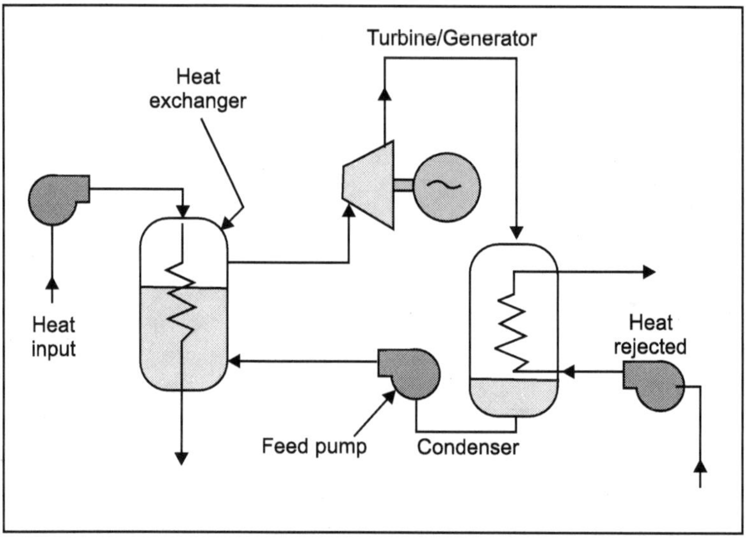

Figure 14.9 Rankine cycle schematic

14.5.4 Flash Binary Combined Cycle

A combination of flash and binary technology, known as the flash/binary combined cycle, has been used effectively to take advantage of the benefits of both technologies. In this type of plant, the flashed steam is first converted to electricity with a steam turbine, and the low-pressure steam exiting the back-pressure turbine is condensed in a binary system. This allows for the effective use of air cooling towers with flash applications and takes advantage of the binary process. The flash/binary system has a higher efficiency where the well-fields produce high pressure steam. This type of system has been operating in Hawaii since 1991 at the Puna Geo Venture facility.

For a high enthalpy water-dominated resource, the most effective power plant configuration may be integration of a combined cycle for the steam and a standard binary unit for the separated brine into one unified plant. In this case, each unit operates with common controls, fluid collection, and reinjection systems. The developer must closely monitor the injection water temperature in combined cycle systems, as declines could occur that lead to scaling. As with any geothermal conversion technology, proper management is critical.

14.5.5 Choosing a conversion technology

Resource characteristics—temperature, pressure, volumes of fluid produced, and chemical properties of the geothermal reservoir—are the primary determinants of the size and type of power conversion equipment. Assuming sufficient volumes of fluid are produced, temperature determines the most efficient conversion design.

While binary plants can utilise any temperature resource, low temperature resources are constrained to the binary model. Medium temperature resources can be economical by using either flash or binary systems. High temperature resources are most economical when steam or flash systems are employed, as these are simpler and therefore less costly. Flash systems are less expensive than binary systems, but may not be as efficient at lower temperatures.

Steam plant equipment costs rise as temperature decreases (as a result of efficiency losses). Despite a more complex design, binary power systems are generally less expensive than steam systems for temperature close to 350°F. The cost of binary systems rises as temperature drops. Binary systems may be preferred in highly sensitive environmental areas, since they operate as closed-loop, virtually emissions-free systems.

14.6 Cooling types

A cooling system, which condenses the working fluid, is essential for the operation of any modern geothermal power plant. A cooling tower provides a greater temperature and pressure differential across the turbine to increase efficiency. The larger this differential, the greater the driving force across the turbine, and the greater the efficiency. Since the earliest days of the industrial revolution, improving the condensing (cooling) process has been a prime concern of scientists and engineers. Advances during the past few years have improved the cooling process. For example, high efficiency fills offer low-cost, compact towers that enhance air-to-water contact. New fills can improve the flow of the geothermal resource, reduce clogging, and assist with cleaning insoluble materials. The use of fibreglass structures rather than wood for cooling towers can improve both cost and fire resistance.

Developers have two basic cooling options: water or air cooling. Hybrid air-water cooled systems have been demonstrated to a limited extent and are considered important for future advancement. Both air and water-cooled systems use cooling fan motors. Some maintenance is required, typically an annual check-up of fan motors and belts as well as system lubrication.

14.6.1 Water cooled

Most power plants, including most geothermal plants, use water-cooled systems—typically in cooling towers. As these are more efficient, they generally require less land than air-cooled systems. Water-cooled systems are less expensive to build and operate if water is readily available and inexpensive to obtain. These systems lose most of the water to the atmosphere by evaporation in the form of water vapour, while the remainder is injected back into the system. Emissions from a wet cooling tower (i.e. water vapour plus dissolved solids or minerals) depend upon the quality of the geothermal liquid injected back through the system.

While today water cooling is mostly used in higher-temperature non-binary facilities due to the use of the geothermal fluid for cooling, a few existing and developing binary facilities in the US utilise water-cooled systems. The binary plants at Heber, East Mesa and Wendel-Amedee, all in California, use water cooling. The Raft River geothermal plant, the first in Idaho, is a binary facility that uses water cooling.

14.6.2 Air cooled

Because the efficiency of power generation is affected by the difference between the temperature of the fluid exiting the turbine and the temperature of the cooling medium, air-cooled systems are influenced by seasonal changes in air temperature. These systems can be extremely efficient in the winter months, but are less efficient in hotter seasons when the contrast between the air and water temperature is reduced. Plant efficiency typically increases by 15 per cent during colder months and decreases by 15 per cent during warmer months. This means that an air-cooled plant is least efficient during summer peak energy demand, which typically takes place during the hottest hours of the day due to air conditioning. The ideal temperature difference between the air and the resource is 200°F (93°C) for an air-cooled system.

Air cooling is beneficial in areas where extremely low emissions are desired, where water resources are limited or where the view of the landscape is particularly sensitive to the effects of vapour plumes (as vapour plumes are only emitted into the air by water cooling towers). While air-cooled systems are only used at binary facilities today, these could theoretically be used with any geothermal conversion technology (Fig. 14.10).

14.6.3 Choosing the right cooling system for a site

Climate and altitude can impact cooling technology. Water cooling is very efficient in hot dry climates; air cooling is most efficient in cool climates; and in hot humid climates where efficiency for both technologies is reduced, either may be applied. Other factors to consider include water and land availability, value of power during hot months, aesthetics, and environmental issues.

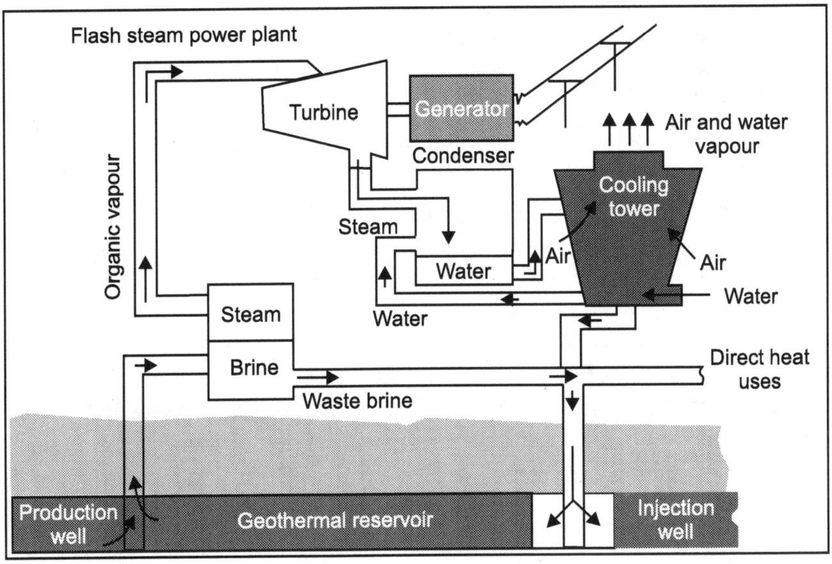

Figure 14.10 Diagram of geothermal power plant with water-cooled system

Because water-cooled systems require biotic and sometimes chemical water treatment to prevent algae blooms or mineral deposition, some developers claim that operation costs of air-cooled systems are lower than those of water-cooled systems. However, the upfront cost of an air-cooled system is higher per kilowatt than a water-cooled system. As is typically the case, a developer must consider upfront versus lifetime costs.

14.7 Structuring power plant to minimise impact

While a geothermal power plant's impact is relatively small compared to that of a fossil fuel plant, geothermal operators still take steps to mitigate any negative affects caused by development.

14.7.1 Noise

A variety of noise muffling techniques and equipment are available for geothermal facilities. During drilling, temporary noise shields can be constructed around portions of drilling rigs. Noise controls can be used on standard construction equipment, impact tools can be shielded, and exhaust muffling equipment can be installed where appropriate. Turbine-generator buildings, designed to accommodate cold temperatures, are typically well-insulated acoustically and thermally, and equipped with noise absorptive interior walls.

14.7.2 Visual impacts

Visual impacts related to geothermal development include night lighting on the power plant, visibility of the transmission line, and the presence of plumes at facilities using water-cooled systems. Fossil fired power plants have all of these visual effects and more. Detailed site planning, facility design, materials selection, landscaping programs, and adjustment to transmission line routing are key aspects of geothermal operations that can reduce impacts. Developers may paint their power facility forest green to blend in with the surrounding landscape.

Additionally, some companies use non-specular conductors, which reduce reflection and glare on transmission lines. Other visual impacts, such as construction equipment, are only of concern on a temporary basis. Construction vehicles, drill rigs, and other heavy equipment impact the visual quality of an area for a limited amount of time.

14.7.3 Mitigation to reduce impact on wildlife and vegetation

Geothermal plants are designed to minimise the potential effect on wildlife and vegetation. Pipes are insulated, which prevents thermal losses and protects animals from burns if they contact the pipes. Spill containment systems are constructed, and areas with sensitive biological or cultural/archeological resources and threatened or endangered species are avoided. Pipelines are built high or low to help minimise impacts to wildlife movement. Geothermal plants do not cause additional disruption from offsite drilling, the construction of pipelines over long distances (as is typically necessary for natural gas transportation), and mining and transportation of coal and uranium ores.

14.7.4 Monitoring activities

Monitoring is a key component of geothermal maintenance and mitigation. Ongoing monitoring activities could include but are not limited to well pressure, water chemistry, surface site, subsidence, biological resources, and deep temperature monitoring. Most mitigation measures are set forth in permitting conditions and environmental documents available for public review.

14.7.5 Maintenance

Geothermal plants are designed to fit the resource of the plant site. Reservoir chemistry can vary dramatically from one resource to another. California's Salton Sea area, for example, has some of the most mineral-rich geothermal water anywhere in the world. This can create challenges caused by corrosion and scaling. The facility at Mammoth Lakes, with a resource comparatively lower in mineral concentration, requires less maintenance.

On average, geothermal plants are available for power generation 97 per cent of the time. Plants are typically off-line near 3 per cent of the time due to routine scheduled maintenance as part of the power cycle management process. By proactively inspecting parts, wear and tear-associated problems are kept to a minimum.

Like any facility, a geothermal plant can be impacted by fire, lightning, a wind storm or other natural disasters. Geothermal developers are prepared for such unlikely occurrences, both in their mitigation and their maintenance techniques. Stop-gap emergency measures are typically put in place.

14.8 Efficiency

Efficiency is broadly defined as the ratio of the output to the input of any system. All thermal power plants have a fraction of 'waste heat'. While efficiency is an important measure of power generating facility performance, comparing efficiency values for geothermal and other renewable technologies, as well as for fossil fuels, poses significant challenges.

The public interest in energy efficiency arose as a fossil fuel issue: that is, the less fuel used per output, the fewer emissions and the greater quantity of depleting fuels conserved. Burning fossil fuels to generate electricity contributes to climate change, health problems, and ecosystem damage. As fossil fuel resources become scarcer, costs skyrocket. That's why efficiency—maximising the energy output from a quantity of burned fossil fuel—is so important for traditional power plants.

The American Council for an Energy Efficient Economy (ACEEE) has pursued increased coal plant efficiency as a means for reducing emissions. According to ACEEE, coal plants grandfathered by the clean air act 'emit 3–5 times as much pollution per unit of power generated as newer, coal-fired power plants and 15–50 times as much nitrogen oxides and particulates as new combined-cycle natural gas power plants'. Older and less efficient plants have 15 per cent higher average heat rates per unit of generation than modern combined-cycle plants, according to ACEEE.

For renewable energy use, in contrast to fossil fuel use, efficiency is primarily an economic concern. Maximising the output per input of available energy is still important, but the public issues are confined primarily to land use, not climate change, health and conservation issues. Unlike geothermal and other renewables, fossil fuel use is not sustainable even if managed properly and used efficiently.

At a geothermal facility, the fuel source is not burned. That means air emissions are substantially lower than at a fossil fuel facility. Because the geothermal resource—the fuel source—doesn't have to be shipped from far-off locations, there is no environmental impact related to transportation as with traditional resources. The geothermal resource is continuously available and highly reliable. Geothermal power plants regularly inject geothermal liquids back through the reservoir, thereby improving the lifetime of the plants. While both conventional plants and geothermal plants must reject heat to the surroundings—a consequence of the Second Law of thermodynamics, discussed in further detail below—geothermal plants result in more heat rejection per unit of useful power output than conventional plants.

Besides more obvious distinctions related to emissions and sustainability, other technology and resource differences must be considered when comparing efficiencies. Resource temperature is one important factor. At fossil fuel facilities, resources can reach temperatures of 1000°F (538°C) or higher. At geothermal power plants, in contrast, temperatures are lower because resources are heated naturally, within their natural confines, rather than through external heating. Because efficiency decreases with lower temperature resources, the quantity of heat input required to produce a given megawatt output increases, and so too does the percentage of that input that must be rejected as waste heat. In geothermal plants, in contrast to fossil fuel plants, more of the energy resource is needed to produce the same output of electricity. That's because more low 'quality' energy sources (e.g. resources at lower temperatures) are needed to produce the same electrical output, and geothermal resource temperatures are lower than coal temperatures.

The geothermal industry has, over the years, worked to define geothermal efficiency in a way that can be easily understood by the general public and compared across technologies, while accurately and fairly characterising the renewable energy resource. The issue is further complicated because experts use a variety of efficiency categorisations, depending upon the context in which an efficiency measurement is needed and the characteristics of the resource and plant. Some of these efficiency measurements cannot be equitably compared. What follows is a selection of some of the ways in which geothermal efficiency can be described. Rather than choosing any one efficiency description over another, this section seeks to identify several of the most popular ways of assessing geothermal efficiency.

14.8.1 First Law of Thermodynamics

The First Law of Thermodynamics states that energy cannot be created or destroyed; it can only be converted from one form to another. The First Law is used to categorise 'the performance of cyclic conversion systems like fossil-fired, steam power cycles or geothermal cycles. This efficiency is a measure of the portion of heat added to a power cycle that is converted to work, i.e. the ratio of network produced to the heat added to the cycle.' The first law is a conservation law (Law of Conservation of Energy). Regarding geothermal power plants, 'the 1st Law requires that any electricity that is generated (energy out) must balance with the energy extracted from the geothermal resource ± any other energy uses and losses to the environment'.

The First Law does not distinguish between the potential type and quality of energy that is received or delivered by a plant. This means that high quality, concentrated energy that can be used to produce electricity is not valued any more than dispersed, low-grade heat energy incapable of producing electricity.

14.8.2 Second Law of Thermodynamics

The Second Law provides direction to the First Law: while energy can neither be created nor destroyed (First Law), important limitations exist in the capacity of energy to do useful work. The Second Law states, in simple, generalised terms, that heat can never be converted completely into work, because some of the energy must flow from high temperatures to a low temperature sink. This means that 100 per cent efficiency is impossible. Second Law efficiency is defined as the ratio of the network to the available energy. Available energy, also referred to as exergy or availability, is the amount of work that can be completed using ideal thermodynamic processes to bring a fluid into equilibrium with a heat reservoir. 'It differs from energy in that it is consumed or dissipated during processes that change a fluid's entropy', or measure of disorder of a system. These processes include heat transfer processes, pressure rises in pumps, compressors and fans, expansions in turbines, pressure drops in piping, etc. The Second Law efficiency measures how 'efficiently the power cycle converts available energy into work'. Second Law efficiencies depend upon a number of factors, including the sink temperature of the power plant. The larger the temperature difference between the heating source and cooling sink, the more efficient a plant will be, assuming all other factors are the same. Plant developers may purposely limit the temperature of the geothermal fluid leaving the plant to prevent mineral precipitation (primarily silica). 'When this temperature limit is imposed, some define Second Law efficiency based on a modified available energy term that uses the minimum temperature instead of the ambient or sink temperature as the reference condition. Second Law efficiencies defined in this manner will be higher in magnitude.'

14.8.3 Comparing First and Second Law efficiencies

Second law efficiencies are generally, but not always, higher than first law because the available energy term (used in Second Law efficiency calculations) tends to be less than the quantity of heat removed from the geothermal fluid (First Law). Other, more subtle differences can sometimes mean that a plant at which Second Law efficiency is made higher through improvements or adjustments will actually result in a lower First Law efficiency.

14.8.4 Carnot efficiency

Carnot efficiency is often used to discuss geothermal power plants and heat engines in general. Binary plants, unique among geothermal conversion types, exist in closed power cycles. As such, power cycle definitions are often considered the most appropriate mechanism through which to categorise binary efficiency.

Carnot cycle

An ideal, frictionless engine in a closed power cycle is known as a Carnot cycle. A Carnot cycle involves four processes and represents the maximum First Law efficiency possible in a specified system. A Carnot Engine is reversible and runs in a cycle, with all of its heat exchanges taking place at a source temperature and a sink temperature. A working engine operating between two heat temperature limits can never exceed Carnot efficiency. Even an ideal, frictionless engine can't convert 100 per cent of its input heat into work.

Some geothermal experts prefer to use other vehicles to set the thermodynamic limit on cycle efficiency for geothermal binary plants. The main reason for this is that geothermal is a variable temperature process—a geothermal liquid enters a plant at a high temperature and cools off as it moves through the plant—while Carnot assumes a heat source operates at a constant temperature.

Triangular cycle

One expert has suggested that the Triangular cycle is the more appropriate ideal cycle upon which to base the maximum binary efficiency. A Triangular cycle recognises that a heating source cools rather than remains at a constant temperature as it transfers heat. An ideal Triangular cycle will more closely mirror a working binary plant without imposing Carnot's 'unreasonably high bar'. Carnot and Triangular cycles are identical in the first two processes. However, the Triangular cycle adjusts for temperature differences in the last half of the cycle: ideal Carnot efficiency assesses the difference between the heat source and heat sink as a fraction of the heat source; ideal Triangular efficiency assesses that same difference (i.e. heat sink subtracted from heat source) as a fraction of the heat source considered along with the heat sink. The distinction is more succinctly represented in the equations below:

Ideal carnot efficiency: $(T_H - T_L)/(T_H)$

Ideal triangular efficiency: $(T_H - T_L)/(T_H + T_L)$

where, T_H = absolute temperature of the heat source and T_L = absolute temperature of the heat sink.

Ideal triangular cycle efficiencies will always be lower than ideal Carnot cycle efficiencies for the same temperature limits.

14.8.5 Efficiency using power and flow measurements

One way to measure efficiency without using the First or Second Law is through power and flow measurements. Such measurements can be classified as either specific power output (SPO) or specific geofluid consumption (SGC). SPO, which can be used for any type of geothermal power system, considers the amount of net power produced per unit flow of geothermal fluid. For given resource conditions, the higher the SPO—typically measured in watt hours per pound or kilogram of geothermal fluid—the more efficient a plant is. The geofluid flow rate needed to produce a certain net power is termed the SGC, which is the inverse of the SPO. SGC measures the flow rate per unit power produced. Some experts consider this the simplest and most effective measurement of efficiency. Dividing the SPO by the available energy is one way to measure the Second Law efficiency.

14.8.6 Turbine efficiency

Rather than considering the efficiency of the entire conversion system, another method for rating geothermal efficiency is to consider certain power plant components. The turbine, for example, provides a useful measurement.

A turbine is a steam-powered machine that causes a shaft—a rotating rod that transmits power or motion from the turbine—to produce electricity through movement. Improvements in turbine design in the past several years have increased geothermal turbine efficiency to over 85 per cent.

14.8.7 Gross versus net efficiency

When parasitic load is reduced at a facility, a plant will operate more efficiently. Gross plant efficiency includes the parasitic load in its assessment, while net plant efficiency only considers the electricity that can produce power (total power minus parasitic load). As a result, gross plant efficiency will always be higher than net efficiency.

14.8.8 Assessing efficiency measurements

While efficiency is important, it is only one characteristic among many that must be considered when choosing the most appropriate energy option for a particular location. Other factors, such as reliability, cost, environmental impact, and sustainability must also be considered. A significant 'energy cost', is associated with producing and transporting fossil fuel for use at a power plant, while the associated costs at a geothermal facility are minimal.

When efficiency assessments must be made, a single number will be meaningless unless the calculations and assumptions used to arrive at that number are made transparent. For example, if a fossil fuel power plant developer cites its 'efficiency' as 40 per cent, and a geothermal developer cites a similar number, these two plants do not necessarily have the same efficiency. To begin to assess efficiency values, an inquiry must be made into any calculations used. In many cases, two efficiency numbers should not even be compared because they measure two different types of efficiencies.

As has been shown, efficiency can be represented in a variety of ways, all of which can be useful and accurate depending upon the situation and the needs of the developer. The point is not to choose one method of calculating efficiency over another. Rather, it is to consider efficiency as one of many factors that can influence power plant development preferences; and then to show which efficiency method is chosen, which calculations and adjustments are made, and why.

14.9 Non-traditional geothermal systems

Most of today's geothermal electricity comes from traditional geothermal conversion technology that integrates no other types of resources into the system. While a significant undeveloped hydrothermal resource base is available, several other technology applications have been considered, or are emerging, that could further expand geothermal potential. Still other applications have already been successfully demonstrated and used commercially. What follows is a rundown of some of these applications, with particular focus upon power plant infrastructure.

14.9.1 Geothermal Hybrid Systems

Hybrid systems, which pair a geothermal hydrothermal resource with another type of resource, offer the flexibility of determining the optimal steam temperature independent of the geothermal source temperature. This adds increased reliability to the system design.

Hybrid systems can increase efficiency, and therefore create more electricity without expanding the use of the geothermal resource. In Fig. 14.11, the energy source for the first two heat exchangers is geothermal; the energy source for the third (labelled 'superheater'), could come from any other source, including biomass, coal or hydropower.

Biomass

Geothermal can be used in conjunction with biomass. The company Infinifuel Biodiesel, for example, has constructed a biodiesel processing facility at a small Nevada geothermal power plant in Wabuska. At this facility, camelina oil seed algae is transformed into diesel fuel. The facility is almost entirely self-contained, largely due to heat supplied by a geothermal plant. The plant works by growing algae, crushing or pressing these materials into vegetable oil and biomass, adding the biomass to alcohol, and finally mixing the biomass/alcohol combination with vegetable oil and heated it using geothermal power for the biodiesel plant. This geothermal facility, which uses 220°F (104°C) water, produces enough power to run the Wabuska facility and sell additional power. Infinifuel plans to expand to other locations in Nevada and beyond.

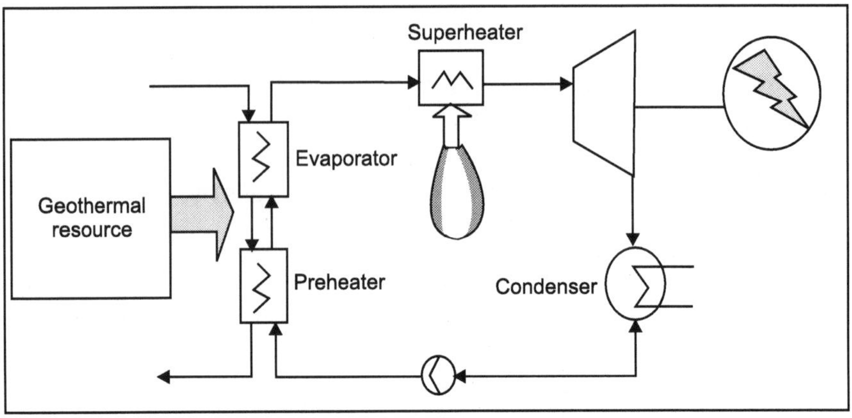

Figure 14.11 Geothermal hybrid power plant system

Combined heat and power

At certain resource locations and under favourable circumstances, geothermal resources can be used both to produce electricity and also for direct use purposes. This hybrid model is known as 'combined heat and power' or CHP. CHP increases net efficiency, improves power plant economics, and creates jobs. CHP essentially takes the 'waste' heat produced by geothermal electric plants and uses it for other useful purposes. Cascading water from a geothermal power plant provides energy for direct use projects such as district heating (and possibly cooling), greenhouse and fish pond heating, industrial applications, and spa and pool heating. CHP has been used on a commercial basis at several sites around the country. Figure 14.12 shows a diagram of a combined heat and power geothermal facility.

Another example of the combined use of geothermal electricity and direct-use heating is highlighted in one of Iceland's top tourist attractions, the blue lagoon. At this location, geothermal water from a working geothermal power plant is piped directly to a large body of water, the blue lagoon. The water is said to offer healing properties due to its unique array of minerals, silica, and blue algae. While the blue lagoon is a great success in Iceland, similar applications have not been constructed at large-scale US power plants.

Solar

The US Department of energy (DOE) has considered solar-geothermal hybrid electric power systems on and off for decades. In 1979, a hybrid concept was considered where wellhead fluid was pressurised, with solar heat added prior to flashing. Such a system was assumed to provide higher quality steam and thermodynamic advantage over a conventional geothermal flash system, but the introduction of hotter than usual geothermal liquid resulted in increased scaling and corrosion.

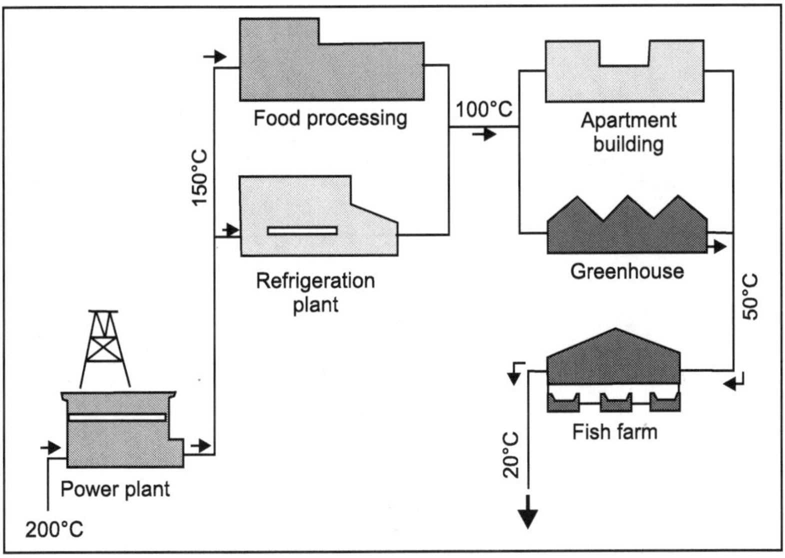

Figure 14.12 Combined heat and power (CHP) diagram

The study conducted by the DOE in 1979 concluded that while geothermal-solar hybrid systems would be comparatively more efficient than stand-alone geothermal plants, hybrids would not offer economic advantages over stand-alone systems. This conclusion has been replicated in subsequent years. Solar-geothermal hybrids may, however, be more cost effective than stand-alone solar facilities. Figure 14.13 shows the solar-hybrid configuration.

Solar has also been proposed as a means to hedge risk associated with geothermal production. In such solar-geothermal hybrid scenarios, a geothermal power plant capacity is proposed that is higher than the 'proven capacity' derived from reservoir engineering assessments of the geothermal resource. A larger power plant size results in higher initial income but also a higher probability of more rapid resource depletion.

To offset the risk that, after 5–10 years of operation, resource depletion might impact generation, the power plant would be constructed so that its heat supply could be supplemented by solar, thereby maintaining generation.

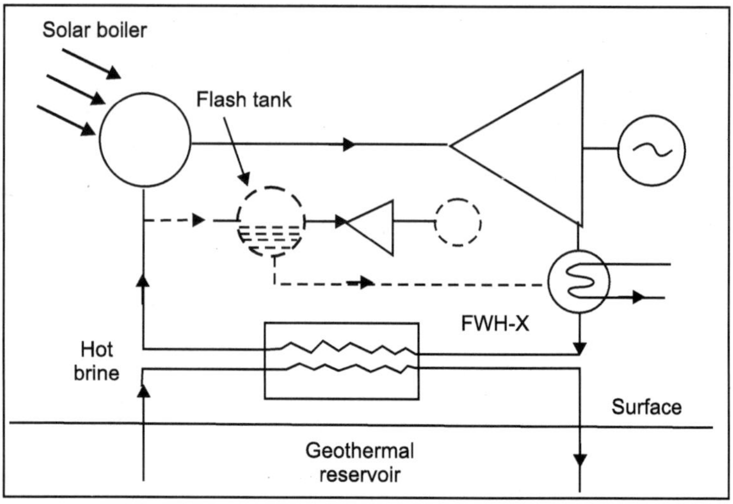

Figure 14.13 Solar-geothermal hybrid facility

One new idea that has been proposed, though not commercially implemented, involves the use of solar and geothermal energy to recover oil from depleting oil and gas fields. Solar augmented geothermal energy (SAGE), according to an abstract presented at the University of Texas, converts depleting oil and gas fields and comparable reservoir strata to 'synthetic geothermal' reservoirs over wider regions. SAGE stores/banks solar energy, using naturally occurring geothermal liquids, for geothermal power generation, while enhancing oil recovery.

Hybrids with heat pumps

A separate application of geothermal energy, geothermal heat pumps, can be used in combination with any other electricity source. Geothermal heat pumps make use of the natural heat trapped below the surface of the earth that averages 65°F (18°C). This heat provides cooling in the winter and warming in the summer. Unlike hydrothermal electricity production, direct use systems do not require a geothermal reservoir. Geothermal heat pumps function like traditional heat pumps: they can heat, cool, and, if so equipped, supply a house with hot water, yet they are significantly more efficient that traditional heat pumps, allowing the user to become less dependent upon the electric grid for heating and cooling. Though any electricity source can be used to power a geothermal heat pump, using renewable power makes the pump 100 per cent green. Figure 14.14 shows a hybrid cycle flow diagram.

Geopressured resources

One particularly promising geothermal-fossil fuel hybrid is known as a 'geopressured' system. A geopressured geothermal facility operates on both natural gas and geothermal fuel. Geopressured geothermal resources have not been tapped since they were successfully demonstrated almost two decades ago, when oil and gas costs were low. Today's increasing oil and gas costs make the economics of geopressured applications look particularly promising.

14.9.2 Enhanced Geothermal Systems (EGS)

Enhanced geothermal systems (EGS) have recently gained much attention as a promising application for geothermal energy. A Massachusetts Institute of Technology (MIT) report discusses possible modifications to the design of a power plant that captures resources through EGS technology. According to the report, EGS technology could be used through existing geothermal power systems, including binary, flash-binary combined cycle, double-flash plant or single flash plant (with several adjustments in power plant components).

EGS resources are similar to traditional hydrothermal resources, except they include one or all of the following: (i) a dense rock reservoir through which liquid cannot easily pass (one that lacks the usual needed porosity and permeability), (ii) insufficient quantities of steam and/or hot water and, (iii) deeper than usual drilling depths.

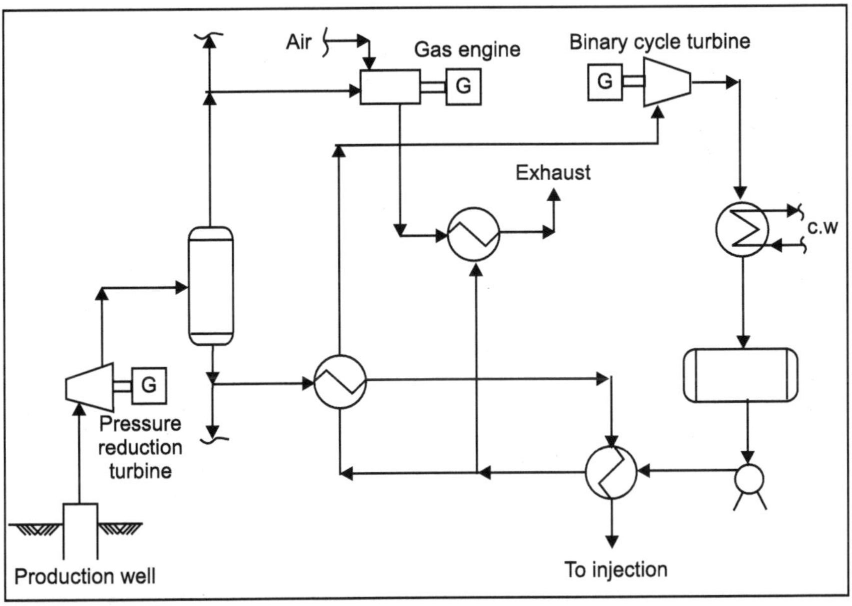

Figure 14.14 Hybrid cycle flow diagram

Highest temperature resources, known as supercritical geothermal systems, would need so-called 'triple-expansion systems'. These are variations on double-flash, with the addition of a 'topping' dense-fluid, back-pressure turbine. The turbine is designed to handle the very high pressures likely to be found with EGS geofluid. However, these systems would require great depths, deeper than 4.7 miles (7.5 km) except for very few areas in the United States. Thus standard EGS systems utilising binary or flash plants are more likely to be developed in the near term. Because EGS has not been successfully demonstrated to date, researchers can only postulate about the power plant that will be needed. However, certain aspects of EGS technology could impact power plant choice. These include:

1. Fluctuations in non-condensable gas content
2. Temperature variations of the heat source
3. Flow variations of the resource
4. Size of the power plant

14.9.3 Outlook for Non-Traditional Geothermal Power Systems

Some of the non-traditional systems discussed in this section, including EGS and oil and gas co-production, show great promise. Some have been demonstrated (geopressured) or even used commercially (CHP) but could be significantly expanded in the future. Still other non-traditional systems, such as solar-hybrids, are not as cost-effective as stand-alone geothermal systems, despite efficiency increases. The commercial and technological viability of each of these non-traditional systems requires not only research and development, but also demonstration projects. Geothermal investors are generally wary of incurring the extra risk associated with trying out a new system before any other developer, especially when the high upfront costs of traditional systems are taken into account. Many developers say they'll be the first in line to construct the second model, but they can't afford to take the risk to construct the first.

14.10 New technology

The value of new technological innovations varies from site to site. At The Geysers, for example, where non-condensable gases influence electricity output, a major improvement for the future could be to upgrade the gas removal equipment for added flexibility, thereby handling a wider range of non-condensable gas concentrations in the steam. At resource locations without non-condensable gases, such innovation would be of little value to developers. What follows is a sample of some promising new developments in geothermal surface technology. It is not an exhaustive list, nor is it an evaluation of which technology is better than another. Rather, this section offers a selection based on discussions with geothermal experts.

14.10.1 Near term versus long term

Incremental technology improvements—those that marginally increase the efficiency of a particular power plant component, for example—might prove most valuable in the near term. Geothermal power plants can always benefit from reduced parasitic load, reduced power expenditures related to cooling fans, improvements to the power substation, and other modest technological advances. Incremental improvements can be commercially implemented more quickly than larger, more revolutionary advances, and can be incorporated into existing designs with comparatively lower risk.

Major surface technology advances, in contrast, typically require considerable research, field experience, and technology trials, along with multiple field applications, before industry and investors are willing to incur the added risk associated with large-scale innovations. Major technology advances could be equally or even more valuable than incremental advances over the long-term, but the former will likely be implemented commercially only after years of research, development, and field experience.

14.10.2 Increasingly standardised, modular geothermal conversion systems

Modular energy conversion systems have already been engineered by existing geothermal companies. Even so, some experts believe modular systems continue to hold promise for further cost reductions. Modular components and subcomponents reduce costs because they can be pulled from off-the-shelf designs that are standardised and mass-produced. Once a plant site is established, a developer can move ahead more rapidly with plant development using modular systems. More costly custom-engineered modules tend therefore to be needed only on a limited basis. Modularity allows developers to more easily add capacity after a reservoir has been found to be capable of additional production.

Increasing modularity does not necessarily indicate small size. Modular units can be in the hundreds of kilowatts or the hundreds of megawatts. Smaller power plants may be more promising in non-traditional power applications, such as oilfield co-production, but larger plants will continue to play a role as new technologies such as EGS become commercially viable.

14.10.3 Mineral recovery

Further research and development could make the separation of minerals from geothermal water, known as mineral recovery, a viable technology. Some geothermal fluids contain significant concentrations of dissolved minerals, while others are virtually mineral free. Mineral recovery offers several benefits, which generally fall into categories of either improving the function of the power plant (reducing scaling, allowing greater power production by lowering the injection temperature) or increasing profits (through the sale of mineral by-products). Often a variety of benefits will result. Minerals found at geothermal power plants include zinc, silica, lithium, manganese, boron, lead, silver, antimony and strontium.

Silica recovery

While silica can be a useful additive in products such as paint, paper, tyres, and toothpaste, it negatively impacts geothermal power plants by clogging pipes, wells, and heat exchangers. DOE has designated silica as one of the most promising minerals suitable for recovery due to its high commercial demand and its potentially negative impact on geothermal systems when not removed or reduced.

Lawrence Livermore National Laboratory (LLNL) has developed a technology for extracting silica from geothermal water. Silica extraction increases the efficiency of a geothermal power plant, provides a marketable silica by-product, and produces freshwater that can be used as a heat exchanger coolant. In the livermore extraction process, geothermal water is separated into freshwater and concentrated brine (heavily salted water) by a reverse-osmosis separation process.

The freshwater is used for evaporative cooling, and the concentrated brine is pumped into a reactor where chemicals are added and silica is extracted. If other minerals are present in the silica-free brine, these can be extracted, and the mineral-mined water is finally pumped to a surface pond and injected into the subsurface (Fig. 14.15).

14.10.4 Case study: Mammoth Pacific

In 2006, LLNL scientists, in collaboration with other agencies, worked with managers at the Mammoth Pacific power plant complex (known as MPLP) in Mammoth Lakes, California to remove silica from the geothermal water. Because the silica content of the geothermal water at MPLP is low compared to typical geothermal water, the co-produced silica is of high, marketable quality. However, the low silica content makes conventional methods for silica recovery less effective than at average geothermal facilities.

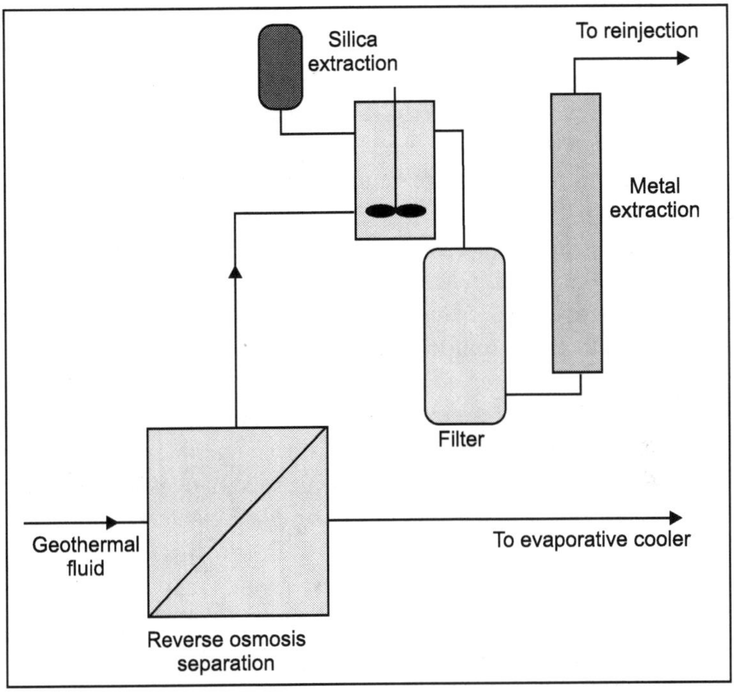

Figure 14.15 Silica extraction diagram

To begin the silica extraction process at Mammoth, the geothermal water is tapped after passing through the heat exchanger and before injection, where it undergoes reverse osmosis. This process concentrates the silica, which then flows through a stirred reactor where salts or polyelectrolytes—synthetic chemicals used to clump solids—are added to induce silica precipitation. The simple silica molecules bond together to form colloids, which are silica particles about 10–100 nanometres in size. These larger molecules cluster to form particles that can be removed by filters downstream from the reactor. Preliminary results suggested in 2006 that silica recovery at Mammoth Lakes could reduce the cost of geothermal electricity production by 1.0′/kWh. The market value of silica that could be produced from the Mammoth Lakes site if silica is removed from all geothermal liquid is estimated to total $11,000,000/year (as on 2006). LLNL is also considering using reverse osmosis to separate lithium, cesium, rubidium and tungsten. However, these activities have not yet been pursued.

14.10.5 Working fluids for Rankine Cycle Power Plants

Studies at National Renewable Energy Laboratory (NREL), Idaho National Laboratory (INL), and elsewhere have shown that mixed working fluids in binary-cycle geothermal power plants can potentially reduce thermodynamic inefficiencies in the boiler and condenser, thereby improving overall plant efficiency. Researchers have investigated various pure and mixed working fluids to optimise power conversion efficiency. One potential working fluid, especially suitable for lower temperature resources, is ammonia-water as used in the Kalina cycle.

The Kalina cycle uses a mixture of 70 per cent ammonia and 30 per cent water as the working fluid. The fluid is vapourised using the geothermal water as the heat source. Because the ammonia-water mixture boils over a range of temperatures, its temperature through the vapouriser aligns closely with the geothermal water temperature, improving the thermodynamic efficiency of the heat transfer process. However, this alignment can result in a smaller temperature differential between the geothermal water and the working fluid, which can require greater heat exchanger surface area. And while Kalina affords a higher thermodynamic efficiency because the cycle reduces so-called thermodynamic irreversibilities, the split ammonia and water streams add complexity to the system and may require additional pumping power. Plant operators and developers offer differing views about the potential for the Kalina model to improve upon Rankine-type power plant performance. However, many experts see the possibility for marketability in the future. One challenge is that the Kalina cycle has not yet been commercially verified in the US, making it difficult for developers to secure investors. A solution may be for the government to provide cost-shared projects with industry, so as to verify the viability of any new power conversion cycle. This process of cost-sharing and verification is important for all emerging geothermal technologies.

14.10.6 Hybrid cooling systems

Experts site hybrid cooling as one of the most important areas for surface improvements. In an increasingly water-constrained world, air cooling will likely become the preferred cooling option. However, the relative inefficiency of air-cooled systems during the summer, when the temperature differential between the air and water is reduced, has proven in some cases to be a liability. On days with extremely hot conditions, some plants put out half the power they'd produce on a cold day.

14.10.7 Coatings

Corrosion and deposition of mineral scale (known as fouling) can occur at geothermal resource areas with high concentrations of dissolved and suspended solids, such as the geothermal water at the Salton Sea in California. When scale accumulates over time, it can clog pipes or vessels and decrease the effectiveness of heat exchangers. Fouling can be controlled by the use of chemical additives to keep scale from forming or by periodic cleaning fouled surfaces, but these practices add to operating costs. Traditionally, expensive materials such as high-alloy steel and titanium have been used to reduce corrosion. However, these do not transfer heat well, and can cost at least three times as much as traditional materials.

In order to reduce the cost of maintaining open flow paths and efficient heat transfer, Brookhaven National Laboratory (BNL) developed durable, scale-resistant polyphenyl sulphide-based coatings for carbon steel. These coatings can be used for heat exchangers—devices which transfer heat through a conducting wall from one fluid to another—as well as for binary cycle power plants, piping, flash vessels, and other plant components. Coatings applied to carbon steel are less expensive than alloys, deliver equivalent corrosion protection, and are easier to clean than stainless steel and titanium.

When carbon steel tubes coated with a thermally-conductive coating variant were field tested by the National Renewable Energy Laboratory and compared to a high-alloy alternative, the coating variant resisted fouling and maintained heat transfer as well or better than the high-alloy. In locations where geothermal fluid is particularly mineral laden or corrosive, coated carbon steel may be a good alternative to more expensive construction materials.

Coatings are also useful with air-cooled systems. Spraying air-cooled condensers with water increases the efficiency of the system, but tends to corrode the heat exchanger fins. Researchers at BNL and NREL have therefore considered different coatings to apply to aluminium fins used in air-cooled heat exchangers to prevent corrosion.

While coatings have been used commercially in military, nuclear, and petroleum facilities, they have not yet been applied commercially to geothermal applications.

14.11 Direct use

Geothermal resources, through geothermal hot springs, have been used directly for centuries. Though hot springs are still used today, geothermal water can now be used directly (known as 'direct use') for an expanded variety of uses, primarily related to heating and cooling. The main utilisation categories are swimming, bathing and balneology; space heating and cooling, including district energy systems; agricultural applications such as greenhouse and soil heating; aquaculture application such as pond and raceway water heating; and industrial applications such as mineral extraction, food and grain drying.

Direct uses work best with temperatures between 70° and 300°F (21° and 149°C). Resources in this range are widespread and exist in at least 80 countries at economic drilling depths. No conversion efficiency losses result when resources are used in this range, and projects can use conventional water-well drilling and off-the-shelf heating and cooling equipment (allowing for the temperature and chemistry of the geothermal water).

Projects can be built on a small scale (mom and pop operations) such as for an individual home, single greenhouse or aquaculture pond, but can also be built on a large scale such as for district heating/cooling, food and lumber drying, and mineral ore extraction.

In modern direct-use systems, a well is drilled into a geothermal reservoir to provide a steady stream of hot water. The water is brought up through the well, and a mechanical system—piping, a heat exchanger, and controls—delivers the heat directly for its intended use.

Care must be taken to prevent oxygen from entering a direct use system, as geothermal water is normally oxygen free, and dissolved gases and minerals such a boron, arsenic, and hydrogen sulphide must be removed or isolated to prevent corrosion, scaling, and harm to plants and animals. Carbon dioxide, which often occurs in geothermal water, can be extracted and used for carbonated beverages or to enhance growth in greenhouses.

The typical equipment for a direct-use system is illustrated in Fig. 14.16, and includes downhole and circulation pumps, heat exchangers (normally the plate type), transmission and distribution lines (normally insulated pipes), heat extraction equipment, peaking or back-up generators (usually fossil fuel fired) to reduce the use of geothermal water and reduce the number of wells required, and water disposal systems (injection wells).

14.12 New technologies: the path forward

Regardless of advances in geothermal technology, geothermal will not provide the silver bullet solution to our energy needs. Nor will any single resource. Instead, geothermal technology advances can help increase our country's share of renewable, sustainable geothermal energy. And that's important—today, only a fraction of the geothermal resource base is tapped. Surface technology developments, in particular, can increase energy production without impacting the resource base. Technology, both surface and subsurface, can be developed a variety of ways.

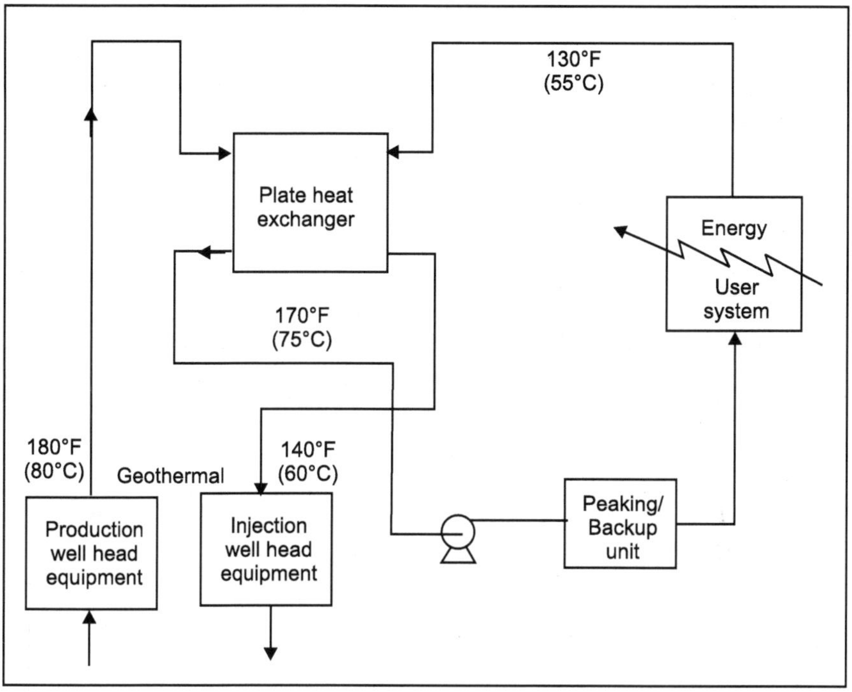

Figure 14.16 Typical direct use geothermal heating system configuration

14.13 Nanotechnology and geothermal energy

Nanotechnology and geothermal energy have been hot topics lately. Currently, geothermal energy accounts for just about 0.5 per cent of energy production, but that number could rise to as much as 10 per cent in the future. Below the surface of the earth, there is a storehouse of heat, waiting to be tapped into for energy production purposes.

14.13.1 Challenges and controversy with commercial systems

Geothermal energy has been a topic of research that is somewhat controversial. By using the heat deep in the earth to produce energy, dependence on fossil fuels can be reduced, but is it really worth it?

Deep drilling for this resource has been shown to cause small earthquakes. Putting earthquakes aside, geothermal energy requires very high temperatures to be effective. There are only a few places in the world that could even benefit from this source of energy and they would face problems such as needing to provide antifreeze for the pipes that run the system in cold times, and the risk of pollution and altering the landscape. Thankfully, these problems have only encouraged scientists to continue research using nanotechnology to solve these issues. Solutions are being developed that uses nanotechnology, where energy can be produced from lower temperature sources. Earthquakes? Not any more. There won't be the need to drill as deeply to produce effective amounts of energy, greatly reducing the risk of earthquakes. Because of lower temperature requirements, nanotechnology and geothermal energy will make it possible to generate power in more places, making the number of people who would benefit much greater.

14.13.2 Solving the challenges

Many worry about the environmental effects and it is wise to evaluate the unintended consequences of any new technology. Nanotechnology has actually helped eliminate many of the effects on the environment from drilling into the earth's crust. In the past, poisonous gases have been released and land sinkage has occurred when too much water has been removed from the reservoir at the bottom of the well. The new technology has the ability to clean the gases out of the air, reducing the amount of pollution. They can also replace the water back into the ground from where it came, reducing land sinkage and reusing the resource more efficiently. What does all of this mean for you? Well, if you live in an area that can benefit from nanotechnology and geothermal energy, this means a more renewable source of energy that is cleaner and much less expensive. Your local utility company can produce cheaper electricity, lower pollution, and rely less on fossil fuels.

14.13.3 Nanotech tool for enhanced geothermal

Enhanced geothermal systems (EGS) could be the Holy Grail of renewable energy sources. If it works, it could potentially supply an almost unlimited amount of energy, cheaper than conventional fuels, from virtually anywhere in the world.

That is, if it works… Unlike conventional geothermal, EGS instead involves drilling much deeper, and fracturing rock kilometres below the surface to create an artificial circulation system.

Geodynamics found out the hard way that this is not so easy. Their pilot project in the Australian Outback was on-line to be the first commercial example of EGS anywhere in the world until they had a drill hole blowout last May. The project remains in limbo until a field investigation is completed. Researchers at Stanford University may have a new tool for companies taking on this very risky venture through the novel use of nanotechnology. One of the many challenges facing EGS pioneers is: how do you determine the porosity of very hot bedrock at the bottom of a very deep and expensive drill hole?

Professor Roland Horne and his grad student Mohammad Alaskar managed to pass silicon nanoparticles through solid rock, which could provide invaluable information for EGS engineers as they move their projects forward. 'Nanotechnology has found a tremendously varied number of uses', said Horne. 'And, this occurred to us to be one of them. By passing the nanoparticles through the fractures we have a way of determining how big they are and potentially other things, such as their temperature, perhaps pressure, chemistry, all of those things.'

Stanford's research is funded through ongoing support of the Department of Energy, which has made enhanced geothermal a strategic priority.

Enhanced geothermal is a potentially game-changing exciting technology. As costs and risks come down, perhaps with the help of nanotechnology, it is definitely worth keeping an eye on.

14.14 Summary

Geothermal surface technology is an important part of geothermal energy development. Geothermal fluid—a hot, sometimes salty, mineral-rich liquid and/or vapour—is the carrier medium that brings geothermal energy up through wells from the subsurface to the surface. It is withdrawn from a deep underground reservoir, isolated from groundwater by thickly encased pipes, making the facility virtually free of water pollution. Once used, the water and condensed steam is injected back into the geothermal reservoir to be reheated. A geothermal resource that uses an existing accumulation of hot water or steam is known as a 'hydrothermal' resource. All geothermal electricity produced today derives from the hydrothermal resource base.

Characteristics of the geothermal fluid, including temperature, chemistry, and non-condensable gas content (NCG), can influence power plant design. Two substances sometimes found in geothermal fluid, hydrogen sulphide (H_2S) and mercury, are regularly abated at geothermal facilities, eliminating 90–99.9 per cent of the substances.

Power plant basics: Like all conventional thermal power plants, a geothermal plant uses a heat source to expand a liquid to vapour/steam. This high pressure vapour/steam is used to mechanically turn a turbine-generator. At a geothermal plant, fuel is geothermal water heated naturally in the earth, so no burning of fuel is required.

Power plant size: Though the size of a power plant is determined primarily by resource characteristics, these are not the only determining factors. Factors that favour the development of larger geothermal plants include:

1. Cost decreases when larger quantities of materials, including steel, concrete, oil, and fuel, are purchased at one time.
2. High transmission costs, regardless of plant size, can include land use and rights-of-way fees.
3. Though some automated facilities require few personnel, a minimum number of people are typically required to run a geothermal power plant.

Factors that favour the development of smaller geothermal plants include:

1. Developers may opt to start small and increase output as they come to understand the potential of the resource through continued use.
2. Smaller plants require less time to permit.
3. The production tax credit (PTC) induces developers to construct smaller plants that can qualify for the short time frame of the PTC.
4. A developer's power purchase agreement may require that he start with a small output and gradually increase production.

Conversion technology: A conversion technology represents the entire process of turning hydrothermal resources into electricity. Four options are available to developers:

1. Dry steam plants, which have been operating for over 100 years, make use of a direct flow of geothermal steam.
2. The most common type of power plant, a flash power plant, uses a mixture of liquid water and steam.
3. Binary geothermal plants function as closed loop systems that make use of resource temperatures as low as 165°F (74°C). A Rankine cycle is the commercial binary cycle used in the United States.
4. A combination of flash and binary technology, known as the flash/binary combined cycle, has been used effectively to take advantage of both technologies.

Cooling system: Most power plants, including most geothermal plants, use water-cooled systems—typically in cooling towers. In areas with scarce or expensive water resources or where the aesthetic impact of steam plumes (produced only in water-cooled systems) are a concern, air cooling may be preferred. However, air-cooled systems are influenced by seasonal changes in air temperature.

Structuring power plant to minimise impact: A geothermal developer mitigates potential impacts in a variety of ways. Developers may use noise muffling equipment, visual mitigation techniques, strategies to reduce potential effects on wildlife and vegetation, monitoring activities, and regular maintenance and upkeep activities.

Efficiency: The public interest in energy efficiency arose as a fossil fuel issue: that is, the less fuel used per output, the fewer emissions and the greater quantity of depleting fuel conserved. For renewable energy use, in contrast to fossil fuel use, efficiency is primarily an economic concern. This is because at renewable sites like geothermal power plants, the fuel source is not burned, and thus few emissions are released. Geothermal developers choose to discuss efficiency in a variety of ways, depending upon the context in which an efficiency measurement is needed and the characteristics of the resource and plant.

Several 'non-traditional' technology applications have been considered or are emerging, that could further expand geothermal potential:

1. *Hybrid systems:* A hybrid system integrates another resource into a hydrothermal geothermal power plant, therefore creating more electricity without expanding the use of the geothermal resource. Geothermal can be used in combination with biomass, combined heat and power or CHP (geothermal electricity plus a geothermal direct use application), geothermal heat pumps, and geopressured resources (those that operate on both natural gas and geothermal fuel).
2. *Enhanced geothermal systems (EGS):* EGS resources could be harnessed using existing geothermal power systems.
3. *Oil and gas co-production:* An oilfield co-produced resource makes use of wells already drilled by oil and gas developers. These wells are either deep enough to encounter hot water, or could be deepened into hot zones.

New technology: Several surface technology applications look particularly promising for the future. These include:

1. *Incremental technology improvements:* Small-scale improvements can be commercially implemented more quickly than larger, more revolutionary advances, and can be incorporated into existing designs with comparatively lower risk.
2. *Increasingly standardised, modular geothermal conversion systems:* Modular components and subcomponents reduce costs because they can be pulled from off-the-shelf designs that are mass-produced. They allow developers to move ahead more rapidly with plant development and, once a plant is established, capacity additions.
3. *Mineral recovery:* Further research and development could make the separation of minerals from geothermal water, known as mineral recovery, a viable technology. Mineral recovery offers benefits such as reduced scaling and increased revenue.
4. *Mixed fluids:* One working fluid especially suitable for lower temperature resources is an ammonia-water mixed fluid system as used in Kalina and other cycles. Studies have shown that mixed

working fluids in binary-cycle geothermal power plants can reduce thermodynamic inefficiencies.

5. *Hybrid cooling:* In an increasingly water-constrained world, air cooling will likely become the preferred cooling option. However, the relative inefficiency of air-cooled systems during the summer has proven in some cases to be a liability. Hybrid cooling systems seek to integrate the best of both systems, increasing seasonal efficiency while also reducing water use and aesthetic impact.
6. *Coatings:* Traditional materials used to reduce corrosion do not transfer heat well and can cost at least three times as much as traditional materials. Researchers have engineered less expensive 'coatings' that can be applied to various power plant components to reduce scaling and fouling.

Direct use: Geothermal resources have been utilised for centuries through 'direct use'. Direct use resources are tapped by drilling wells and bringing hot water to the surface directly for a varicty of uscs, primarily for space heating, but also for drying farm and timber products, aquaculture and industrial uses. Geothermal heat pumps: According to the DOE, geothermal heat pumps (GHPs) use 25–50 per cent less electricity than conventional heating or cooling systems. Geothermal heat pumps can reduce energy consumption—and corresponding emissions—from 45 to 70 per cent when compared to traditional systems. They also improve humidity control. Because heat pumps do not require a geothermal reservoir, they can be used anywhere in the world.

14.15 Environmental monitoring of geothermal power plants

During the operation phase of geothermal power plants, environmental engineers and scientist are entrusted with the task of monitoring their impacts on the environment. With a well designed monitoring plan, they are kept informed of the conditions of the system which they are responsible for maintaining and a sound environmental management plan helps ensure that the monitoring is carried out in an effective manner within the boundaries of their observation space, in accordance with regulation and best practices. As is the case with the power plant engineer, the environmental engineer should strive to apply preventive measures, to ensure continued sustenance of the natural environment.

As all other anthropic projects, geothermal power plants interact with their surrounding environment in various ways over the course of their lifetime. They thus become a part of the complex planetary environment without any obvious demarcated boundaries. The impacts can vary in nature, severity and scope over the different phases of development, which according to Steingrímsson consist of a preliminary study, an appraisal study, project design and construction, commissioning and operation, and finally shutdown and abandonment. Early impacts during surface exploration are usually minimal, but they become gradually more pronounced as the project moves through exploratory drilling and on to the appraisal study phase (Fig. 14.17). The greatest impacts are brought about during the design and construction phase, when the local environment in the geothermal field and at the power plant site may change significantly with the clearing of land and the construction of man-made structures, when wells are being flow tested and the economic and social effects of the power plant are felt most profoundly in the neighbouring communities. After commissioning, the power plant usually falls into a rather steady interaction with the environment if maintenance and operation activities are carried out according to best practices. This interaction finally stops when the plant is shut down and abandoned, although there will be lingering local effects, depending on the reversibility of impacts and the degree of restoration.

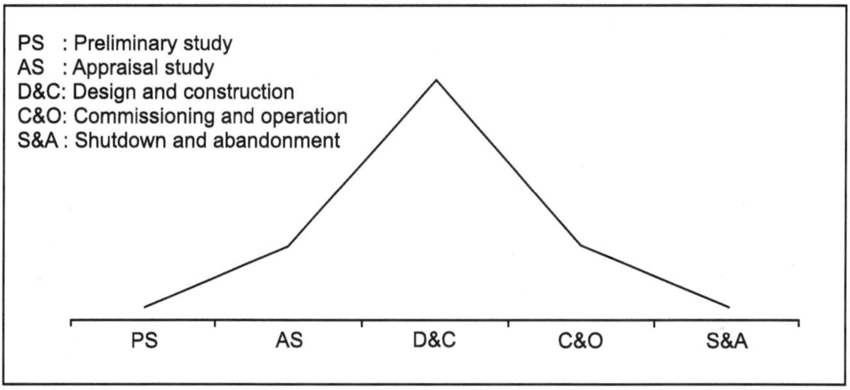

Figure 14.17 Hypothetical relative degrees of the environmental impacts of the different phases of geothermal power plant development

The intent of this chapter is to present a broad overview of environmental monitoring of geothermal power plants in operation and to examine to a limited extent the role of environmental engineers and scientists in that context.

14.16 Geothermal power plants in the environment

Geothermal power plants in operation cannot be viewed as isolated systems. They rely upon high enthalpy geothermal fluids that may draw their heat content from rocks several kilometres beneath the surface of the earth—heat that may have been locked in the deep interior of the planet since its formation around 4.5 billion years ago or released from radioactive heavy elements that captured and locked away some of the tremendous energy of supernovae long before the formation of the solar system. The plants process these fluids and re-emit them to the environment in different states and compositions: water and dissolved substances are reinjected or discharged to the surface and steam and non-condensable gases are emitted to the atmosphere, where they may be advected and dispersed over long distances while undergoing chemical reactions with other atmospheric components. One of these gases is CO_2, which has an estimated 200 years combined lifetime in the atmosphere, biosphere, and upper ocean. Over this period, it can contribute to radiative forcing of the climate and acidification of the oceans. Geothermal power plants therefore interact with the environment over vast distance and time scales.

Geothermal power plants need cooling fluids to discard waste heat to the environment. These can be fresh or salt water that makes a one-time pass through a condenser, carrying the waste heat into a nearby river, lake or ocean. The cooling fluid can also be water that carries waste heat into the environment in latent form through an evaporative process in a cooling tower or air that cools a dry condenser. The first method increases the temperature of the receiving river, lake or ocean, which may affect the biosphere, the second increases the temperature and humidity of the local atmosphere and the third increases the temperature of the local atmosphere.

In addition to geothermal fluids and cooling fluids, a geothermal power plant in operation needs to discard various substances, materials and equipment and new ones are required for maintenance purposes. A pump made by workers in China may be bought for maintenance in a power plant in El Salvador, Iceland or Kenya, and transformer oil or cooling water treatment chemicals may likewise be shipped between continents. Last, but not least, geothermal power plants may provide electricity to the public and industry on a country-wide scale and hot water to district heating systems on a municipal scale, thereby affecting the social and economic fabric in these spaces. It is thus clear that geothermal power plants in operation affect the environment and socio-economic sphere near and far.

A common practice in environmental engineering, fluid mechanics and thermodynamics is the application of control volumes to systems of interest. They can range from the infinitesimal to the very large depending on the problem at hand and are used for simplification to draw an arbitrary border between the system of choice and its surroundings. The goal is usually to bundle the various processes of the system within the control volume together without regard to the details of system components. The interaction of such systems with the surrounding environment can be evaluated on the basis of mass, energy or information exchange through the imaginary borders of the control volume.

One way of selecting the control volume, or observation space, of a geothermal power plant is to imagine the borders as consisting of the land surface and the borders of man-made structures and equipment, or perhaps extending them below the surface to the casings of wells (Fig. 14.18). This may be the preferred observation space of power plant engineers who are concerned with daily operations and maintenance of their power plants. In order to keep the plants in optimum shape for future operations, they may apply preventive maintenance philosophies within this space, since the key to reliable operation is constant monitoring and condition-based maintenance. In such an observation space, the environment may be a concern primarily as an outside agent of influence on the system of relevance (the power plant), as weather conditions and the chemistry of surrounding fluids may detrimentally affect equipment and materials.

The environmental engineer may take the stance of looking at a bounded or unbounded observation space surrounding the power plant to look at the effects of the plant on the environment. In this case, the power plant has become the outside agent of influence on the system of relevance, which is the environment, often extended to the socio-economic sphere. The environmental engineer is concerned with sustaining the environment, so that it is passed on to the future in the same or better shape than at present. As one synonym of the verb to 'sustain' is to 'maintain', the goals of the power plant engineer who seeks to maintain his system in optimum working condition is analogous to that of the environmental engineer who seeks to sustain the system which he or she is responsible for monitoring.

In this respect, it is important to note that the environmental engineer may choose to look at details or overall effects within the observation space. An example is the clearing of a patch of land and the associated loss of vegetation and wildlife, which may or may not be considered acceptable depending on the scarcity of the affected flora and fauna and the magnitude of the effects, but which may possibly be compensated for by restoring another patch of land to similar conditions. Therefore sustained balance of the system on a predetermined scale is sometimes what should be sought rather than simply keeping individual small scale parts of the system in sustained condition.

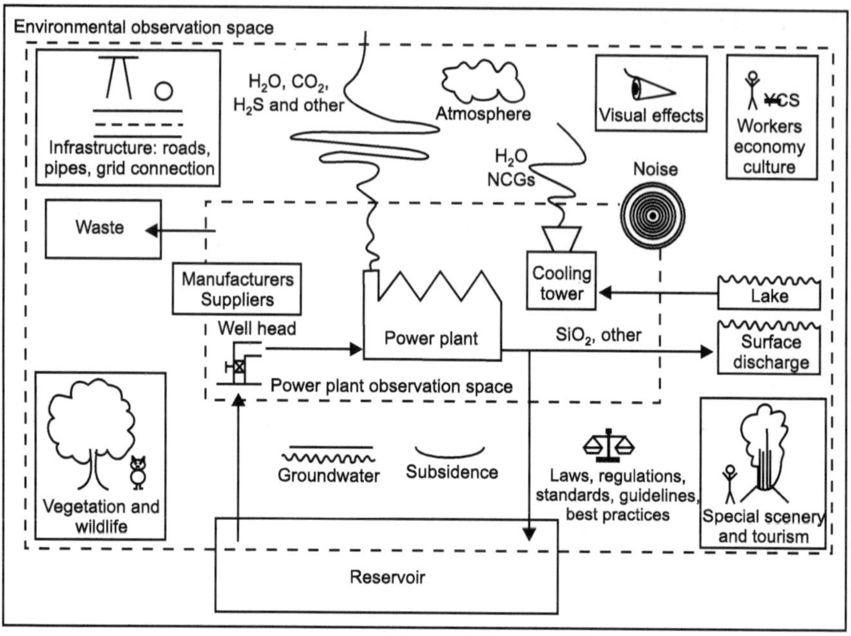

Figure 14.18 A simple presentation of one possible demarcation of the environmental and power plant observation spaces.

While arbitrary compartmentalisation into observation spaces can be useful, it should be remembered that the power plant and its surrounding environment can also be viewed as a single system of interacting parts and perhaps such an approach is most fit.

14.16.1 Interaction with the environment

The interaction between geothermal power plants and the surrounding environment can be divided into the impacts of the power plants on the environment and the impacts of the environment on the power plants. Kristmannsdóttir and Ármannsson have classified the former into the following categories:

1. Surface disturbances
2. Physical effects of fluid withdrawal
3. Noise
4. Thermal effects
5. Chemical pollution
6. Biological effects
7. Impact on (protected) natural features

Surface disturbances include the clearing of land, changes in landscape and the introduction of man-made structures where none existed before. Physical effects of fluid withdrawal include subsidence, lowering of the groundwater table and induced seismicity as earth layers consolidate due to the removal of fluids from matrix pore spaces or when increased pressure due to injection causes the relief of accumulated geological stresses. Thermal effects include elevated temperatures of rivers, lakes and groundwater due to thermal fluid discharges and changes in cloud formation and local weather due to steam emissions. Chemical pollution can be caused by steam and non-condensible gas emissions to the atmosphere and the discharge of brine to surface or subsurface water bodies. All of these have the potential of impacting wildlife, vegetation and the socio-economic sphere around a geothermal power plant, as well as altering natural features for the short-term or permanently. Monitoring and controlling sources and their impacts is therefore of utmost importance.

On the other hand, the impacts of the environment on geothermal power plants include:

1. The application of external dynamic forces to power plant structures. These include seismic forces changing over short time scales and deforming forces that may change over much longer time scales due to subsidence. Wind (hurricanes), floods, snow and ice are also examples of dynamic environmental forces that may follow a seasonal pattern and affect power plant structures.
2. Degradation of equipment and materials due to environmental chemistry. These effects include corrosion and scaling.

All of these effects need to be monitored to maintain and sustain the power plant and its environment.

14.16.2 Environmental monitoring

The goal of environmental monitoring is to keep the environmental engineer informed of the conditions of the system which he/she is responsible for maintaining. Ideally, information about all possible factors of concern should be available at all times to give instantaneous snapshots of the system when needed. The environmental engineer, however, does not have the same elaborate monitoring and control system at his/her disposal as the power plant engineer, where the most important parameters of power plant operations are constantly monitored and recorded, and can be viewed on a display in a central location. While such frequent information may be obtained by chemical, meteorological and seismic sensors in the environment, the environmental engineer will in most cases have to make his/her own observations and measurements. As this can be costly, it is important to design a well structured monitoring program that fulfills the demands of regulations and best practices, while minimising cost. Quality of data should be emphasised over quantity, since data volume by itself is not an indicator of meaningful information, but rather the context in which the data is collected and how it is related to processes of interest.

In 2007, the World Bank Group (WBG) listed the common components of a pollution monitoring program in the following manner:

The elements of a monitoring plan normally include selection of the parameters of concern; the method of collection and handling of samples (specifying the location, the frequency, type, and quantity of samples, and sampling equipment); sample analysis (or alternatively, online monitoring); and a format for reporting the results.

While pollution monitoring is essentially the monitoring of various chemical species and other potentially harmful agents (such as particulate matter) in the environment and unmistakably important in the context of geothermal power plants, it is only a part of a thorough and well rounded geothermal environmental monitoring program. The WBG description is therefore not complete in the context of geothermal power plants, but can nevertheless serve as a basis that can be extended to the other factors of concern to design a sound environmental monitoring program.

According to the WBG, 'an environmental management system is a structured program of continuous environmental improvement that follows procedures drawn from established business management practices'. One such system is the ISO 14000 series, which helps delineate well organised management programs. As it provides structure more than content, it leaves the responsibility of identifying environmental concerns to the users of the system. The ISO 14001 environmental management system has for example been implemented by Instituto Costarricense de Electricidad for the Miravalles power plant in Costa Rica and has allowed continuous improvement in controlling the environmental impacts of the power plant.

In order to clarify the goals of monitoring and establish the content of an environmental management plan, it can be helpful to keep the following basic questions in mind:

1. What aspects need to be monitored?
2. Why are they a concern?
3. Where should the monitoring take place?
4. How should monitoring and analysis be carried out?
5. When is monitoring needed?

While environmental monitoring programs can be designed meticulously, they are meaningless without references to compare observations against, and these mainly take two forms:

1. The natural state before alteration.
2. Laws, regulations, standards, codes, guidelines, and best practices.

14.16.3 Baseline establishment

It is necessary to know the natural state of the environment before it is changed by exploration, testing or utilisation. Baseline data collection involves collecting background information for this purpose on the physical, chemical, biological, social and economic settings in the vicinity of a proposed geothermal power plant and is usually carried out as part of scoping or an environmental impact assessment. The information may be obtained from secondary sources or gathered through measurements, field samplings, surveys, interviews and consultations. A well-established baseline allows the environmental engineer to assess how significantly specific environmental conditions deviate from the natural state during the power plant operation phase and to what extent they are caused by the plant.

14.16.4 Role of laws, regulations, standards, codes, guidelines and best practices

Most countries have established specific laws that address the environment, and environmental regulations may be issued by relevant ministries. The particular aspects addressed and the detail in which they are covered can however vary from country to country. Power plant designers and operators may be obliged to follow certain standards, norms and codes that may be relevant to the environment, and financing institutions may condition the financing of projects to the observation of environmental guidelines that they have established. Companies and developers can also decide on their own accord to follow best practices without being obliged to do so by outside parties.

Environmental laws are important in channelling and placing constraints on human interaction with nature. Governmental and municipal regulations are an extension of the law, which may address specific topics in more detail than the laws themselves and dictate limits, such as the allowable concentration of pollutants in the environment. They may also provide a framework for the development of geothermal resources and address issues as diverse as planning, the acquisition of land, the application of concessions, exploration, exploitation, rights, permits, performance guarantees, taxation, general procedures and obligations of actors.

Engineers may be constrained by standards and codes in their design work that are pertinent to the environment. Civil engineers who design concrete containment structures for conveying, storing or treating liquid or solid wastes may for example have to take account of specific codes in structural design and materials selection in order to ensure that those wastes will not find their way into the environment.

Health guidelines are important to the environmental engineer as they often provide advice on maximum allowable concentrations of particular pollutants over a set time period. Such guidelines are issued by national agencies or international organisations such as the World Health Organisation (WHO) and are often specific to the fluid medium carrying the pollutants, such as the atmosphere or water bodies. One of the tasks of the environmental engineer is to assess measured concentrations against such guidelines and to devise mitigation schemes if they are surpassed.

14.16.5 Environmental engineering and mitigation

It is important that environmental considerations be taken into account from the very first stages of geothermal resources development and that these considerations are noted in the design of a power plant. To this end, it is important to start gathering environmental baseline information as soon as possible and that an environmental impact assessment be carried out—the results of which will be of use in the power plant design. If environmental constraints are likely to be broken based upon available information about the geothermal fluid properties and composition and initial proposals for utilisation, it is the task of the environmental engineer to devise schemes to prevent such scenarios from unfolding. A geothermal power plant that is well designed with regard to the environment and well managed is not likely to cause environmental harm. However, as regulations and guidelines may change and conditions become more stringent with time, the environmental engineer must keep abreast of developments in the regulatory sphere that may increase environmental demands on a power plant, even if the operation remains in a steady phase.